T0327381

Charge-based MOS Transistor Modeling

Charge-based MOS Transistor Modeling

The EKV model for low-power and RF IC design

Christian C. Enz
Eric A. Vittoz

John Wiley & Sons, Ltd

Other Wiley Editorial Offices

John Wiley & Sons Inc., 111 River Street, Hoboken, NJ 07030, USA

Jossey-Bass, 989 Market Street, San Francisco, CA 94103-1741, USA

Wiley-VCH Verlag GmbH, Boschstr. 12, D-69469 Weinheim, Germany

John Wiley & Sons Australia Ltd, 42 McDougall Street, Milton, Queensland 4064, Australia

John Wiley & Sons (Asia) Pte Ltd, 2 Clementi Loop #02-01, Jin Xing Distripark, Singapore 129809

John Wiley & Sons Canada Ltd, 22 Worcester Road, Etobicoke, Ontario, Canada M9W 1L1

Wiley also publishes its books in a variety of electronic formats. Some content that appears
in print may not be available in electronic books.

Library of Congress Cataloging-in-Publication Data

Enz, Christian.
    Charge-based MOS transistor modelling : the EKV model for low-power and RF IC
design / Christian Enz and Eric Vittoz.
      p. cm.
Includes bibliographical references and index.
ISBN-13: 978-0-470-85541-6 (alk. paper)
ISBN-10: 0-470-85541-X (alk. paper)
1. Metal oxide semiconductors–Mathematical models.   2. Metal oxide semiconductor
field-effect transistors–Mathematical models.   I. Vittoz, Eric A., 1938–   II. Title.

TK7871.99.M44E59 2006
621.3815′284–dc22   2006041744

A catalogue record for this book is available from the British Library

ISBN 13 978-0-470-85541-6
ISBN 10 0-470-85541-X

Typeset in 10/12pt Times by TechBooks, New Delhi, India

This book is printed on acid-free paper responsibly manufactured from sustainable forestry
in which at least two trees are planted for each one used for paper production.

To our families

Dominique, Adrien, Mathilde and Simon
and
Monique, Nathalie and Didier

Contents

Foreword

Modern electronic technology is largely based on MOS integrated circuits containing both analog and digital parts. In designing such circuits with high performance, a correct MOS transistor model is a must. The designer needs a model he or she can rely on, which correctly describes the numerous physical phenomena in MOS transistors, allowing the performance of a circuit composed of such devices to be predicted with accuracy during circuit simulation. In addition, for preliminary "hand" analysis and design, it is desirable to have a simple model that makes evident the inter-relations between the various parameters, and allows the designer to correctly identify the trade-offs involved. The EKV approach to MOS transistor modeling combines both of these attributes.

The EKV model is the result of a large body of work by Drs C. C. Enz, F. Krummenacher, and E. A. Vittoz, and several of their students and colleagues. The work has its origins in the pioneering work at CEH (now CSEM), on micropower devices and circuits for watches in the late sixties. This has given the EKV model development a unique aspect: it originated with highly competent circuit designers, notably analog ones, and was developed by them, or at least with constant feedback from them, every step of the way. Thus, it not only describes the physics of the MOS transistor, but takes into account carefully what circuit designers need. The result is a model that is accurate and predictive, correctly treats the MOSFET as a four-terminal, nominally symmetric device, has smooth behavior without discontinuities in all regions of operation, and correctly predicts small-signal parameters. In addition, the basic part of the model consists of a simple set of equations that are intuitively appealing, which makes it possible for the circuit designer to have a feel for the model and its parameters, rather than treating the model simply as a black box in which no designer dares to tread. This helps make circuit design a systematic process, and less a cut-and-try approach.

Drs. Enz and Vittoz, well-known for their contributions to MOS devices and circuits, have done a great job putting together a streamlined presentation of the EKV model. The book covers every aspect of the model, from DC large-signal I-V equations, to charge modeling, nonquasi-static effects, small-signal modeling, noise, small-channel effects, and matching. I have followed the work of the authors and their colleagues for many years with appreciation, and I am delighted to see their results presented in this unified manner. This book will help spread the understanding and use of the EKV model, as the latter certainly deserves.

PROF. Y. TSIVIDIS
Columbia University, New-York

Preface

The aggressive downscaling of CMOS technologies that has been going on for more than 25 years has led to an increase in the number of transistors per chip and hence extend the functionality while at the same time dramatically pushing the speed performance. Although these tremendous speed improvements have been mainly driven by the requirements of VLSI digital chips, they have also been exploited for analog and RF circuits. Today, ultra deep-submicron (UDSM) technologies have caught up and even surpassed the transit frequencies achieved by bipolar transistors. This has clearly opened the door to full CMOS highly integrated solutions for wireless applications. Of course, in addition to high transit frequency, good noise performance and low-power consumption are required as well. Since the noise figure also decreases as the transit frequency is increased, it has also clearly taken advantage of the downscaling of the transistor length. At the same time, the supply voltage has had to be decreased progressively in order to limit high electric fields within the device and hence avoid the related high-field effects. The threshold voltage could unfortunately not be scaled in the same proportion without strongly increasing the drain leakage current, which is now seriously affecting the static power consumption of digital chips. This has resulted in a decrease of the overdrive voltage which in turn has moved the operating points of analog transistors more and more from strong inversion to moderate inversion and even into weak inversion. From this perspective, it is important to have a model that accurately predicts the behavior of the MOS transistor in all regions of operation, from weak to strong inversion, through moderate inversion, in a consistent way. This was the primary motivation for developing what today is known as the EKV model.

The purpose of this book is to assemble and explain in a coherent manner all the know-how and all the publications related to the particular MOS transistor modeling approach embodied by the EKV model. This model borrows from the work of a long line of researchers, starting in the early times of semiconductor physics. It has its roots in the search of early designers of very low-power and low-voltage integrated circuits for a description of the transistor behavior fulfilling their specific needs. This book focuses on this particular line of research, with no intention to present all alternative ways of modeling the transistor. Being written by analog circuit designers, it is clearly design-oriented with the purpose of describing the transistor as the basic component of integrated circuits, rather than the result of a sequence of physical processing steps. It gives to emphasis highlighting the properties of the device that can be used by designers to build new robust circuits, or to understand existing circuits and assess their robustness. The book is organized in three hierarchically structured parts. It firstly describes the basic behavior of the generic MOS transistor, then focuses on additional effects essentially due to scaling down the device dimensions, and finally discusses the transistors to be used in RF circuits.

Based on the charge in the channel, the EKV model describes in a continuous manner the static, dynamic and noise characteristics of the transistor down to very low current levels. The basic model requires a very limited set of parameters, all of them directly related to basic independent physical parameters. Intended for analog designers, it conserves the intrinsic source-drain symmetry of the transistor by using the substrate as the voltage reference and by introducing the concept of forward and reverse components of the drain current. This symmetrical approach makes it easier to understand the various modes of static operation of the device, and to describe them by a single uncomplicated equation. The charge-based approach lends itself naturally to a coherent description of the dynamic and noise behavior of the transistor.

The authors want to acknowledge the numerous persons who contributed directly and indirectly to this book. We are grateful to Dr François Krummenacher for his invaluable contribution to the EKV model and for his many inputs and suggestions that greatly helped us to write this book. We have benefited from the many discussions we had with Jean-Michel Sallese and Ananda Roy who helped us to clarify many fine points along the process of writing this book. We also would like to acknowledge the contribution of all the other members of the EKV development team, who each brought their own contribution to the EKV model: Matthias Bucher, Christophe Lallement, Alain-Serge Porret, Wladek Grabinski. Our gratitude also goes to Henri Oguey and Stephan Cserveny who pioneered the work for a continuous model and paved the way for the current EKV model.

Finally, we would like to give special thanks to our families – Dominique, Adrien, Mathilde and Simon Enz, and Monique, Nathalie and Didier Vittoz – for their support and understanding during this seemingly endless task.

CHRISTIAN C. ENZ, ERIC A. VITTOZ
St-Aubin-Sauges, Switzerland
Cernier, Switzerland

List of Symbols

Table 0.1 Symbols and their definitions

Symbol	Description	Reference
	Physical parameters	
q	Electron charge	(2.1)
k	Boltzmann's constant	(2.1)
T	Absolute temperature in degree Kelvin	(2.1)
T_n	Noise temperature in degree Kelvin	(9.141)
T_L	Lattice temperature in degree Kelvin	(9.143)
T_C	Carrier temperature in degree Kelvin	(9.142)
ϵ_0	Permittivity of free space	
ϵ_{si}	Permittivity of silicon	3.1
ϵ_{ox}	Permittivity of SiO_2	3.2
n_i	Intrinsic carrier concentration	3.1
μ	Mobility of current carriers	(4.1)
μ_0	Low-field surface mobility	(8.1)
μ_z	Mobility including the effect of the vertical field	(8.1)
μ_{eff}	Effective mobility including the effects of the vertical and longitudinal fields	(9.2)
	Process parameters	
n_p	Electron concentration (in P-type Si)	(3.1)
p_p	Hole concentration (in P-type Si)	(3.1)
N_b	Doping concentration of the substrate	3.1
N_g	Doping concentration of the polysilicon gate	8.4
N_{diff}	Doping concentration of the source and drain diffusions	4.6.1
Γ_b	Substrate modulation factor	(3.30)
Γ_g	Depletion factor in the polysilicon gate	(8.54)
v_{drift}	Drift velocity	9.1
v_{sat}	Saturated drift velocity	9.1

continued on next page

continued from previous page

Symbol	Description	Reference
	Geometry	
W	Channel width	Figure 2.1
L	Channel length	Figure 2.1
L_{SD}	Distance between the source and drain metallurgical junctions	Figure 4.12
L_{eff}	Effective channel length	Figure 9.26
L_{ov}	Gate overlap length	Figure 10.10
L_f	Length of a single finger	Figure 11.1
W_f	Width of a single finger	Figure 11.1
ΔL_S	Channel length reduction at the source	Figure 4.12
ΔL_D	Channel length reduction at the drain	Figure 4.12
t_{ox}	Oxide thickness	Figure 2.1
x	Distance from source along the channel	Figure 2.1
y	Distance across the channel	Figure 2.1
z	Distance in direction perpendicular to the surface into the bulk	Figure 2.1
	Voltages and potentials	
U_T	Thermodynamic voltage	(2.1)
Ψ	Electrostatic potential	2.2
Ψ_s	Surface potential $[\Psi_s \triangleq \Psi(z=0)]$	2.2
Ψ_{sS}	Surface potential at the source	
Ψ_{sD}	Surface potential at the drain	
Ψ_P	Pinch-off surface potential	(3.37)
Ψ_0	Approximation of Ψ_s in strong inversion at equilibrium ($V=0$)	(3.56)
Φ_{ms}	Difference between the work functions of the gate and the substrate	2.2
Φ_F	Fermi potential of silicon substrate	3.1
Φ_{Fn}	Quasi-Fermi potential of electrons	3.1
Φ_B	Potential barrier of source and drain junctions at equilibrium	(4.55)
V_{FB}	Flat-band voltage	(3.22)
V_{TB}	Threshold function	(3.33)
V_{T0}	Equilibrium threshold voltage	(3.58)
V	Channel voltage	Figure 2.1
V_G	DC gate-to-bulk voltage	Figure 2.1
V_S	DC source-to-bulk voltage	Figure 2.1
V_D	DC drain-to-source voltage	Figure 2.1
ΔV_G	Incremental gate-to-bulk voltage	5.1.1
ΔV_S	Incremental source-to-bulk voltage	5.1.1
ΔV_D	Incremental drain-to-bulk voltage	5.1.1
V_P	Pinch-off voltage	(3.46)
V_{sh}	Channel voltage shift	(3.44), (8.100)
$V_{DS\ sat}$	Drain-to-source saturation voltage	(4.12)
V_M	Channel-length modulation voltage (Early voltage)	(5.22)
V_{G0}	Extrapolated band gap voltage	Figure 7.1

continued on next page

continued from previous page

Symbol	Description	Reference
	Electric fields	
E_c	Critical longitudinal electric field	(9.1)
E_{ox}	Electric field in the oxide	Figure 2.2
E_x	Electric field along the longitudinal direction	3.1
E_y	Electric field along the lateral direction	3.1
E_z	Electric field along the vertical direction	3.1
E_{zs}	Electric field along the vertical direction at the surface $[E_{zs} \triangleq E_z(z = 0)]$	3.1
	Currents	
I_D	Static drain current flowing into the drain terminal	Figure 2.1
I_S	Static source current flowing into the source terminal	5.3
I_B	Static bulk current flowing into the bulk terminal	
I_G	Static gate current flowing into the gate terminal	(8.107)
I_{spec}	Specific current	(4.14)
I_F	Static forward current	(4.9)
I_R	Static reverse current	(4.9)
I_{D0}	Off drain current	(4.38)
ΔI_D	Incremental drain current	5.1.1
ΔI_S	Incremental source current	
ΔI_G	Incremental gate current	
ΔI_B	Incremental bulk current	
	Charges	
Q_i	Inversion mobile charge density	Figure 2.2
Q_{iS}	Inversion mobile charge density at the source	Figure 2.2
Q_{iD}	Inversion mobile charge density at the drain	Figure 2.2
Q_b	Depletion charge density	Figure 2.2
Q_g	Gate charge density	Figure 2.2
Q_{fc}	Fixed charge density	Figure 2.2
Q_{spec}	Specific charge density	(3.42)
Q_{si}	Semiconductor total charge density	2.2
Q_I	Total channel charge	(6.16), (6.19)
	Resistances, conductances, and transconductances	
R_S	Source series resistance	Figure 10.1
R_D	Drain series resistance	Figure 10.1
R_G	Gate series resistance	Figure 10.1
R_B	Bulk series resistance	Figure 10.1
R_{sde}	Source and drain extension resistance	Figure 10.2(b)

continued on next page

continued from previous page

Symbol	Description	Reference
R_{con}	Source and drain contact resistance	Figure 10.2(b)
R_{sal}	Source and drain salicide resistance	Figure 10.2(b)
R_{via}	Source and drain via resistance	Figure 10.2(b)
R_{DSB}	Source-to-drain substrate resistance	Figure 11.9(c)
R_{BS}	Source-to-bulk substrate resistance	Figure 11.9(c)
R_{BD}	Drain-to-bulk substrate resistance	Figure 11.9(c)
G_{ch}	Channel conductance	(9.116)
G_{ds}	Residual output conductance in saturation	(5.22)
G_{spec}	Specific conductance	(5.6)
G_m	Gate transconductance	(5.2c)
G_{ms}	Source transconductance	(5.2a)
G_{md}	Drain transconductance	(5.2b)
G_{mb}	Bulk transconductance	13.3.2

Capacitances and transcapacitances

Symbol	Description	Reference
C_{ox}	Oxide capacitance per unit area	3.2
C_{OX}	Total oxide capacitance	5.2
C_{si}	Silicon capacitance per unit area	(3.23)
C_g	Gate capacitance per unit area	(3.25)
C_d	Depletion capacitance per unit area	3.3
C_{OX}	Total oxide capacitance	(5.38)
C_{GSi}	Intrinsic gate-to-source capacitance	Figure 5.14
C_{GDi}	Intrinsic gate-to-drain capacitance	Figure 5.14
C_{GBi}	Intrinsic gate-to-bulk capacitance	Figure 5.14
C_{BSi}	Intrinsic bulk-to-source capacitance	Figure 5.14
C_{BDi}	Intrinsic bulk-to-drain capacitance	Figure 5.14
C_{GGi}	Total intrinsic gate capacitance	(11.4)
C_m	Intrinsic gate transcapacitance	(5.58)
C_{ms}	Intrinsic source transcapacitance	(5.56)
C_{md}	Intrinsic drain transcapacitance	(5.57)
C_{GSo}	Gate-to-source overlap capacitance	Figure 10.1
C_{GDo}	Gate-to-drain overlap capacitance	Figure 10.1
C_{GBo}	Gate-to-bulk overlap capacitance	Figure 10.1
C_{GGo}	Total gate overlap capacitance	Figure 10.1
C_{BSj}	Source-to-bulk junction capacitance	10.4
C_{BDj}	Drain-to-bulk junction capacitance	10.4
C_{GS}	Total gate-to-source capacitance	(12.1)
C_{GD}	Total gate-to-drain capacitance	(12.1)
C_{GB}	Total gate-to-bulk capacitance	(12.1)
C_{BS}	Total bulk-to-source capacitance	(12.1)
C_{BD}	Total bulk-to-drain capacitance	(12.1)
C_G	Total gate capacitance	(12.6)
C_g	Local gate capacitance per unit area	3.3

continued on next page

continued from previous page

Symbol	Description	Reference
Admittances and transadmittances		
Y_{GSi}	Intrinsic gate-to-source admittance	Figure 5.9
Y_{GDi}	Intrinsic gate-to-drain admittance	Figure 5.9
Y_{GBi}	Intrinsic gate-to-bulk admittance	Figure 5.9
Y_{BSi}	Intrinsic bulk-to-source admittance	Figure 5.9
Y_{BDi}	Intrinsic bulk-to-drain admittance	Figure 5.9
Y_m	Intrinsic gate transadmittance	(5.36)
Y_{ms}	Intrinsic source transadmittance	(5.36)
Y_{md}	Intrinsic drain transadmittance	(5.36)
Y_{sub}	Substrate admittance	Figure 12.4
Frequency and time constants		
ω_t	Transit frequency	(11.4)
τ_{qs}	Intrinsic channel time constant	(5.32)
ω_{qs}	Intrinsic channel transit frequency (also limit between quasi-static and non-quasi static operation)	(5.32)
ω_{max}	Extrapolated maximum frequency of oscillation	(11.18)
ω_{spec}	Specific (or critical) frequency	(5.33)
τ_{spec}	Specific time constant	(5.33)
Noise		
$S_{\Delta I_{nD}^2}$	Thermal noise power spectral density at the drain	(6.4), (6.14)
$S_{\Delta I_{nS}^2}$	Thermal noise power spectral density at the source	(13.42)
$S_{\Delta I_{nG}^2}$	Thermal noise power spectral density at the gate (induced gate noise power spectral density)	(13.42)
$S_{\Delta I_{nB}^2}$	Thermal noise power spectral density at the bulk	(13.42)
$S_{\Delta I_{nG} \Delta I_{nD}^*}$	Thermal noise gate-drain cross-power spectral density	(13.42)
G_{nD}	Drain thermal noise conductance	(6.15)
G_{nG}	Gate thermal noise conductance (induced gate noise thermal conductance)	(13.49)
δ_{nD}	Thermal noise parameter at the drain	(6.26)
δ_{nG}	Thermal noise parameter at the gate	(13.49)
γ_{nD}	Thermal noise excess factor at the drain	(6.30)
γ_{nG}	Thermal noise excess factor at the gate	(13.49)
ρ_{GD}	Gate-drain thermal noise correlation factor	(13.71)
S_v	Input-referred thermal noise voltage power spectral density	(13.13)
S_i	Input-referred thermal noise current power spectral density	(13.13)
R_v	Input-referred thermal noise voltage resistance	(13.13)
G_i	Input-referred thermal noise current conductance	(13.13)
G_{iu}	Uncorrelated part of G_i	(13.14)

continued on next page

continued from previous page

Symbol	Description	Reference
G_{ic}	Correlated part of G_i	(13.14)
Y_c	Noise correlation admittance	(13.8)
G_c	Noise correlation conductance	(13.16)
B_c	Noise correlation susceptance	(13.16)
F	Noise factor	(13.17), (13.21), (13.26)
NF	Noise figure	(13.17)
F_{min}	Minimum noise factor	(13.25)
NF_{min}	Minimum noise figure	(13.25)
Y_{opt}	Optimum source admittance for $F = F_{min}$	(13.24)
G_{opt}	Optimum source conductance	(13.24)
B_{opt}	Optimum source susceptance	(13.24)
	Other	
ρ	Charge concentration	(3.1)
L_D	Extrinsic Debye length	(3.15)
L_{c0}	Characteristic length for DIBL	(9.100)
t_d	Thickness of the depletion layer	(3.26)
n	Slope factor	(3.34)
n_w	Slope factor evaluated at pinch-off	(3.68)
n_0	Slope factor evaluated at $V = 0$	(3.73)
n_q	Charge slope factor	(8.60)
n_v	Voltage slope factor	(8.60)
β	Transconductance factor or transfer parameter	(4.7)
D_S	Source-to-bulk diode	Figure 10.1
D_D	Drain-to-bulk diode	Figure 10.1
$A_{v\,max}$	Maximum voltage gain in common gate	(5.24), (9.115)
ΔP	Mismatch of parameter P	(7.52)
A_P	Area proportionality constant of parameter P	(7.52)
θ	Parameter of field-dependent mobility	(8.5)
z_c	Characteristic depth	Figure 8.7
v_{drift}	Drift velocity of carriers	Figure 9.1
v_{sat}	Saturation value of v_{drift}	Figure 9.1
λ_c	Velocity saturation parameter	(9.19)

Table 0.2 Normalization factor definition

Symbol	Description	Reference
L	Transistor length for normalizing distance along the x-axis	2.1
$U_T \triangleq \frac{kT}{q}$	Thermodynamic voltage for normalizing voltages and potentials	2.1
$I_{spec} \triangleq 2n\beta U_T^2$	Specific current for normalizing currents	(4.14)
$Q_{spec} \triangleq -2nC_{ox}U_T$	Specific charge density for normalizing charge densities	3.6.1
$C_{OX} \triangleq WLC_{ox}$	Total oxide capacitance for normalizing capacitances	5.2
$G_{spec} \triangleq \frac{I_{spec}}{U_T}$	Specific admittance for normalizing admittances	(5.6)
$\omega_{spec} \triangleq \frac{\mu_n U_T}{L^2}$	Specific angular frequency for normalizing angular frequency	5.2

Table 0.3 Normalized symbols

Symbol	Description	Reference
$\xi \triangleq x/L$	Normalized position along the x-axis	(4.21)
ζ_c	Normalized characteristic depth	(8.24)
ν	Doping ratio	(8.32)
$\lambda_0 \triangleq L/L_{c0}$	Channel length normalized to characteristic length	(9.100)
$v_x \triangleq V_X/U_T$	Normalized voltage	(3.43)
$\gamma_j \triangleq (\Gamma_j/U_T)^2$	Normalized modulation factor	(3.43)
$\phi_f \triangleq \Phi_F/U_T$	Normalized Fermi potential of silicon substrate	3.1
$\psi_p \triangleq \Psi_P/U_T$	Normalized pinch-off surface potential	(3.37)
$\psi_0 \triangleq \Psi_0/U_T$	Normalized approximation of Ψ_s in strong inversion	(3.66)
$i_x \triangleq I_X/I_{spec}$	Normalized drain current	(4.15)
IC	Inversion coefficient or factor	(4.26)
$g_x \triangleq G_x/G_{spec}$	Normalized conductance or transconductance	
$q_x \triangleq Q_x/Q_{spec}$	Normalized charge density	(3.41)
$q_s \triangleq Q_{iS}/Q_{spec}$	Normalized inversion charge density at the source	(5.7a)
$q_d \triangleq Q_{iD}/Q_{spec}$	Normalized inversion charge density at the drain	(5.7b)
$q_X \triangleq Q_X/Q_{spec}$	Normalized total charge	
$\Omega \triangleq \omega/\omega_{spec}$	Normalized frequency	13.2.2
$\Omega_{qs} \triangleq \omega_{qs}/\omega_{spec}$	Normalized QS frequency	(5.32)
$c_j \triangleq C_j/C_{ox}$	Normalized capacitance per unit area	(5.50)
$c_J \triangleq C_J/C_{OX}$	Normalized total capacitance	(5.50)

1 Introduction

This chapter explains the basic motivations for developing MOS transistor models that can be used for the design of complementary MOS (CMOS) integrated circuits. It then gives a short history of the EKV MOS transistor model starting from the early development, motivated by the design of micropower circuits for watch applications, to the most recent developments. Finally, the structure of the book is highlighted in order to help the reader organizing his reading.

1.1 THE IMPORTANCE OF DEVICE MODELING FOR IC DESIGN

Modern large-scale integrated circuits are essentially composed of MOS transistors and their interconnections. Therefore, the design of such circuits requires some kind of a model for the transistors.

For noncritical digital circuits, this model may in principle be very simple. Indeed, modeling each transistor as an on–off switch would be sufficient to design purely logic circuits. However, as soon as there are critical races among transitions, the model must be extended to describe the dynamic behavior of the device, in order to obtain the rise and fall time of these transitions. This dynamic behavior is also needed when the frequency of operation approaches its maximum limit. With the reduction of supply voltage, more details must be introduced, such as the residual current of blocked transistors, the importance of which is increased.

Analog circuits contain usually a smaller number of transistors, but they are even more dependent on the exact behavior of each transistor. The design of high-performance analog circuits therefore requires a very detailed model of the transistor. This model must include a precise description of the voltage–current relationships, including the effect of the source that is often not grounded, and of the dynamic behavior of the device. Its behavior with respect to noise and to temperature variations must also be accounted for.

A transistor model intended for circuit design should serve two essential purposes:

It should first provide a good understanding of the various properties of the device to facilitate the synthesis of optimum circuit architectures. Indeed, in order to build robust circuits, the physical properties of the transistor must be exploited in a way that is minimally dependent on temperature and process variations. For this purpose, the model should be explicit. It should "speak to the mind," using no complicated or chained equations. Clarity should supersede precision and can be enhanced by means of graphical representations. This important aspect of a model is often underestimated and overlooked. It will be emphasized in this book, in particular in Part I, which essentially describes what can be called the core of the model.

Second, the model should be adapted to numerical simulations on a computer, embedded in a circuit simulator. For this purpose, precision supersedes clarity, and second-order effects must be accounted for. This can be obtained by predistorting variables, by chaining equations and/or by providing additional layers around a core model. The model does not need to be fully explicit, but it should be compact: it should use sufficiently simple expressions with minimum need for numerical iterations, in order to limit the computation time.

A transistor model should include a minimum number of process-dependent device parameters. This is to facilitate the very heavy task of extracting and following-up the value of these parameters, with their statistical distribution and temperature dependency.

Now, the correlation between these parameters (with process and temperature variations) must be known, in order to avoid designing circuits for irrelevant worst cases. For this reason, the device parameters should be explicitly based on independent and measurable process parameters. This is essential to be able to ascertain their amount of correlations while avoiding the almost impossible task of measuring all these correlations. It also makes the model predictive, allowing to foresee the characteristics of the transistor and hence the performance of the resulting circuits even before measuring the device.

The EKV model described in this book is believed to meet all the above expectations. It serves the two main purposes in a coherent manner. Its core requires just a few parameters to describe all the basic properties of the long-channel intrinsic device in an explicit manner. Layers are added to this core to account for short-channel and secondary effects.

1.2 A SHORT HISTORY OF THE EKV MOS TRANSISTOR MODEL

The model presented in this book results from a series of direct and indirect contributions along several decades. Its origins can be traced back to the early developments of electronic watches at CEH (French acronym for Watchmakers Electronic Center) in Switzerland [2].

The total power consumption had to be extremely low, less than 1μW, to ensure a few years of life to the single button-size cell battery. After the very first versions based on bipolar transistors [3], the CMOS technology was soon identified as the best approach to implement the digital electronic circuitry needed in a watch using a crystal resonator as the time reference. Supply voltage had to be very low, compatible with the 1.3 V delivered by the cell, so the development of low-threshold CMOS was a major challenge in the late 1960's [4].

The digital circuitry was essentially an asynchronous chain of divide-by-two stages. The main design problem was to minimize the number of node transitions in order to minimize the dynamic power. Another one was the elimination of logic hazards to improve the robustness against large local variations of the small gate voltage overhead, and this led to the first single clock circuits [5–7]. For these digital circuits, MOS transistors could be considered just as switches and hence no special model was required.

The problem was very different for the few analog subcircuits. Most important was the circuitry needed to sustain the oscillation of the quartz crystal resonator (the quartz oscillator). Each transistor had to be biased at a drain current much below $1\mu A$. Early measurements carried out in 1967 showed that the transistor behaved in a very strange manner at these very low current levels. Indeed, the well-known square-law transfer characteristics were replaced by an exponential over more than 5 order of magnitude of the drain current, very similar to bipolar transistors. This is how *weak inversion* popped out to the attention of micropower circuit designers in the late 1960s.

At that time, no transistor model was available for weak inversion, but they started coming out in subsequent years, mainly to account for what appears in digital circuits as a leakage current of blocked transistors. In 1972, M. B. Barron published a model for the grounded source device showing the exponential dependencies on drain voltage and on surface potential, with a rather complex expression relating the surface potential to the gate voltage [8]. The same year, R. M. Swanson and J. D. Meindl [9] showed that this relation could be accounted for by means of an almost constant factor, which became the *slope factor n* of our model. The following year, R. R. Troutman and S. N. Chakravarti [10] treated the case of nonzero source voltage. Then T. Mashuhara *et al.* [11] showed that the current depends on a difference of exponential functions of source and drain voltages. In the mean time, micropower analog circuit blocks were developed at CEH. They were first published in 1976 [12, 13], together with a model applicable for weak inversion circuit design, which was based on the previously mentioned work. This model already included two important features of the EKV model: *reference to the (local) substrate* (and not to the source) for all voltages and full *source–drain symmetry*. The related small-signal model including noise was also presented [14].

A symmetrical model of the MOS transistor in strong inversion was first published by P. Jespers in 1977 [15, 16]. Based on an idea of O. Memelink, this graphical model uses the approximately linear relationship between the local mobile charge density and the local "non-equilibrium" voltage in the channel. This charge-based approach has been adopted and generalized to all levels of current in the EKV model.

Another ingredient of EKV is the representation of the drain current as the difference between a *forward* and a *reverse* component. This idea was first introduced in 1979 by J.-D. Châtelain [17], by similarity with the Ebers–Moll model of bipolar transistors [18]. However, his definition of these two components was different from that adopted later, and was not applicable to weak inversion.

Even in micropower analog circuits, not all transistors should be biased in weak inversion. There was therefore a need for a good continuous model from weak to strong inversion. Such a model was developed at CEH by H. Oguey and S. Cserveny, and was first published in French in 1982 [19]. The only publication in English was at a Summer Course given in 1983 [20].

This model embodied most of the basic features that were retained later. It introduced a function of the gate voltage called control voltage, later renamed *pinch-off voltage V_P*. A single function of this control voltage and of either the source voltage or the drain voltage defined two components of the drain current (which became the forward and reverse components). This function was continuous from weak to strong inversion, using a mathematical interpolation to best fit moderate inversion.

In the mid-1980s, the model of Oguey and Cserveny was simplified by the second author for his undergraduate teaching at EPFL (Swiss federal Institute of Technology, Lausanne, Switzerland), and most further developments were carried out there. They started with the Ph.D. Thesis of the first author [21], in collaboration with F. Krummenacher. The model was formulated more explicitly. Noise and dynamic behavior were introduced by exploiting the fundamental source–drain symmetry. The status of the model was presented at various Summer Courses [22–24] and a full paper was finally published in 1995 [25]. This publication gave its name to the model, but many important extensions were added later.

Probably the most important extension was the replacement of the current and transconductance interpolation functions between weak and strong inversion presented in [25] by a more physical based one, derived from an explicit linearization of the inversion charge versus the surface potential. The incremental linear relationship between inversion charge and surface potential was first considered by M. Bagheri and C. Turchetti [26], but the linearization of the inversion charge versus surface potential was originally proposed in 1987 by M. Maher and C. Mead [27, 28]. Several years later, different groups looked at this problem. B. Iñiguez and E. G. Moreno [29, 30] derived an approximate explicit relation between inversion charges and surface potential which included a fitting parameter. While their first linearization was done at the source [29], they later obtained a substrate referenced model based on the original EKV MOSFET model approach [25], which also included some short-channel effects. A similar approach was also proposed by Cunha *et al.* [31–34] who obtained an interpolated expression of the charges versus the potentials that used the basic EKV model definitions[1] [25] and was closely inspired from our approach. We also adopted the inversion charge linearization approach, since it offers physical expressions for both the transconductance-to-current ratio and the current that are valid from weak to strong inversion [35–38]. This gave rise to the *charge-based EKV model* which is discussed in this book. The inversion charge linearization principle was rediscovered once more in 2001 by H. K. Gummel and K. Singhal [39, 40]. Finally, a formal detailed analysis of the inversion charge linearization process and a rigorous derivation of the EKV model was finally published by J.-M. Sallese *et al.* in [41].

Note that this approach actually provides voltages versus currents expressions that cannot be explicitly inverted. It can nevertheless be easily inverted by using a straightforward Newton-Raphson technique or by an appropriate approximation. Both these techniques have been used in the final model implementation.

The basic long-channel charge-based EKV model was further developed by the EKV team to include the following additional effects:

Nonuniform doping: Nonuniform doping in the vertical direction was proposed by C. Lallement *et al.* in [42, 43].

Non-quasi-static model: A small-signal charge-based non-quasi-static model was presented by J.-M. Sallese and A.-S. Porret in [38, 44].

Polysilicon depletion and quantum effects: Polysilicon depletion and quantum effects were also added [45–47].

RF modeling: The EKV model was extended by the first author to also cover high-frequency operation for the design of RF CMOS integrated circuits [48–52].

Thermal noise: An accurate thermal noise model accounting for short-channel effects was developed by A. S. Roy and C. C. Enz [53–55].

---

[1] Unfortunately, Cunha *et al.* did not use the same definition of the specific current we have been using. Their specific current is actually four times smaller.

Extrinsic components: An accurate model of the parasitic capacitances was developed by F. Prégaldiny *et al.* [56].

EKV compact model: A model of the MOS transistor would be almost useless if it could not be used by circuit designers with a circuit simulator. To this purpose, the model has to be carefully implemented in the simulator so that it can run efficiently avoiding any convergence problems. The early EKV model (version 2.7) was implemented by M. Bucher as a compact model in many circuit simulators [37]. All the more recent developments were implemented in the version 3.0 of the EKV MOS transistor compact model [57, 58].

Parameter extraction: A compact model cannot be used without an efficient parameter extraction methodology. The EKV model uses an original parameter extraction methodology presented in [59–62]. (Reference [61] can be found on line at the EKV Web site [63].)

More recently, the research of the EKV team is more oriented toward the modeling of multigate MOS devices and more particularly on double-gate devices [64, 65].

Further parts of the model were derived by members of the team of researchers and Ph.D. students that developed its implementation as a CAD tool at EPFL [63].

1.3 THE BOOK STRUCTURE

This book is organized in three parts, which are briefly described below:

Part I describes the basic long-channel charge-based MOS transistor model. It is the core of the model around which all the other parts are built in a hierarchic manner following the basic structure of the EKV MOS transistor model. This part is self-contained and the reader can stop after it while still having a strong background in all the fundamental aspects of the EKV MOS transistor model. It includes all the most important aspects such as basic large-signal static model, small-signal dynamic model, noise model, and a discussion of temperature effects and matching properties. The other parts complete the basic model by adding more detailed descriptions of advanced aspects.

Part II presents more advanced aspects which are of utmost importance for understanding the operation of deep-submicron devices. It starts with the modeling of several nonideal effects that already affects long-channel devices before concentrating on short-channel effects. The model is then extended to also include the extrinsic part of the device.

Part III discusses additional aspects which become important when increasing the operating frequency. It presents a complete MOS transistor model, built on top of the two first parts, which is required for designing RF CMOS integrated circuits.

Part I

The Basic Long-Channel Intrinsic Charge-Based Model

The first part models the intrinsic part of the most basic MOS transistor. The channel is assumed to be sufficiently long to avoid all short-channel effects. The doping concentration in the substrate and the carrier mobility are constant. The gate is a perfect equipotential conductor, and the gate oxide is thick enough to prevent quantum effects and tunneling current. Some basic definitions are introduced in Chapter 2 in order to preserve the symmetry of the device. Chapter 3 models the density of mobile charge as a function of the gate voltage and of the local channel voltage. This function is used in Chapter 4 to obtain the drain current in function of the source and drain voltages, and introduces the unusual concept of forward and reverse current components. Chapter 5 then establishes the corresponding small-signal DC and AC models. Chapter 6 is dedicated to modeling the noise, whereas Chapter 7 investigates the effect of temperature and the problem of mismatch between devices.

Charge-Based MOS Transistor Modeling: The EKV Model for Low-Power and RF IC Design   C. Enz and E. Vittoz
© 2006 John Wiley & Sons, Ltd.

2 Definitions

This short introductory chapter starts by describing the generic structure of the N-channel MOS transistor that will be analyzed in Part I. It defines the essential geometrical dimensions and voltages, including the "channel voltage" V that will play an important role in the model. Section 2.2 introduces the definition of additional important variables, in particular the density of mobile inverted charge Q_i that underlies the whole charge-based modeling approach. Symbols for the four-terminal N- and P-channel transistors are proposed in Section 2.3, together with sign conventions that will render all results derived for the N-channel transistor applicable to the P-channel device.

2.1 THE N-CHANNEL TRANSISTOR STRUCTURE

The schematic cross section of a generic N-channel MOS transistor is shown in Figure 2.1. The source S and drain D are highly doped N-type islands (N+) diffused in a P-type local silicon substrate (or bulk) B.

In between, the active part of the transistor of length L and width W is controlled by the gate electrode G, separated by a dielectric layer called gate oxide, since it is normally made of silicon dioxide. The P+ diffusion is needed to ensure good ohmic contact with the lightly doped P-type local substrate. The position along the channel is defined by x, whereas the distance from the silicon surface is given by z. The y-axis is perpendicular to the plane of the cross section. We shall assume for the time being that the net doping concentration N_b of the local substrate and the oxide thickness t_{ox} are both constant along the channel. Hence, this four-terminal device has a symmetrical structure with respect to source and drain. In order to keep this symmetry in the model, the source voltage V_S, the drain voltage V_D, and the gate voltage V_G are all defined with respect to the local substrate. By definition, V_S and V_D are positive when they block the corresponding junction.

As shown in Figure 2.1, the active region of the transistor located between source and drain is limited to the gate-to-surface capacitor plus a thin layer of silicon in which the potential and the charge distribution are modified by the effect of the gate. It is called the *intrinsic part* of the transistor. All the rest is the *extrinsic part*. It includes the source and drain diodes, series access

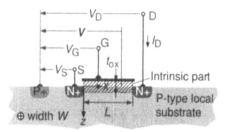

Figure 2.1 Schematic cross section of a MOS transistor

resistors or inductors to the four terminals, and all external parasitic capacitors, in particular those of the D and S junctions and the direct overlap capacitors from gate electrode to source and drain islands. This extrinsic part of the device will be discussed in Chapter 10.

Application of a voltage across the source-to-substrate and/or the drain-to-substrate junctions forces electrons and holes out of equilibrium, splitting their respective quasi-Fermi potential by V_S at the source end of the channel and V_D at the drain end. This splitting propagates along the channel, and can be characterized by a *channel voltage* V that varies monotonically from V_S at $x = 0$ (source end) to V_D at $x = L$ (drain end).

Now, for an N-channel device in normal operation (potential increased at the surface by the voltage applied to the gate), the quasi-Fermi potential of holes can be assumed to be constant throughout the structure [1,66,67]. Thus V is (within a constant) the quasi-Fermi potential of electrons in the channel.

Another important voltage is the thermodynamic voltage

$$U_T \triangleq kT/q, \tag{2.1}$$

where k is the Bolzmann constant and q is the elementary charge. Proportional to the absolute temperature T, it is a measure of the thermal energy of electrons. Since it appears ubiquitously in MOS modeling equations, it is a more natural unit of voltage for devices and circuits than the standard unit of 1 V. Its value is 25.8 mV at 300 K or 27 °C.

2.2 DEFINITION OF CHARGES, CURRENT, POTENTIAL, AND ELECTRIC FIELDS

For zero electric field at the silicon surface, the source to drain structure of Figure 2.1 corresponds to two back-to-back diodes connected in series; thus, no current other than the junction leakage current can flow as long as V_S and V_D are positive. The situation remains qualitatively the same when more holes are attracted at the surface by applying a negative gate voltage V_G.

On the contrary, if a positive voltage is applied to the gate, the holes are repelled from the surface, leaving the negatively charged P-doping atoms. As shown schematically in Figure 2.2, this corresponds to a negative charge of density Q_b per unit area. This charge is fixed and therefore cannot carry any current.

By further increasing V_G, negative electrons are attracted to the surface thereby forming an N-type channel. It is this negative mobile inversion charge, of density Q_i per unit area, which will carry the drain-to-source current by a combination of drift and diffusion mechanisms of

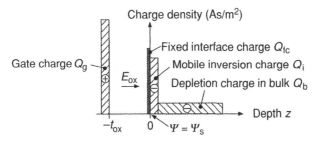

Figure 2.2 Schematic representation of various local charge densities

electrons. For the N-channel device, this current I_D will be defined positive if it enters the drain terminal.

The *charge-based model* presented in this book will first calculate the dependency of the density Q_i of induced mobile charge on the voltages applied to the transistor. Then, it will *rely on* Q_i, and on its particular values Q_{iS} and Q_{iD} at the source and drain ends of the channel, to calculate the drain current and *to model all aspects* of the device behavior.

The total net charge induced underneath the surface of silicon per unit area of channel is given by

$$Q_{si} \overset{\triangle}{=} Q_b + Q_i. \qquad (2.2)$$

As depicted in Figure 2.2, an additional component of charge Q_{fc} is present at the silicon-oxide interface. This is a fixed charge that includes the effect of charges trapped inside the oxide and weighted by their relative distance to the interface. This charge will be assumed to be independent of the gate voltage, although it might change very slowly in time at very high values of gate voltage. Additional voltage-dependent charges due to fast surface state will not be considered, since they are negligible in modern processes.

The 0-reference of electrostatic potential ψ is that of the bulk of silicon, at a distance from the surface where it is not affected by the gate voltage. At the silicon surface ($z = 0$), Ψ takes the particular value Ψ_s called the *surface potential*. The electric field E_{ox} in the oxide depends on $V_G - \Psi_s$, but is modified by Φ_{ms}, the *difference between the extraction potentials* of gate and channel materials. It corresponds to the barrier of potential that would be created at their interface if the oxide thickness t_{ox} would be zero. The electric field in the oxide is therefore given by

$$E_{ox} = \frac{V_G - \Phi_{ms} - \Psi_s}{t_{ox}}. \qquad (2.3)$$

2.3 TRANSISTOR SYMBOL AND P-CHANNEL TRANSISTOR

In order to reflect the symmetrical structure of the MOS transistor, the symbols to be used in circuit schematics should also be symmetrical, as shown in Figure 2.3. Figure 2.3(a) shows the symbol of an N-channel transistor, with the definitions of voltages and current already introduced in Figure 2.1. The arrow on the bulk (B) terminal symbolizes the bulk-to-channel "junction," whereas the source S and drain D terminals are symmetrical.

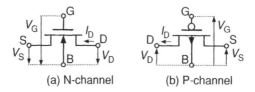

(a) N-channel   (b) P-channel

Figure 2.3   Symbols and definitions for N-channel and P-channel MOS transistors

The generic structure of a P-channel transistor is similar to that of the N-channel, but the doping types are opposite, with P+ source and drain diffusions inside an N-type local substrate (which is usually an N-well in modern processes). Hence a negative voltage must be applied to the gate to obtain a positive inverted charge of holes Q_i, after creating a positive depletion charge Q_b.

In spite of this sign difference, all equations that will be derived for modeling the N-channel transistor will be applicable to the P-channel device, provided the *definitions of positive voltages and currents are inverted*, as shown in Figure 2.3(b). The arrow in the bulk connection is inverted, which might be sufficient to distinguish it from the N-channel. However, since very often the bulk connection is the same for all transistors of the same type and is therefore not represented, a small circle is added at the gate of the P-channel device.

It should be pointed out that, except for the gate electrode, the generic structure illustrated in Figure 2.1 is also that of an NPN lateral bipolar transistor. Indeed, this parasitic transistor will be activated as soon as one of the two junctions is sufficiently forward biased ($-V_S$ and/or $-V_D$ larger than a few hundred millivolts). This device may be exploited as a true bipolar transistor, provided it is in a separate well (which becomes the base of the transistor) and the MOS current is canceled by a negative gate voltage [68]. This special mode of operation will not be further discussed in this book.

3 The Basic Charge Model

This chapter is dedicated to the calculation and the modeling of the density of induced mobile charge Q_i as a function of the various voltages applied to the transistor. Section 3.1 is a repetition of the classical one-dimensional analysis of the total charge density Q_{si} induced at the surface of a long transistor channel by a nonzero surface potential. It already shows the fundamental difference between weak and strong inversion. This difference is further highlighted by the dependency of the surface potential on the gate voltage, which is derived in Section 3.2. Section 3.3 takes advantage of the results obtained so far to calculate the variation of the local gate capacitance per unit area as a function of the gate voltage and the local channel voltage. Section 3.4 introduces and justifies the charge sheet approximation, which will be used throughout the rest of the book. Using this approximation, Q_i is then obtained in Section 3.5 by the difference between the gate voltage and a threshold function V_{TB} of the surface potential Ψ_s, and it cancels at a value called pinch-off surface potential Ψ_P. The slope n of $V_{TB}(\Psi_s)$ will become one of the few basic parameters of the model. In Section 3.6, an important simplification of the model is introduced by exploiting the fact that this slope n may be considered constant, which corresponds to a linearization of the charge versus potential relationship. Based on this linearization, an explicit expression is obtained that relates Q_i to the channel voltage V, within a constant called pinch-off (channel) voltage V_P. This expression is normalized and depends only on process parameters through a specific charge Q_{spec}. By using an approximation of this expression in strong inversion, a threshold voltage (at equilibrium) V_{T0} is defined, by which V_P can then be directly related to the gate voltage.

3.1 POISSON'S EQUATION AND GRADUAL CHANNEL APPROXIMATION

In the P-type substrate of an N-channel MOS transistor, the total charge concentration ρ is the net effect of the concentrations N_b of doping atoms, of holes p_p and n_p of electrons:

$$\rho = q(p_p - n_p - N_b). \tag{3.1}$$

Charge-Based MOS Transistor Modeling: The EKV Model for Low-Power and RF IC Design   C. Enz and E. Vittoz
© 2006 John Wiley & Sons, Ltd.

Far from the surface, the semiconductor is neutral, and $\rho = 0$. Closer to the surface, the spatial variations of potential Ψ due to the electric field produced by the gate result in a nonzero charge concentration according to the three-dimensional Poisson's equation

$$\frac{\partial^2 \Psi}{\partial x^2} + \frac{\partial^2 \Psi}{\partial y^2} + \frac{\partial^2 \Psi}{\partial z^2} = -\frac{\rho}{\epsilon_{si}}, \tag{3.2}$$

where ϵ_{si} is the dielectric constant of silicon.

The channel is long and wide compared to the oxide thickness t_{ox}, allowing the *gradual channel approximation*, stating that the electric field variation in the z-direction (perpendicular to the surface) is much larger than that in the x- and y-directions. Therefore, the second derivative of potential Ψ in the directions parallel to the surface can be neglected in (3.2), and this three-dimensional equation can be reduced to the one-dimensional equation in z:

$$\frac{d^2 \Psi}{dz^2} = \frac{q}{\epsilon_{si}} (n_p - p_p + N_b). \tag{3.3}$$

For an N-channel transistor, the quasi-Fermi potential of holes can be assumed to be constant for $\Psi \geq 0$ [1, 66, 67]. The hole concentration can therefore be expressed as

$$p_p = n_i \exp \frac{\Phi_F - \Psi}{U_T}, \tag{3.4}$$

where n_i is the *intrinsic carrier concentration* and Φ_F is the *Fermi potential* of the silicon substrate. At $T = 300$ K, $n_i = 1.45 \times 10^{10}$ cm^{-3}.

Application of a source voltage V_S and/or a drain voltage V_D brings the electrons in the channel out of equilibrium, which is characterized by a *quasi-Fermi potential* Φ_{Fn} different from Φ_F. As explained in the introduction, this difference is called channel voltage V, and $\Phi_{Fn} = \Phi_F + V$. The concentration of electrons can thus be expressed as

$$n_p = n_i \exp \frac{\Psi - \Phi_{Fn}}{U_T} = n_i \exp \frac{\Psi - \Phi_F - V}{U_T}. \tag{3.5}$$

The doping concentration N_b is assumed to be constant in the channel region. Far from the surface, the effect of V_S, V_D, and E_{ox} vanishes; thus $V = \Psi = 0$, and the silicon is neutral with $\rho = 0$. Combining equations (3.1), (3.4), and (3.5) then yields

$$N_b = n_i \left(\exp \frac{\Phi_F}{U_T} - \exp \frac{-\Phi_F}{U_T} \right) \cong n_i \exp \frac{\Phi_F}{U_T} \tag{3.6}$$

since $\Phi_F \gg U_T$, resulting in

$$n_i = N_b \exp \frac{-\Phi_F}{U_T}. \tag{3.7}$$

The expression relating the Fermi potential Φ_F to the doping concentration N_b is then given by

$$\Phi_F = U_T \ln\frac{N_b}{n_i}. \tag{3.8}$$

Introducing (3.4), (3.5), and (3.7) in (3.3) yields

$$\frac{d^2\Psi}{dz^2} = \frac{qN_b}{\epsilon_{si}} \left(\underbrace{\exp\frac{\Psi - 2\Phi_F - V}{U_T}}_{\text{electrons}} - \underbrace{\exp\frac{-\Psi}{U_T}}_{\text{holes}} \underbrace{+1}_{\text{fixed charge}} \right)$$

$$\triangleq \frac{qN_b}{\epsilon_{si}} G(\Psi, 2\Phi_F + V). \tag{3.9}$$

The first term in the parentheses is the contribution of electrons, the second that of holes, and the third that of the fixed depletion charge.

Now, the vertical field is given by

$$E_z = -\frac{d\Psi}{dz}; \tag{3.10}$$

hence,

$$\frac{d^2\Psi}{dz^2} = -\frac{dE_z}{dz} = -\frac{dE_z}{d\Psi}\frac{d\Psi}{dz} = E_z\frac{dE_z}{d\Psi} \tag{3.11}$$

and Poisson's equation (3.9) becomes

$$E_z\, dE_z = \frac{qN_b}{\epsilon_{si}} G(\Psi,\, 2\Phi_F + V)\, d\Psi. \tag{3.12}$$

Both sides of this equation can now be integrated from far below the surface, where $E_z = 0$ and $\Psi = 0$, to closer to the surface where they become nonzero:

$$\int_0^{E_z} E_z\, dE_z = \frac{E_z^2}{2} = \frac{qN_b}{\epsilon_{si}} \int_0^{\Psi} G(\Psi,\, 2\Phi_F + V)\, d\Psi. \tag{3.13}$$

This finally yields the vertical field E_z as a function of Ψ and V:

$$E_z = \text{sgn}(\Psi) \frac{U_T}{L_D} F(\Psi,\, 2\Phi_F + V), \tag{3.14}$$

where L_D is a combination of constants called the extrinsic Debye length

$$L_D \triangleq \sqrt{\frac{\epsilon_{si}U_T}{2qN_b}} \tag{3.15}$$

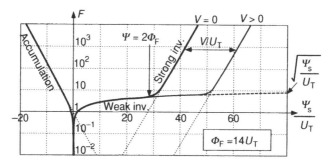

Figure 3.1   Function F relating E_z to Ψ, and E_{zs} and Q_{si} to Ψ_s

and

$$F(\Psi, \, 2\Phi_F + V) \triangleq$$

$$\sqrt{\left(\exp \frac{\Psi}{U_T} - 1\right) \underbrace{\exp \frac{-(2\Phi_F + V)}{U_T}}_{\text{contribution of electrons}} + \underbrace{\left(\exp \frac{-\Psi}{U_T} - 1\right)}_{\text{of holes}} + \underbrace{\frac{\Psi}{U_T}}_{\text{of fixed charge}}} \,. \quad (3.16)$$

This function is represented in Figure 3.1 for $\Phi_F = 14U_T$, which, according to (3.8), corresponds at ambient temperature to a doping concentration N_b of 1.7×10^{16} cm^{-3}. At the surface of silicon, the potential Ψ takes the value Ψ_s, and the vertical field has the value E_{zs} given by $F(\Psi_s, \, 2\Phi_F + V)$. Knowing the surface field, the local silicon charge density Q_{si} can be obtained by applying Gauss' law to a short section of channel, as explained by Figure 3.2.

A parallelepiped volume of channel starts at the surface of silicon where $E_z = E_{zs}$ and ends deep in the substrate where $E_z = 0$; according to the gradual channel approximation, the variation of horizontal field along a short section of channel is negligible; thus, $E_{x+\Delta x} = E_x$. Finally, the lateral field $E_y = 0$ so that the only side contributing to the electrical flux leaving the volume is the surface. According to Gauss' law, this flux is equal to the net charge inside

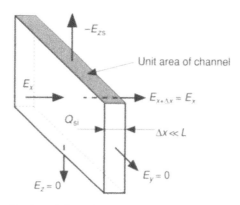

Figure 3.2   Application of Gauss' law to calculate Q_{si} from the surface field E_{zs}

the volume, which is equal to Q_{si} for a unit area of channel. Thus, for $\Psi_s > 0$,

$$Q_{si} = -\epsilon_{si} E_{zs} = -\frac{\epsilon_{si} U_T}{L_D} F(\Psi_s, 2\Phi_F + V). \tag{3.17}$$

For a given value of $2\Phi_F + V$, the function F plotted in Figure 3.1 thus represents $E_z(\Psi)$, $E_{zs}(\Psi_s)$, and $Q_{si}(\Psi_s)$. It results from the contribution of the three types of charge, which can still be identified inside the square root of (3.16): the first term is the contribution of electrons, the second that of holes (which become negligible for $\Psi \gg U_T$), and the third term that of the fixed depletion charge.

For negative values of Ψ_s, the second term dominates and holes accumulate exponentially. If $\Psi_s = 0$ then $F = 0$; the silicon is neutral up to the surface. This is called the *flat-band* situation.

For $0 < \Psi_s \ll 2\Phi_F + V$, the last term due to the fixed depletion charge dominates. The *small quantity of electrons* corresponding to the second term *is negligible in the calculation of field and total charge*, but will be the only charge available to transport current. This situation is called *weak inversion*.

When Ψ_s exceeds $2\Phi_F + V$, the first term becomes nonnegligible and would keep its exponential growth if Ψ_s could be increased much above. This situation is called *strong inversion*.

It must be pointed out that F is a nonlinear function of the three kinds of charge, and so neither Q_i nor Q_b can be identified separately in this function, except if one of them strongly dominates. This is the case in weak inversion where, from (3.17) and (3.16),

$$Q_{si} \cong Q_b = -\frac{\epsilon_{si} U_T}{L_D} \sqrt{\frac{\Psi_s}{U_T}} \quad \text{(weak inversion).} \tag{3.18}$$

3.2 SURFACE POTENTIAL AS A FUNCTION OF GATE VOLTAGE

The surface potential Ψ_s increases with the gate voltage V_G. This dependency can be obtained by again applying Gauss' law as illustrated in Figure 3.2, but with the upper part of the volume ending inside the oxide layer, and thus including the fixed charge density Q_{fc}. The electric field at the upper face is the oxide field E_{ox} given by (2.3), and thus

$$Q_{si} + Q_{fc} = -\epsilon_{ox} E_{ox} = -C_{ox}(V_G - \Phi_{ms} - \Psi_s), \tag{3.19}$$

where

$$C_{ox} = \epsilon_{ox}/t_{ox} \tag{3.20}$$

is the gate oxide capacitance per unit area. Introducing expression (3.17) of Q_{si} and solving for V_G yields

$$\frac{V_G - V_{FB}}{U_T} = \frac{\Psi_s}{U_T} + \frac{\epsilon_{si}}{C_{ox} L_D} F(\Psi_s, 2\Phi_F + V), \tag{3.21}$$

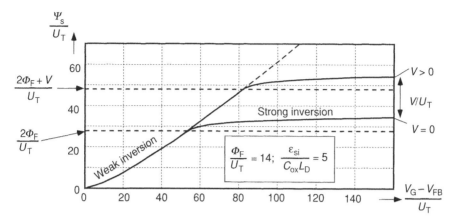

Figure 3.3 Surface potential Ψ_s as a function of gate voltage V_G

where V_{FB} is the *flat-band voltage* given by

$$V_{FB} \triangleq \Phi_{ms} + \frac{-Q_{fc}}{C_{ox}}. \tag{3.22}$$

It is the value of the gate voltage V_G needed to obtain the flat-band situation, for which $\Psi_s = 0$ and $Q_{si} = 0$.

Relation (3.21) is represented in Figure 3.3 for realistic particular values of Φ_F and $\epsilon_{si}/(C_{ox}L_D)$.

As long as $\Psi_s \ll 2\Phi_F + V$ (corresponding to weak inversion), $Q_{si} = Q_b$ increases only with the square root of Ψ_s. Thus Ψ_s increases with V_G. But as soon as Ψ_s reaches $2\Phi_F + V$, Q_{si} starts increasing much more rapidly with Ψ_s due to the important contribution of mobile charge. This is not compatible with the limited field, and thus the surface potential only increases very slowly to become *almost constant in strong inversion*.

3.3 GATE CAPACITANCE

Since the function F represents the variation of the charge in silicon Q_{si} with surface potential Ψ_s, its derivative gives the corresponding silicon capacitance C_{si}:

$$C_{si} \triangleq \frac{d(-Q_{si})}{d\Psi_s} = \frac{\epsilon_{si}U_T}{L_D}\frac{dF}{d\Psi_s} = \frac{\epsilon_{si}}{2L_D}\frac{\exp\frac{\Psi_s - 2\Phi_F - V}{U_T} - \exp\frac{-\Psi_s}{U_T} + 1}{F(\Psi_s, 2\Phi_F + V)}. \tag{3.23}$$

This equation and equation (3.21) of $V_G(\Psi_s)$ can be used as parametric equations of $C_{si}(V_G)$, with Ψ_s as parameter. The *local* gate capacitance per unit area C_g is then obtained by the series connection of C_{si} and C_{ox}:

$$\frac{C_g}{C_{ox}} = \frac{C_{si}}{C_{si} + C_{ox}}. \tag{3.24}$$

The resulting $C_g(V_G)$ curve is plotted in Figure 3.4.

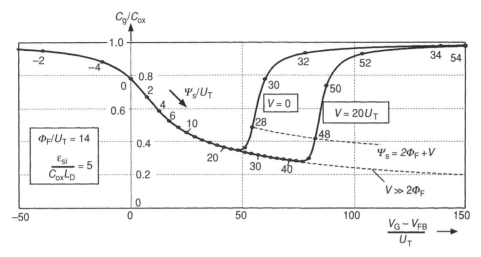

Figure 3.4 Local gate capacitance C_g as a function of gate voltage V_G

For negative gate voltages, holes are attracted to the surface. C_{si} becomes much larger than C_{ox} and thus C_g tends to C_{ox}. The same is true for large positive gate voltages that attract a large number of electrons corresponding to channel inversion.

In between, holes are repelled from the surface, leaving the depletion charge Q_b, while the electron charge remains negligible. The silicon capacitance is thus reduced to the depletion capacitance C_d given by

$$C_{si} = C_d = \frac{\epsilon_{si}}{2L_D\sqrt{\Psi_s/U_T}} = \sqrt{\frac{\epsilon_{si}q N_b}{2\Psi_s}}. \tag{3.25}$$

It corresponds to a thickness t_d of the dielectric depletion layer given by

$$t_d = 2L_D\sqrt{\frac{\Psi_s}{U_T}} = \sqrt{\frac{2\Psi_s\epsilon_{si}}{q N_b}} \tag{3.26}$$

that increases with the square root of the surface potential. Hence the gate capacitance C_g slowly decreases, until electrons are no longer negligible and rapidly dominate in strong inversion, causing an abrupt increase of C_g. If the channel voltage V is increased, it increases the value of Ψ_s required for strong inversion (see Figure 3.3), letting the gate capacitance keep descending further.

It must be remembered that C_g is the *local* gate capacitance per unit area. Thus, since there is only a single gate electrode, the curve of Figure 3.4 can only be measured when the whole channel is at the same voltage $V = V_S = V_D$. Indeed, if $V_D \neq V_S$, then V changes along the channel, and the total gate capacitance is a combination of different local values.

These very nonlinear characteristics of gate capacitance must be taken into consideration when a MOS transistor is used to implement a capacitor. To obtain a value as constant as possible, the device must be biased in accumulation or in strong inversion. A voltage-dependent capacitor (varicap) is obtained by exploiting the slow decay of C_G in weak inversion, or its rapid increase at the verge of strong inversion [69].

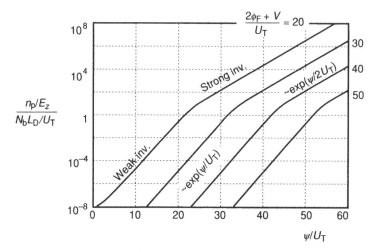

Figure 3.5   Integrand of equation (3.27)

3.4 CHARGE SHEET APPROXIMATION

The mobile inversion charge Q_i defined in Section 2.2 is obtained by integrating the electron concentration n_p below the surface of silicon:

$$Q_i = -q \int_0^\infty n_p \, dz = -q \int_0^{\Psi_s} \frac{n_p}{E_z} \, d\Psi. \tag{3.27}$$

The integrand n_p/E_z can be expressed from (3.5), (3.7), and (3.14) [70]:

$$\frac{n_p}{E_z} = \frac{N_b L_D}{U_T} \frac{\exp \frac{\Psi - 2\Phi_F - V}{U_T}}{F(\Psi_s, \, 2\Phi_F + V)}. \tag{3.28}$$

This expression, represented in the semilog plot of Figure 3.5 for several values of $2\Phi_F + V$, unfortunately cannot be integrated explicitly. As can be seen, it is essentially an exponential function $\exp[\Psi/(mU_T)]$, with a slope factor m changing from 1 in weak inversion to 2 in strong inversion. Thus, as illustrated by the linear plot on Figure 3.6, 95% of charge Q_i is at a potential within $3U_T$ to $6U_T$ below Ψ_s.

On this basis, the *charge sheet approximation* [71] illustrated in the same figure *will be used in the rest of the book*. It assumes that *the whole charge Q_i is at the surface potential*

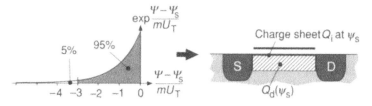

Figure 3.6   Charge sheet approximation

Ψ_s. Thus, since there is no voltage drop across this thin charge sheet, *the depletion charge Q_b is controlled by Ψ_s*. Hence, it can be approximated by expression (3.18) for weak inversion (where Q_i is really negligible):

$$-Q_b = \frac{\epsilon_{si} U_T}{L_D} \sqrt{\frac{\Psi_s}{U_T}} = \Gamma_b C_{ox} \sqrt{\Psi_s}, \qquad (3.29)$$

where Γ_b is the *substrate modulation factor*, given by

$$\Gamma_b \triangleq \frac{\epsilon_{si}}{L_D C_{ox}} \sqrt{U_T} = \frac{\sqrt{2q N_b \epsilon_{si}}}{C_{ox}}. \qquad (3.30)$$

3.5 DENSITY OF MOBILE INVERTED CHARGE

3.5.1 Mobile Charge as a Function of Gate Voltage and Surface Potential

The total charge density Q_{si} can be calculated from the field in the oxide by using relations (3.19) and (3.22),

$$Q_{si} = -C_{ox}(V_G - V_{FB} - \Psi_s), \qquad (3.31)$$

and, with the charge sheet approximation, the depletion charge Q_b is given by (3.29). Although the inversion charge Q_i cannot be explicitly obtained directly from (3.27), it can be expressed as their difference

$$Q_i = Q_{si} - Q_b = -C_{ox}(V_G - V_{FB} - \Psi_s - \Gamma_b \sqrt{\Psi_s}) = -C_{ox}(V_G - V_{TB}), \qquad (3.32)$$

where

$$V_{TB} \triangleq V_{FB} + \Psi_s + \Gamma_b \sqrt{\Psi_s} \qquad (3.33)$$

is the *threshold function*. This function of Ψ_s that depends on the process through parameters V_{FB} and Γ_b is represented in Figure 3.7 for two extreme values of Γ_b.

This function is nonlinear due to the contribution of Q_b and its slope $n > 1$ is obtained by differentiation of (3.33):

$$n \triangleq \frac{dV_{TB}}{d\Psi_s} = 1 + \frac{\Gamma_b}{2\sqrt{\Psi_s}}. \qquad (3.34)$$

It is represented in Figure 3.8.

As can be seen, n is a very slow function of Ψ_s, especially since in normal situations, Ψ_s is not lower than about $20 U_T$. Its maximum value thus ranges between 1.2 and 1.7, and it tends to 1 for very large values of Ψ_s.

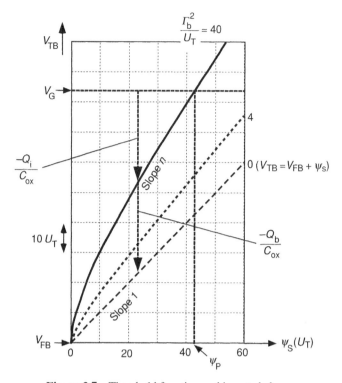

Figure 3.7   Threshold function and inverted charge

Now according to (3.32), the density of inverted charge Q_i is proportional to the difference between the gate voltage and the threshold function, as illustrated in Figure 3.7. Hence, for V_G constant,

$$\frac{dQ_i/C_{ox}}{d\Psi_s} = n. \tag{3.35}$$

For a given value of the gate voltage, the inverted charge becomes zero for a particular value Ψ_P of the surface potential called *pinch-off (surface) potential*. It is directly related to the gate

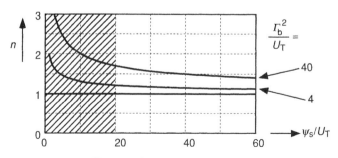

Figure 3.8   Slope n of $V_{TB}(\Psi_s)$

voltage, as can be obtained by (3.32) for $Q_i = 0$:

$$V_G = V_{FB} + \Psi_P + \Gamma_b\sqrt{\Psi_P}. \qquad (3.36)$$

This function $V_G(\Psi_P)$ is identical to $V_{TB}(\Psi_s)$ given by (3.33), as can be confirmed by inspection of Figure 3.7.

Solving (3.36) for Ψ_P yields

$$\Psi_P \triangleq \Psi_s(Q_i = 0) = V_G - V_{FB} - \Gamma_b^2\left(\sqrt{\frac{V_G - V_{FB}}{\Gamma_b^2} + \frac{1}{4}} - \frac{1}{2}\right). \qquad (3.37)$$

3.5.2 Mobile Charge as a Function of Channel Voltage and Surface Potential

The total charge in silicon, Q_{si}, can also be expressed by (3.17), in which L_D can be eliminated using (3.30). Furthermore, the function $F(\Psi, 2\Phi_F + V)$ given by (3.16) can be simplified if the channel potential is sufficiently positive ($\Psi_s \gg U_T$), the contribution of holes being negligible.

With the charge sheet approximation, the depletion charge Q_b is again given by (3.29). The inverted charge can then be expressed as

$$Q_i = Q_{si} - Q_b = -\Gamma_b C_{ox}\sqrt{U_T}\left(\sqrt{\frac{\Psi_s}{U_T} + \exp\frac{\Psi_s - 2\Phi_F - V}{U_T}} - \sqrt{\frac{\Psi_s}{U_T}}\right). \qquad (3.38)$$

3.6 CHARGE-POTENTIAL LINEARIZATION

3.6.1 Linearization of $Q_i(\Psi_s)$

Since n is only slightly dependent on Ψ_s, it will be *considered constant*, which amounts to a *linearization of the mobile inverted charge Q_i in function of the surface potential Ψ_s* [27].

Thus, although it is a (slow) function of Ψ_s, n will become a *fixed device parameter called the slope factor*. It should be evaluated for the best coverage of the device operation range, as will be discussed in Section 3.6.4.

The inverted charge Q_i cannot be calculated from expression (3.32) or (3.38), since the surface potential Ψ_s cannot be expressed analytically as a function of the gate voltage V_G. Indeed, Figure 3.3 was obtained from equation (3.21) which cannot be inverted. However, the linearization of $Q_i(\Psi_s)$ by means of the constant slope factor n can be exploited to obtain an explicit solution, as will be explained below.

This linearization is illustrated in Figure 3.9, which is just a qualitative replication of Figure 3.7. With the slope factor n constant, the surface potential is related to its pinch-off value Ψ_P by

$$Q_i = nC_{ox}(\Psi_s - \Psi_P) \quad \text{or} \quad \Psi_s = \Psi_P + \frac{Q_i}{nC_{ox}}. \qquad (3.39)$$

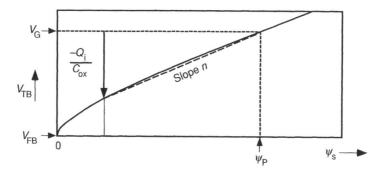

Figure 3.9   Linearization of mobile inverted charge Q_i with surface potential Ψ_s

This expression can be introduced in (3.38) to eliminate Ψ_s [41]. Arranging the result to extract $\Psi_P - 2\Phi_F - V$ yields

$$\frac{\Psi_P - 2\Phi_F - V}{U_T} = \frac{-Q_i}{nC_{ox}U_T} + \ln\left[\frac{-Q_i}{\Gamma_b C_{ox}\sqrt{U_T}}\left(\frac{-Q_i}{\Gamma_b C_{ox}\sqrt{U_T}} + 2\sqrt{\frac{Q_i}{nC_{ox}U_T} + \frac{\Psi_P}{U_T}}\right)\right].$$

(3.40)

This equation expresses a general relation between the density of inversion charge Q_i and voltages V and V_G (since Ψ_P is a direct function of V_G given by (3.37)).

The inversion charge can be normalized by introducing a specific charge Q_{spec}

$$q_i = Q_i/Q_{spec}$$

(3.41)

with

$$Q_{spec} \triangleq -2nU_T C_{ox}.$$

(3.42)

The negative sign in Q_{spec} takes care of the negative charge of electrons, whereas reasons for introducing the factor 2 will be explained later. Furthermore, voltages normalized to U_T will be represented by lowercase letters:

$$\frac{V}{v} = \frac{\Psi}{\psi} = \frac{\Phi}{\phi} = \left(\frac{\Gamma_b}{\gamma_b}\right)^2 = U_T.$$

(3.43)

Equation (3.40) then becomes

$$2q_i + \ln q_i + \underbrace{\ln\left[\frac{2n}{\gamma_b}\left(q_i\frac{2n}{\gamma_b} + 2\sqrt{\psi_p - 2q_i}\right)\right]}_{v_{sh}} = \psi_p - 2\phi_f - v.$$

(3.44)

Due to the simplifications of the function $F(\Psi, 2\Phi_F + V)$ in (3.38), this expression is valid only for $\psi_p \gg 2q_i$ ($\psi_p = 2q_i$ corresponds to $\psi_s = 0$). Figure 3.10 shows the variation with q_i

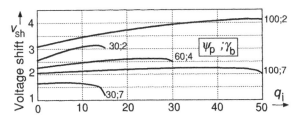

Figure 3.10   Voltage shift v_{sh} in (3.44).

of the term labeled v_{sh} (voltage shift) in (3.44) for various values of ψ_p and γ_b, and with the slope factor n defined by (3.34) evaluated at $\Psi_s = \Psi_p$.

As can be seen, even for extreme values of ψ_p and γ_b the variation of v_{sh} with q_i never exceeds unity. The terms in q_i can therefore be neglected, and (3.44) simplifies to

$$2q_i + \ln q_i + \underbrace{\ln\left(\frac{4n}{\gamma_b}\sqrt{\psi_p}\right)}_{v_{sh}} = \psi_p - 2\phi_f - v. \tag{3.45}$$

The remaining dependency of v_{sh} on ψ_p is very weak; hence, v_{sh} can be considered *constant* (evaluated for example at $\psi_p = 2\phi_f$).

Equation (3.45) constitutes a general normalized relation between the pinch-off potential Ψ_P, the local channel voltage V, and the resulting local inversion charge density Q_i. Since Ψ_P is a direct function of the gate voltage V_G given by (3.37), $Q_i(V_G, V)$ can be obtained by numerical computation.

But Ψ_P is a particular value of the surface potential that does not appear explicitly in the measurable characteristics of the transistor. It is therefore very useful for circuit applications to relate the inversion charge (and later the drain current) to a particular value of the channel voltage V called *pinch-off voltage* V_P [72] and defined by

$$v_p \triangleq \frac{V_P}{U_T} \triangleq v(2q_i + \ln q_i = 0) = v(q_i = 0.4263) \tag{3.46}$$

or, by using (3.44),

$$v_p = \psi_p - (2\phi_f + v_{sh}). \tag{3.47}$$

Introducing this definition in (3.44) results in

$$2q_i + \ln q_i = v_p - v. \tag{3.48}$$

This expression provides $v(q_i)$, but it cannot be inverted analytically to provide a general expression of $q_i(v)$. However, the two axes can be exchanged to represent $q_i(v)$, as shown in Figure 3.11.

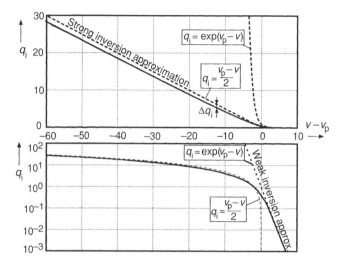

Figure 3.11 Normalized inverted charge vs. channel voltage

In weak inversion, $q_i \ll 1$; thus the linear term becomes negligible. The mobile inverted charge can be approximated by

$$q_i = \exp(v_p - v) \quad \text{(weak inversion)}. \tag{3.49}$$

In strong inversion, $q_i \gg 1$; thus the logarithmic term becomes negligible. The mobile inverted charge can be approximated by

$$q_i = \frac{v_p - v}{2} \quad \text{(strong inversion)}. \tag{3.50}$$

This approximation will be further discussed in Section 3.6.3.

In moderate inversion, both terms contribute to the variation of mobile charge and neither approximation is valid.

As expressed by (3.47) and shown by Figure 3.10, $\Psi_P - V_P$ is almost exactly constant. Hence by inspection of Figure 3.7,

$$\frac{dV_G}{dV_P} = \frac{dV_G}{d\Psi_P} = n. \tag{3.51}$$

3.6.2 Linearized Bulk Depletion Charge Q_b

Differentiating (3.29) and introducing (3.34) give

$$\frac{dQ_b/C_{ox}}{d\Psi_s} = 1 - n \tag{3.52}$$

as can be confirmed by inspection of Figure 3.7. Furthermore, its value Q_{bP} at $\Psi_s = \Psi_P$ obtained from (3.29) is

$$Q_{bP} = -\Gamma_b C_{ox} \sqrt{\Psi_P}. \tag{3.53}$$

Using the charge-potential linearization introduced in Section 3.6.1,

$$Q_b = Q_{bP} + (1-n)(\Psi_s - \Psi_P)C_{ox} = -\Gamma_b C_{ox} \sqrt{\Psi_P} + (1-n)(\Psi_s - \Psi_P)C_{ox} \tag{3.54}$$

where the linearized expression (3.39) of Q_i can be introduced to obtain in normalized values

$$q_b \triangleq \frac{Q_b}{Q_{spec}} = \frac{\gamma_b \sqrt{\psi_P}}{2n} - \frac{n-1}{n}q_i. \tag{3.55}$$

3.6.3 Strong Inversion Approximation

As was pointed out at the end of Section 3.2, in strong inversion the surface potential Ψ_s increases very slowly with the gate voltage V_G, due to the very rapid increase of the total charge in silicon Q_{si}. Thus, Ψ_s *can be assumed to be independent of V_G*, and approximated, in view of Figure 3.3, by

$$\Psi_s = \Psi_0 + V, \tag{3.56}$$

where Ψ_0 is a constant slightly larger than $2\Phi_F$.

Expression (3.32) of the inverted charge can then be rewritten as

$$Q_i = -C_{ox}[V_G - \underbrace{(V_{FB} + \Psi_0 + V + \Gamma_b\sqrt{\Psi_0 + V})}_{V_{TB}}]. \tag{3.57}$$

The threshold function $V_{TB}(V)$ is identical to $V_{TB}(\Psi_s)$ of Figure 3.7, but shifted by $-\Psi_0$, as shown in Figure 3.12(a).

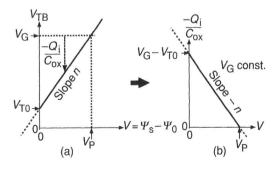

Figure 3.12   Strong inversion approximation of inverted charge

For $V = 0$, V_{TB} has the particular value V_{T0} called *equilibrium threshold voltage*, or in short *threshold voltage*:

$$V_{T0} \triangleq V_{TB}(V = 0) = V_{FB} + \Psi_0 + \Gamma_b\sqrt{\Psi_0}. \tag{3.58}$$

This *bias-independent* device parameter corresponds to the threshold voltage V_T for $V_S = 0$ used in other models. Its precise value depends on the value chosen for Ψ_0.

Introducing (3.58) in (3.57) yields

$$V_{TB} = V_{T0} + V + \Gamma_b\left(\sqrt{\Psi_0 + V} - \sqrt{\Psi_0}\right). \tag{3.59}$$

According to its strong inversion approximation (3.50) derived from the general expression (3.48), the inverted charge would be zero when the channel voltage V reaches its pinch-off value V_P defined by (3.46). Hence, V_P can be obtained as the value of V corresponding to $V_{TB} = V_G$, as illustrated in Figure 3.12(a). Indeed, inspection of this figure shows that V_G depends on V_P exactly as V_{TB} depends on V. Thus according to (3.59),

$$V_G = V_{T0} + V_P + \Gamma_b\left(\sqrt{\Psi_0 + V_P} - \sqrt{\Psi_0}\right), \tag{3.60}$$

which can be inverted to provide

$$V_P = V_G - V_{T0} - \Gamma_b\left[\sqrt{V_G - V_{T0} + \left(\frac{\Gamma_b}{2} + \sqrt{\Psi_0}\right)^2} - \left(\frac{\Gamma_b}{2} + \sqrt{\Psi_0}\right)\right]. \tag{3.61}$$

The slope of $V_G(V_P)$ is still n defined by (3.34) since

$$n = \frac{dV_{TB}}{d\Psi_s} = \frac{dV_{TB}}{dV} = \frac{dV_G}{dV_P} = 1 + \frac{\Gamma_b}{2\sqrt{\Psi_0 + V}}. \tag{3.62}$$

It should be evaluated at a value of V that best covers the range of operation.

Since n is almost constant with V, inspection of Figure 3.12(a) shows that instead of using equation (3.61) the pinch-off voltage can be approximated by

$$V_P \cong \frac{V_G - V_{T0}}{n}. \tag{3.63}$$

It also shows that, instead of using (3.57), the inverted charge can be approximated by

$$\frac{-Q_i}{C_{ox}} = n(V_P - V), \tag{3.64}$$

which corresponds to approximation (3.50).

For a given value of the gate voltage V_G, the inverted charge can be plotted directly by first moving the V-axis to this value in the plot of Figure 3.12(a), and then vertically flipping around this axis to produce Figure 3.12(b). Changing V_G *shifts the curve vertically*, resulting in a change of V_P.

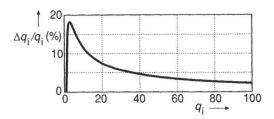

Figure 3.13 Relative error of inverted charge in strong inversion approximation

By introducing expression (3.58) of V_{T0} into (3.61) and comparing the result with expression (3.37) of Ψ_P, we obtain

$$V_P = \Psi_P - \Psi_0. \tag{3.65}$$

Hence, according to (3.47)

$$\psi_0 = \psi_p - v_p = 2\phi_f + v_{sh} \quad \text{or} \quad \Psi_0 = \Psi_p - V_p = 2\Phi_F + V_{sh}. \tag{3.66}$$

Inspection of Figure 3.10 shows that v_{sh} is almost constant, whereas Figure 3.3 shows that the surface potential keeps increasing slowly in strong inversion. This difference is due to the logarithmic term in (3.48), which has been neglected in approximation (3.50). It results in a difference Δq_i of the inverted charge q_i that is represented in relative values in Figure 3.13. As can be seen, the excess of charge obtained in the strong inversion approximation never exceeds 18%.

Notice that although the threshold voltage V_{T0} was introduced in the framework of this strong inversion approximation, it can *always* be used to relate V_P to V_G by equation (3.61) or by its approximation (3.63).

3.6.4 Evaluation of the Slope Factor

As already stated in Section 3.6.1, the slope factor n defined by (3.34) should be evaluated to best fit the range of operation of the transistor.

Inspection of Figure 3.9 shows that n would be best evaluated as the slope of the secant of the threshold function V_{TB} between a particular value of surface potential Ψ_s and its pinch-off value Ψ_P. Indeed, the calculation of Q_i by the linear relation (3.39) is then exact for the selected value of Ψ_s. This slope can be obtained from expression (3.33) of V_{TB}:

$$n = n_{opt} = \frac{V_{TB}(\Psi_P) - V_{TB}(\Psi_s)}{\Psi_P - \Psi_s} = 1 + \frac{\Gamma_b}{\sqrt{\Psi_P} + \sqrt{\Psi_s}}. \tag{3.67}$$

In weak inversion, Q_i is very small; therefore, $\Psi_s \cong \Psi_P$, as can be seen in Figure 3.7. The secant merges with the tangent and slope n is obtained by introducing $\Psi_s = \Psi_P = \Psi_0 + V_P$ in (3.34)

$$n = n_w = 1 + \frac{\Gamma_b}{2\sqrt{\Psi_P}} = 1 + \frac{\Gamma_b}{2\sqrt{\Psi_0 + V_P}}. \tag{3.68}$$

Hence, the slope factor in weak inversion *depends only on the gate voltage* through V_P. Moreover, it has a particular physical meaning. Indeed, since the inverted charge Q_i is negligible with respect to the depletion charge Q_b, inspection of Figure 3.7 shows that the slope factor can be expressed as

$$n = \frac{dV_G}{d\Psi_s} \qquad \text{(weak inversion only)}. \tag{3.69}$$

The surface potential follows almost linearly the gate voltage variations, but with an attenuation by factor n. This attenuation is produced by a capacitive divider made of the oxide capacitance C_{ox} and the surface depletion capacitance $C_d = -dQ_b/d\Psi_s$. Hence,

$$n = n_w = \frac{C_{ox} + C_d}{C_{ox}} \qquad \text{(weak inversion only)}. \tag{3.70}$$

This series connection of C_{ox} and C_d reduces the gate capacitance to

$$\frac{C_g}{C_{ox}} = \frac{C_d}{C_{ox} + C_d} = 1 - \frac{1}{n_w}. \tag{3.71}$$

Hence, $1/n_w$ can be identified on Figure 3.4 as the depth of the dip of the $C_g(V_G)$ curve, as shown in Figure 3.14.

If the whole channel is in weak inversion (see Section 4.4.5), then the surface potential Ψ_s is constant along the channel, since it depends only on V_G, as shown by Figure 3.3. It is smaller than the lowest value of $2\Phi_F + V$, which is normally $2\Phi_F + V_S$. Thus the minimum value of the slope factor (max of $1/n$), which occurs at the minimum of the $C_g(V_G)$ curve, can be expressed as

$$n = n_{w\,min} \cong 1 + \frac{\Gamma_b}{2\sqrt{2\Phi_F + V_S}}. \tag{3.72}$$

The surface potential cannot be much lower than this upper limit, since the inversion charge decreases exponentially and would rapidly become so small that the transistor would be blocked. Hence, n_w is usually not much larger than its minimum value $n_{w\,min}$ given above.

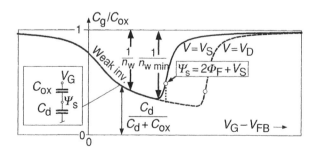

Figure 3.14   Value of the slope factor n in weak inversion

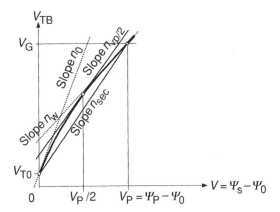

Figure 3.15 Possible evaluations of slope n in strong inversion (nonlinearity of $V_{TB}(V)$ strongly exaggerated)

In strong inversion, evaluating the slope at pinch-off ($n = n_w$) results in a too small value, as illustrated in Figure 3.15, where the nonlinearity has been strongly exaggerated.

Another possibility is to evaluate it at $V = 0$ ($\Psi_s = \Psi_0$), resulting in [17]

$$n = n_0 = 1 + \frac{\Gamma_b}{2\sqrt{\Psi_0}}. \tag{3.73}$$

This evaluation is independent of any bias voltage and can thus be used as a first approximation, but it is overestimated as shown in Figure 3.15.

The same figure shows that a better solution would be to use the slope of the secant from $V = 0$ to $V = V_P$, obtained by introducing $\Psi_P = \Psi_0 + V_P$ and $\Psi_s = \Psi_0$ in (3.67),

$$n = n_{\text{sec}} = 1 + \frac{\Gamma_b}{\sqrt{\Psi_0 + V_P} + \sqrt{\Psi_0}}, \tag{3.74}$$

or to evaluate the slope (3.34) at $\Psi_s = \Psi_0 + V_P/2$,

$$n = n_s = 1 + \frac{\Gamma_b}{\sqrt{\Psi_0 + V_P/2}}, \tag{3.75}$$

which gives a slightly lower value.

According to (3.66), the value of $\Psi_0 - 2\Phi_F$ is just a few U_T. Hence, Ψ_0 can be replaced by $2\Phi_F$ in the above evaluations of n, without much affecting the result.

It must be pointed out again that the dependency of n on bias voltages is very weak, the main dependency being on V_G. Thus the error produced by a nonoptimum evaluation is probably within the spreading range of the various process parameters.

However, this slight difference might be a cause of mismatch between devices at different bias conditions, especially at different values of the gate voltage.

3.6.5 Compact Model Parameters

Equation (3.48) describes the general charge–voltage relationship for a long-channel transistor. This equation is dimensionless and uses normalized variables. Only *three model parameters* and one physical parameter are needed to obtain from it the relation between applied voltages V_G and V, and the resulting inverted charge density Q_i.

The physical parameter, used to normalize all voltages in the dimensionless equation is U_T. The following are the three device parameters:

1. *The slope factor n* defined in Figure 3.7 and by equation (3.34). This parameter was further discussed in Section 3.6.4.

2. *The threshold voltage V_{T0}* defined in Figure 3.12 and by equations (3.58) and (3.66). It is very slightly dependent on V_G through Ψ_p (see Figure 3.10), but can be considered bias independent in practice.

   These first two parameters relate the gate voltage V_G to the pinch-off voltage V_P according to (3.63).

3. *The oxide capacitance per unit area C_{ox}.* It is combined with n and U_T to obtain the specific charge Q_{spec} defined by (3.42) and is used to normalize the charge density.

Introducing these parameters in (3.48) provides the nonnormalized general charge–voltage relation:

$$\frac{-Q_i}{C_{ox}} + nU_T \ln \frac{-Q_i}{2nC_{ox}U_T} = V_G - V_{T0} - nV. \tag{3.76}$$

We can see that U_T plays an important role in weak inversion where the logarithmic term dominates. This role disappears in strong inversion when the logarithmic term becomes negligible.

4 Static Drain Current

In this chapter, the model of the static current–voltage characteristics of the transistor is derived. In Section 4.1, the drain current I_D is shown to be directly related to the charge–voltage relation established in the previous chapter. Section 4.2 introduces the important concept of forward and reverse components of I_D, and Section 4.3 defines the various modes of operation of the transistor. Section 4.4 details the derivation of the drain current for all current levels from the linearized charge model, with its approximations in strong and weak inversion. Section 4.5 is dedicated to a fundamental property of the transistor that facilitates its modeling and opens interesting circuit approaches through the concept of pseudo-resistor. Finally, Section 4.6 introduces a first approximation of the channel length modulation phenomenon, which limits the output resistance in saturation.

4.1 DRAIN CURRENT EXPRESSION

When the source and drain voltages are different, electrons of density n_p forming the mobile inverted charge Q_i move by a combination of drift and diffusion, resulting in a drain current I_D defined in Figure 2.1. For the long channel considered here, all elementary flows of current are along the x-axis with a local current density in this direction given by [67]

$$J_n = \mu q \left(\underbrace{-n_p \frac{d\Psi}{dx}}_{\text{drift}} + \underbrace{U_T \frac{dn_p}{dx}}_{\text{diffusion}} \right), \tag{4.1}$$

where μ is the equivalent mobility of electrons in the channel. The first term is the drift component of the current, proportional to the longitudinal electric field $-d\Psi/dx$. The second term is the diffusion component, proportional to the gradient of charge concentration dn_p/dx.

With the charge sheet approximation, integration in the vertical direction (z-axis) is obtained by replacing qn_p by $-Q_i$. Moreover, if the channel is sufficiently wide, integration along the

Charge-Based MOS Transistor Modeling: The EKV Model for Low-Power and RF IC Design C. Enz and E. Vittoz
© 2006 John Wiley & Sons, Ltd.

y-axis is simply a multiplication by width W. Thus [1],

$$I_D = \mu W \left(\underbrace{-Q_i \frac{d\Psi_s}{dx}}_{\text{drift}} + \underbrace{U_T \frac{dQ_i}{dx}}_{\text{diffusion}} \right), \tag{4.2}$$

Since the whole charge Q_i of electrons is assumed to be concentrated at the surface, the concentration of electrons (3.5) can be replaced by

$$Q_i \propto \exp\frac{\Psi_s - \Phi_F - V}{U_T}. \tag{4.3}$$

Therefore

$$\frac{dQ_i}{dx} = \frac{Q_i}{U_T}\left(\frac{d\Psi_s}{dx} - \frac{dV}{dx} \right), \tag{4.4}$$

which when introduced in (4.2) yields

$$I_D = \mu W(-Q_i)\frac{dV}{dx}. \tag{4.5}$$

This expression *includes the drift and the diffusion* components and shows that the overall current is proportional to the gradient of channel voltage V (which is a quasi-Fermi potential, as discussed in Section 2.1). Now, since the current is constant along the channel, (4.5) can be integrated from source to drain:

$$I_D \int_0^L dx = \int_{V_S}^{V_D} \mu W(-Q_i)dV. \tag{4.6}$$

In this basic model, *mobility μ and channel width W are assumed to be constant*. The dependency of μ on the vertical field E_{zs} and the horizontal field E_x will be introduced in Part II. The effect of a possible variable channel width will be discussed in Section 4.5.2. Since only Q_i changes along the channel, equation (4.6) becomes

$$I_D = \beta \int_{V_S}^{V_D} \frac{-Q_i}{C_{ox}} dV, \tag{4.7}$$

where

$$\beta \triangleq \mu C_{ox}\frac{W}{L} \tag{4.8}$$

is the *transfer parameter* of the transistor depending on the width over length ratio of the channel.

Equation (4.7) is a very interesting result, since it shows that the drain current can be obtained directly from the $Q_i(V)$ as is illustrated in Figure 4.1. This result is *independent of*

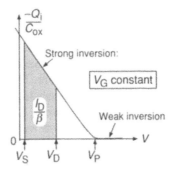

Figure 4.1   Drain current according to (4.7)

the shape of $Q_i(V)$. It is valid for all values of the source and drain voltages, including those larger than the pinch-off voltage V_P, for which the inverted charge is very small.

As established in Chapter 3, a variation of the gate voltage V_G shifts vertically the whole strong inversion part of the curve by the same amount, whereas the weak inversion part is shifted horizontally to follow the variation of V_P.

4.2 FORWARD AND REVERSE CURRENT COMPONENTS

Since the mobile charge Q_i tends to zero for V tending to infinity, the integral (4.7) can be rewritten as

$$I_D = \beta \underbrace{\int_{V_S}^{\infty} \frac{-Q_i}{C_{ox}} \, dV}_{\text{forward current } I_F} - \beta \underbrace{\int_{V_D}^{\infty} \frac{-Q_i}{C_{ox}} \, dV}_{\text{reverse current } I_R} = I_F - I_R. \qquad (4.9)$$

The drain current can thus be expressed as the difference between a *forward current* I_F and a *reverse current* I_R, as illustrated in Figure 4.2. I_F depends on V_G and V_S, but *not on* V_D, whereas I_R depends on V_G and V_D, but *not on* V_S. Furthermore, according to (4.9), $I_F(V_S) \equiv I_R(V_D)$: I_F and I_R are indeed two values of the same function of V.

Thus, the drain current is the *superposition* of *independent* and *symmetrical* effects of the source and drain voltages. This is a *property* of MOS transistors that is *independent of the shape of* $Q_i(V)$ [73]. Its limits of validity will be discussed in Section 4.5.

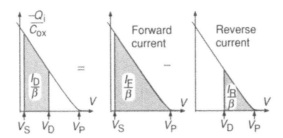

Figure 4.2   Decomposition of the drain current into forward and reverse components

4.3 MODES OF OPERATION

With the bias situation illustrated in Figure 4.1, both V_S and V_D are in the section of the $Q_i(V)$ characteristics corresponding to strong inversion. The channel is strongly inverted from source to drain and the transistor is said to be in *linear* mode. This terminology refers to the fact that the drain current is a linear function of the gate voltage, as will be shown in Section 4.4.4. Alternative appellations are *nonsaturation mode* [1] (meaning that the drain current keeps increasing with the drain voltage) and *triode mode* (referring to the nonsaturating output characteristics of the old triode vacuum tube). In French, this mode is called *conduction* [17], and this terminology has been previously used in some of the authors' publications.

If the drain voltage V_D is increased above the pinch-off voltage V_P, then the current does not increase significantly anymore: the transistor is still in strong inversion, but in *forward saturation*. Forward saturation can be characterized by the fact that the reverse current becomes negligible compared to the forward current:

$$\text{in (forward) saturation}: \quad I_R \ll I_F \qquad \text{thus} \quad I_D = I_F. \qquad (4.10)$$

It can be noticed that the negligible reverse component is in fact in weak inversion, as is the drain end of the channel.

Now, if both V_D and V_S are larger than V_P, then the whole channel is weakly inverted, and the transistor is said to be in *weak inversion* mode.

These various modes of operation are summarized in Figure 4.3 that represents the V_S, V_D plane for a given positive value of V_P (thus for a given value of $V_G > V_{T0}$).

The part of the plane above the $V_D = V_S$ line corresponds to the forward modes described above, for which $I_D > 0$. The other half of the plane corresponds to the reverse modes, with $V_S > V_D$; therefore, $I_D < 0$. It includes reverse saturation with

$$\text{in reverse saturation}: \quad I_F \ll I_R \qquad \text{thus} \quad I_D = -I_R. \qquad (4.11)$$

In forward strong inversion, the transistor enters saturation for $V_D > V_P$; thus, the source to drain voltage necessary for saturation is

$$V_{DSsat} = V_P - V_S \quad \text{(in strong inversion)}. \qquad (4.12)$$

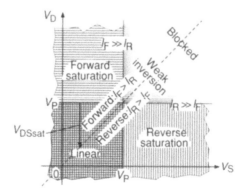

Figure 4.3   Modes of operation of a MOS transistor

This saturation voltage is also represented in the figure. It decreases when V_S approaches V_P. However, it never reaches zero since (4.12) is no longer valid close to or in weak inversion.

In weak inversion, the mobile charge, and thus the two components of current, decrease exponentially with V_S/U_T and V_D/U_T. This mode can be obtained even for zero value of source voltage if the pinch-off voltage V_P is made negative. According to (3.63), this is obtained when $V_G < V_{T0}$, hence the alternative appellation of *subthreshold* mode of operation.

When the larger of I_F or I_R becomes sufficiently small, the transistor is considered to be *blocked*.

It must be reminded that, although V_P is used as the limit between strong and weak inversion in Figure 4.3, the transition is progressive through a range of *moderate inversion*, where all terms of relation (3.48) are significant.

According to the definitions of positive voltages in Figure 2.1, the source and drain junctions are reverse biased in the first quadrant represented in Figure 4.3. Both V_S and V_D can be slightly negative without qualitatively changing the modes of operation described above. However, if these negative values exceed a few hundreds of millivolts, the forward-biased junctions inject minority carriers (electrons for the N-channel transistor) in the local substrate. A parasitic bipolar transistor is superimposed on the MOS transistor. This bipolar mode of operation can be usefully exploited if MOS operation is blocked by applying a negative gate voltage [68,74].

4.4 MODEL OF DRAIN CURRENT BASED ON CHARGE LINEARIZATION

4.4.1 Expression Valid for All Levels of Inversion

By normalizing the charge and voltages according to (3.41) and (3.43), the general drain current (4.7) expression becomes

$$I_D = I_{spec} \int_{v_s}^{v_d} q_i \, dv, \tag{4.13}$$

where I_{spec} is the *specific current* of the transistor defined by

$$I_{spec} \triangleq \mu U_T \frac{W}{L} (-Q_{spec}) = 2n\mu C_{ox} \frac{W}{L} U_T^2 = 2n\beta U_T^2. \tag{4.14}$$

All currents can be normalized to this specific current according to

$$\frac{I_D}{i_d} = \frac{I_F}{i_f} = \frac{I_R}{i_r} = I_{spec}. \tag{4.15}$$

The normalized values of drain current I_D, forward current I_F, and reverse current I_R defined by (4.9) can then be expressed as

$$i_d = \int_{v_s}^{v_d} q_i \, dv, \quad i_f = \int_{v_s}^{\infty} q_i \, dv, \quad \text{and} \quad i_r = \int_{v_d}^{\infty} q_i \, dv. \tag{4.16}$$

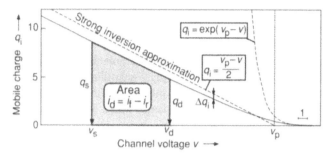

Figure 4.4   Normalized charge after linearization with surface potential, and resulting drain current

An analytic expression relating the channel voltage and the inverted charge was derived in Chapter 3 (equation (3.48); Figure 3.11) based on the linearization of $Q_i(\Psi_s)$ by means of the slope factor n. It is represented again in Figure 4.4 with the drain current shown according to (4.16). The particular values of the source and drain voltages correspond to the linear mode. From (3.48), the element of channel voltage dv can be expressed as

$$dv = -(2 + 1/q_i)\, dq_i,\tag{4.17}$$

which when introduced in (4.16) yields

$$i_{f,r} = \int_0^{q_{s,d}} (2q_i + 1)\, dq_i = q_{s,d}^2 + q_{s,d}\tag{4.18}$$

where $q_{s,d}$ is the value of normalized charge density q_i at the source or at the drain end of the channel.

The normalized forward or reverse component of drain current is thus given by

$$i_{f,r} = q_{s,d}^2 + q_{s,d}.\tag{4.19}$$

It should be reminded that this result was obtained by integrating the $Q_i(V)$ function for its particular expression obtained by linearizing $Q_i(\Psi_s)$ according to (3.35). It is a very simple expression, thanks to the factor 2 included in the definitions of Q_{spec} (3.42) and I_{spec} (4.14).

An alternative approach is possible by introducing the same linearization to eliminate the surface potential Ψ_s from the original drain current expression (4.2), resulting in

$$I_D = \mu W \left(\frac{-Q_i}{nC_{ox}} + U_T \right) \frac{dQ_i}{dx},\tag{4.20}$$

or, by normalizing the charge and the current according to (3.41) and (4.15), and the position x along the channel by $\xi = x/L$,

$$i_d = -(2q_i + 1)\frac{dq_i}{d\xi}.\tag{4.21}$$

Integration along the channel ($\xi = 0$ to 1) then yields

$$i_d = \int_{q_s}^{q_d} -(2q_i + 1)\, dq_i = (q_s^2 + q_s) - (q_d^2 + q_d) = i_f - i_r.\tag{4.22}$$

where i_f and i_r are given by (4.19).

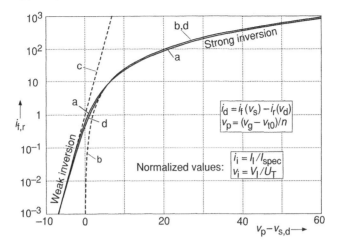

Figure 4.5 Normalized forward or reverse current: (a) from charge model (4.25); (b) strong inversion approximation (4.29); (c) weak inversion approximation (4.33); (d) from the interpolation formula (4.39) between weak and strong inversion approximations

Now, equation (4.19) can be associated with (3.48) applied at the source or at the drain end of the channel

$$v_p - v_{s,d} = 2q_{s,d} + \ln q_{s,d} \tag{4.23}$$

to obtain the relation between drain current components $i_{f,r}$ and bias voltages $v_p - v_{s,d}$ in a parametric form. The parameter $(q_{s,d})$ can be expressed by inverting (4.19):

$$q_{s,d} = \frac{\sqrt{1 + 4i_{f,r}} - 1}{2}. \tag{4.24}$$

It can then be inserted in (4.23), which yields

$$v_p - v_{s,d} = \sqrt{1 + 4i_{f,r}} + \ln\left(\sqrt{1 + 4i_{f,r}} - 1\right) - (1 + \ln 2). \tag{4.25}$$

This general expression cannot be inverted to provide $i_{f,r}$ as a function of $v_{s,d}$, but it can be plotted as shown in Figure 4.5 (curve a).

4.4.2 Compact Model Parameters

Equation (4.25) describes the general current–voltage relationship for a long-channel transistor. It is continuously valid from weak to strong inversion.

Only *three model parameters* and one physical parameter are needed to obtain from this dimensionless equation the relation between bias voltages V_G and $V_{S,D}$, and the resulting current component $I_{F,R}$.

The physical parameter is U_T. It is used to normalize all voltages in the dimensionless equation.

The three device parameters are:

1. *The slope factor n* defined in Figure 3.7 and by equation (3.34). This parameter was further discussed in Section 3.6.4.

2. *The threshold voltage V_{T0}* defined in Fig (3.12) and by equations (3.58) and (3.66). It is very slightly dependent on V_G through ψ_p (see Figure 3.10), but can be considered bias independent in practice.

    These first two parameters relate the gate voltage V_G to the pinch-off voltage V_P according to (3.63). They were already introduced to obtain the charge–voltage relationship in Chapter 3.

3. *The transfer parameter β.* It is combined with n and U_T to obtain the specific current I_{spec} defined by equation (4.14) and is used to normalize components I_F and I_R of the drain current.

The drain current in all the modes of operation of the transistor identified in Figure 4.3 can be obtained by subtracting $I_R(V_D, V_G)$ from $I_F(V_S, V_G)$ according to (4.9). However, these two components of the drain current cannot be obtained analytically from (4.25) since this expression cannot be inverted. This is the reason for the approximative curve d of Figure 4.5, which will be introduced in Section 4.4.6.

4.4.3 Inversion Coefficient

In equation (4.25), the first term that corresponds to strong inversion dominates for $i_{f,r} \gg 1$, whereas the second term corresponding to weak inversion dominates for $i_{f,r} \ll 1$. Thus, the specific current I_{spec} can be used to characterize the current level at which I_F or I_R changes from weak to strong inversion. The level of inversion of the whole transistor can then be characterized by an *inversion coefficient* IC defined by

$$IC = \max (i_f = I_F/I_{spec}, i_r = I_R/I_{spec}), \qquad (4.26)$$

and the diagram of Figure 4.3 can be replaced by that of Figure 4.6.

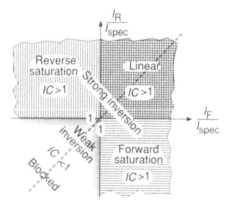

Figure 4.6   Modes of operation characterized by current levels

Again, although $IC = 1$ is used in this diagram as the limit between weak and strong inversion, the transition is progressive through a zone of moderate inversion. Hence the transistor operates in

- weak inversion for $IC \ll 1$;

- strong inversion for $IC \gg 1$; and

- moderate inversion for $IC \cong 1$. The width of this zone is not precisely defined. It must be extended until the weak or strong inversion approximation is sufficient to reach the expected accuracy (see Figures 4.7 and 4.10).

The notion of inversion coefficient is qualitatively equivalent to that of *gate voltage overhead*

$$V_G - V_{T0} - nV_S = V_{GS} - (V_{T0} + (n-1)V_S) = n(V_P - V_S) \qquad (4.27)$$

to characterize the level of inversion. However the latter is not very convenient in moderate or weak inversion where a small variation of voltage produces a large variation of current. Moreover, the gate voltage overhead becomes negative in weak inversion.

4.4.4 Approximation of the Drain Current in Strong Inversion

As established in Section 3.6.3, the inverted charge in strong inversion can be approximated by the linear function of $V_P - V$ described by equation (3.64). The corresponding forward and reverse components of the drain current are obtained by introducing this expression in integral (4.9) as illustrated in Figure 4.7(a), which yields

$$I_{F,R} = \frac{\beta n}{2}(V_P - V_{S,D})^2 \quad \text{for} \quad V_P - V_{S,D} \gg U_T \qquad (4.28)$$

or, with normalized voltages and currents

$$i_{f,r} = \left(\frac{v_p - v_{s,d}}{2}\right)^2 \quad \text{or} \quad v_p - v_{s,d} = 2\sqrt{i_{f,r}}. \qquad (4.29)$$

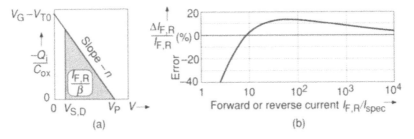

Figure 4.7 Strong inversion approximation: (a) calculation of current; (b) relative error

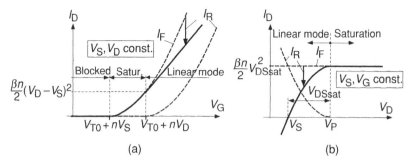

(a)   (b)

Figure 4.8   Characteristics in strong inversion: (a) gate-to-drain transfer characteristics; (b) output characteristics

It can be verified that, for $i_{f,r} \gg 1$, the general expression (4.25) of $v_p - v_{s,d}$ tends toward (4.29). However, since approximation (3.50) of the mobile charge was resulting in an excess of charge, (4.28) and (4.29) result in an excess of current that is hardly visible in Figure 4.5. This error is explicitly represented in Figure 4.7(b), which shows that the excess of current in strong inversion never exceeds 14%.

The current in *linear mode* can be expressed by introducing the approximation (3.63) of $V_P(V_G)$ in (4.28) and by subtracting I_R from I_F:

$$I_D = \frac{\beta}{2n}[\underbrace{(V_G - V_{T0} - nV_S)^2}_{\text{forward}} - \underbrace{(V_G - V_{T0} - nV_D)^2}_{\text{reverse}}]$$

$$= \beta(V_D - V_S)[V_G - V_{T0} - \frac{n}{2}(V_D + V_S)]. \tag{4.30}$$

The drain current is indeed a linear function of the gate voltage (with an offset $V_{T0} + n(V_D + V_S)/2$), because it is the difference of two identical square laws shifted by $n(V_D - V_S)$, as illustrated in Figure 4.8(a).

If the gate voltage is reduced below $V_{T0} + nV_D$ (corresponding to $V_D > V_P$), then the reverse current becomes negligible and the transistor is in *forward saturation* with $I_D = I_F$ given by

$$I_D = \frac{\beta n}{2}(V_P - V_S)^2 = \frac{\beta n}{2}V_{\text{DSsat}}^2 = \frac{\beta}{2n}(V_G - V_{T0} - nV_S)^2. \tag{4.31}$$

The drain current is a square law function of the gate voltage and is, as expected, independent of the drain voltage.

Combining the second expression of this saturation current with definitions (4.26) and (4.14) provides the relation between saturation voltage and inversion coefficient IC in strong inversion:

$$IC = \left(\frac{V_P - V_S}{2U_T}\right)^2 = \left(\frac{V_{\text{DSsat}}}{2U_T}\right)^2 \quad \text{(strong inversion only).} \tag{4.32}$$

If the gate voltage is further reduced below $V_{T0} + nV_S$ (corresponding to $V_S > V_P$), then the forward mode becomes zero and the transistor is blocked in this approximation.

4.4.5 Approximation of the Drain Current in Weak Inversion

As established in Section 3.5, the inverted charge in weak inversion can be approximated by the exponential function of $V_P - V$ described by equation (3.49). The corresponding forward and reverse components of drain current are obtained by introducing this expression in integral (4.16) which yields

$$i_{f,r} = \exp(v_p - v_{s,d}) \quad \text{or} \quad v_p - v_{s,d} = \ln i_{f,r}. \qquad (4.33)$$

It can be verified that the second form of (4.33) is the asymptotic value of $v_p - v_{s,d}$ given by the general current expression (4.25) for $i_{f,r} \ll 1$.

For nonnormalized variables, (4.33) becomes

$$I_{F,R} = I_{spec} \exp \frac{V_P - V_{S,D}}{U_T}. \qquad (4.34)$$

It should be pointed out that (unlike what is suggested by Figure 4.1) weak inversion is usually obtained by applying a value of gate voltage V_G smaller than the threshold V_{T0}, hence the alternative appellation of "*subthreshold*" for this mode of operation. The pinch-off voltage V_P then becomes negative, as illustrated in Figure 4.9, and weak inversion is already reached for $V_S = 0$.

The same figure shows that the approximation is valid only for $V - V_P \gg U_T$, and hence for $I_{F,R} \ll I_{spec}$. It yields an excess of current, which can be calculated by comparing (4.33) with (4.25). This error represented in Figure 4.10 is already about 10% for $I_{F,R}/I_{spec} = 0.1$.

The drain current equation is obtained by introducing approximation (3.63) of $V_P(V_G)$ in (4.34) and by subtracting I_R from I_F:

$$I_D = I_{spec} \exp \frac{V_G - V_{T0}}{nU_T} \left(\exp \frac{-V_S}{U_T} - \exp \frac{-V_D}{U_T} \right) \quad \text{for} \quad IC \ll 1, \qquad (4.35)$$

where the first and the second term in the parentheses are the distinctive parts of I_F and I_R, respectively. This equation is valid as long as the inversion coefficient IC defined by (4.26) is sufficiently smaller than unity.

This expression can also be obtained from expression (4.19) of the components of the drain current, where the square term becomes negligible for $q_{s,d} \ll 1$. Returning to denormalized

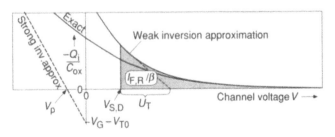

Figure 4.9 Weak inversion approximation

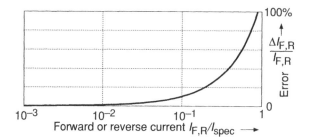

Figure 4.10   Relative excess of current in weak inversion approximation

charge Q_{iS} and Q_{iD} at the source and drain ends of the channel defined by (3.41) results in

$$I_D = I_F - I_R = \frac{I_{spec}}{Q_{spec}}(Q_{iS} - Q_{iD}) = \mu W U_T \underbrace{\frac{Q_{iD} - Q_{iS}}{L}}_{dQ_i/dx}. \tag{4.36}$$

in which expression (3.49) of the mobile charge can be used to replace Q_{iS} and Q_{iD}, and ends up with (4.35).

The comparison of equation (4.36) with (4.2) shows that it is a current carried only by diffusion. Indeed, according to Figure 3.3, the surface potential in weak inversion depends only on the gate voltage and is therefore constant along the channel. Thus, the current in weak inversion can only be carried by diffusion. The current carriers (in this case electrons) are *locally majority carriers*, since the holes have been repelled away from the surface. The channel length can thus be much longer than the diffusion length of minority carriers deep in the substrate.

As discussed in Section 3.6.4, in weak inversion the slope factor n represents the attenuation of the capacitive divider formed by C_{ox} and the surface depletion capacitor C_d. Its value can be evaluated as $n_{w\,min}$ given by (3.72).

The dependency of the drain current on the three control voltages expressed by equation (4.35) is illustrated in Figure 4.11.

The drain current saturates to its forward value I_F as soon as the drain voltage exceeds the source voltage by $3U_T$ to $5U_T$, as illustrated by the output characteristics. The output saturation voltage $V_{DS\,sat}$, which is given by equation (4.12) in strong inversion, is progressively reduced as the inversion coefficient is decreased by reducing $V_P - V_S$, but it is limited to the minimum

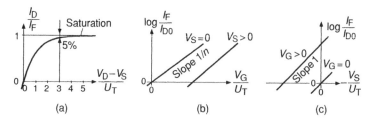

Figure 4.11   Characteristics in weak inversion: (a) output characteristics; (b) transfer from gate; (c) transfer from source

obtained in weak inversion:

$$V_{\text{DS sat}} = V_{\text{DS sat min}} = 3\,U_T \text{ to } 5\,U_T \quad \text{(weak inversion).} \tag{4.37}$$

If $V_G = V_S = 0$, the saturation current (I_F in forward mode) is reduced to

$$I_{D0} \triangleq I_{\text{spec}} \exp \frac{-V_{T0}}{n U_T}. \tag{4.38}$$

In digital CMOS circuits, this is the residual channel current of "off" transistors, which is responsible for their DC current consumption.

According to equation (4.35), the transfer characteristics from the gate are exponential, corresponding to straight lines with slope $1/n$ in the normalized values used in the semilog plot of Figure 4.11(b). In saturation, if V_S is increased by some amount ΔV_S, V_G must be increased by $n\Delta V_S$ to recover the same drain current, corresponding to a right shift of the characteristics. Furthermore, since according to (3.72) n decreases slightly for V_S increasing, the shifted line is slightly steeper.

The transfer characteristics from the source are also exponential in saturation, but without slope factor n. Thus, they correspond to straight lines of slope 1 in the normalized semilog plot of Figure 4.11(c). This exponential behavior is very similar to that of a bipolar transistor in active mode, the base–emitter voltage V_{BE} being replaced by $-V_S$. Indeed, according to equation (3.49), the mobile charge Q_{iS} at the source end of the channel depends exponentially on the source voltage $-V_S$, and the saturation current I_F is a linear function of this charge. Similarly, in a bipolar transistor, the density of minority carriers at the emitter side of the base depends exponentially on V_{BE}, and the collector current is a linear function of this density.

But contrary to the bipolar, the junction is normally reverse biased. Hence, there is no carrier injection into the local substrate.

4.4.6 Alternative Continuous Models

As already pointed out, the general expression (4.25) relating the control voltages (v_p, v_s, and v_d) and the two components of the drain current (i_f and i_r) cannot be analytically inverted to calculate currents from voltages. It is therefore useful to introduce an approximative expression that continuously interpolates the current behavior between weak and strong inversion. One possibility is the simple following expression [19]:

$$i_{f,r} = \ln^2 \left[1 + \exp \frac{v_p - v_{s,d}}{2} \right] \quad \text{or} \quad v_p - v_{s,d} = 2\ln\left(e^{\sqrt{i_{f,r}}} - 1\right). \tag{4.39}$$

This expression is plotted in Figure 4.5 (curve d) for comparison with the exact expression (4.25).

It can be verified that this continuous approximation tends asymptotically to the strong inversion approximation (4.29) for $v_p - v_{s,d} \gg 1$, and to the weak inversion approximation (4.33) for $v_p - v_{s,d} \ll 1$.

It coincides with the exact model (4.25) at $i_{f,r} = 6.48$, giving a value of currents slightly higher above and slightly lower below this limit.

4.5 FUNDAMENTAL PROPERTY: VALIDITY AND APPLICATION

4.5.1 Generalization of Drain Current Expression

The fundamental property of long-channel MOS transistor obtained in Section 4.2 can be generalized [73] if equation (4.5) of the drain current can be written in the form

$$I_D = \frac{F_V(V, V_G)\, dV}{F_x(x, V_G)\, dx} \tag{4.40}$$

where $F_V(V, V_G)$ is a function of V and V_G *but not of x*, and $F_x(x, V_G)$ is a function of x and V_G *but not of V*; I_D is then a separable function of position x and voltage V in the channel. As long as the channel length L is independent of V_D and V_S, $F_x(x, V_G)$ and $F_V(V, V_G)$ can then be integrated separately:

$$I_D \int_0^L F_x(x, V_G)\, dx = \int_{V_S}^{V_D} F_V(V, V_G)\, dV. \tag{4.41}$$

Now, since $F_V(V, V_G)$ tends to zero for large V, this expression can be written as

$$I_D = \frac{1}{\int_0^L F_x(x, V_G)\, dx} \left[\int_{V_S}^{\infty} F_V(V, V_G)\, dV - \int_{V_D}^{\infty} F_V(V, V_G)\, dV \right] \tag{4.42}$$

or

$$I_D = I(V_S, V_G) - I(V_D, V_G) = I_F - I_R, \tag{4.43}$$

where

$$I(V, V_G) = \frac{\int_V^{\infty} F_V(V, V_G)\, dV}{\int_0^L F_x(x, V_G)\, dx}. \tag{4.44}$$

This result is a generalization of (4.9), expressing the fact that the drain current is the *superposition* of *independent* and *symmetrical* (same function I) effects of source and drain voltages.

It is interesting to point out that this property is similar to that of bipolar transistors as expressed by the Ebers–Moll model [18].

4.5.2 Domain of Validity

Let us examine the necessary and sufficient conditions for which equation (4.5) has the required form (4.40).

The channel width W does not depend on V. It may thus depend on position x along the channel, and therefore can be included in $F_x(x, V_G)$. For example, in a concentric circular transistor,

$$W(x) = 2\pi(R_S + x) = \frac{1}{F_x(x, V_G)}, \tag{4.45}$$

where R_S is the radius of the source. Thus

$$\int_0^L F_x(x, V_G)\,\mathrm{d}x = \frac{1}{2\pi}\ln\left(1 + \frac{L}{R_S}\right), \tag{4.46}$$

which replaces L/W in equation (4.8) of β.

In a more general case, the device can be split into several (or an infinity of) transistors of different lengths and variable widths, all connected in parallel. As long as each transistor i fulfills equation (4.40) with

$$I_{Di} = I_i(V_S, V_G) - I_i(V_D, V_G), \tag{4.47}$$

the sum of I_{Di} fulfills it as well.

Equation (3.32) of the mobile inverted charge per unit area Q_i can be rewritten by introducing expression (3.30) of Γ_b:

$$-Q_i = C_{ox}(V_G - V_{FB} - \Psi_s) - \sqrt{2q N_b \epsilon_{si} \Psi_s}, \tag{4.48}$$

which is a function of Ψ_s. Now, Figure 3.3 shows that Ψ_s is possibly a function of V, but not of x. Hence if (and only if) $V_G - V_{FB}$, N_b and C_{ox} are all independent of x (homogeneous channel), then Q_i is a function of V only. It can then be included in $F_V(V, V_G)$ and the property is not affected.

The property is conserved if any other term in expression (4.48) of Q_i also depends on V (or on Ψ_s, but not on x). This includes the effect of gate polysilicon depletion, which is equivalent to a value of C_{ox} function of Ψ_s.

If the doping concentration N_b is a function $N_b(z)$ of the depth z in the substrate, then the last term of (4.48) that represents the depletion charge density Q_b becomes a different function Ψ_s, as will be discussed in Section 8.3. However, the property is not affected as long as this function remains independent of x (homogeneous channel).

If the channel is nonhomogeneous along its lateral dimension (y-axis), the device may again be split into several transistors i connected in parallel, each of them fulfilling (4.47). This includes the possible difference of side structures of a narrow channel transistor, for which the fundamental property is therefore not affected.

As will be discussed in Section 8.2, the value of mobility μ depends on the local vertical surface field E_{zs}. Now, combining (3.17), (3.19), and (3.22) yields

$$E_{zs} = \frac{C_{ox}}{\epsilon_{si}}(V_G - V_{FB} - \Psi_s), \tag{4.49}$$

which depends only on Ψ_s (thus possibly on V) for a homogeneous channel. The variation of mobility with the vertical field can therefore be included in $F_V(V, V_G)$ and does not affect the property.

But *the mobility should be independent of the drain current* I_D. Indeed, such a dependency could be included neither in $F_V(V, V_G)$ nor in $F_x(x, V_G)$, hence, the property would not be conserved.

As another necessary condition, the effective value of the *channel length L* along which $F_x(x, V_G)$ is integrated in (4.41) should be *constant*. It should not depend on the drain current I_D, or on the drain or source voltage V_D or V_S.

In summary, the fundamental property of MOS transistors expressed by equation (4.40) is valid if (and only if) the channel is homogeneous along its source–drain dimension (x-axis) with a fixed effective length, and if (and only if) the mobility is independent of the drain current. The property depends neither on the shape and width of the channel nor on the doping profile of the substrate. It remains valid for large gate voltages, in spite of the mobility reduction due to the vertical field.

4.5.3 Causes of Degradation

4.5.3.1 Finite length of channel

When the channel is not very long, several independent mechanisms degrade the fundamental property. This is the most important reason why this property is never perfectly valid in practice.

Channel length modulation. As will be discussed in Section 4.6, when the drain (or source) voltage is increased, the effective channel length is slightly reduced by the extension of the depleted region surrounding the drain (or the source). As expressed by (4.59) or (4.61), the forward current I_F (or reverse current I_R) is therefore slightly dependent on the drain voltage V_D (or the source voltage V_S), which is not compatible with (4.43). Since this effect is inversely proportional to L, the property is progressively degraded when the channel is shortened.

Short-channel effects. If the channel length is reduced more than proportionally to the gate and drain voltages, the longitudinal field in the channel is increased. Hence, the velocity of mobile carriers is increased, resulting in an increase of drain current. However, at high values, this velocity starts increasing less than proportionally to the field, to finally reach a saturation limit. Thus, mobility μ becomes a function of the field (or of the current), as will be discussed in Section 9.1.

This variation cannot be included in $F_x(x, V_G)$ or $F_V(V, V_G)$ as required by (4.40), and the property is degraded.

For very short channel, additional effects such as drain-induced barrier lowering discussed in Section 9.3 and other two-dimensional effects further degrade and possibly destroy the property.

4.5.3.2 Nonhomogeneous channel

Referring to equation (4.48), if any term of its right-hand side depends on position x along the channel, it makes Q_i a (nonseparable) function of both x and Ψ_s (hence of V); relation (4.40) is then no longer valid, and the property is lost. This is true even if the nonhomogeneous channel remains symmetrical with respect to source and drain: indeed, the effects of source and drain voltages on I_D remain symmetrical, but they are neither independent nor linearly superimposed.

Since ϵ_{si} and V_G are normally constant along the channel, three terms remain to be examined in (4.48), namely N_b, V_{FB}, and C_{ox}.

Variations of substrate doping N_b can be due to some intentional channel engineering such as lightly doped drain (LDD) and "halo" implants, or to some artifact of the process like the piling-up of impurities at both ends of the channel. Whatever the process, variations of N_b always occur at the very ends of the channel, due to the presence of the source and drain diffusions. This is yet another reason why the fundamental property is lost in very short channel devices.

Since the flat-band voltage V_{FB} depends on the Fermi-level of silicon in the channel, it is variable as long as the doping concentration is itself variable. Further variations of V_{FB} could be due to variations of the fixed interface charge, as a consequence of non source–drain symmetrical channel engineering.

There is no reason to intentionally change the value of the oxide capacitance C_{ox} along the channel. However, local variations at both ends of the channel are unavoidable, which further contributes to the degradation of the property for short-channel devices.

Weak inversion represents a special case. It is characterized by the fact that, all along the channel, the mobile charge Q_i is negligible with respect to the depletion charge Q_b. As a consequence, the function F defined by equation (3.16), which relates the vertical surface field to the surface potential, becomes *independent of the channel voltage V*, as illustrated in Figure 3.1. Using equations (3.27) and (3.28), the mobile charge density can thus be expressed as

$$Q_i = F_q \exp \frac{-V}{U_T} \tag{4.50}$$

where function F_q is *independent of V*. It can therefore depend on x and be included in F_x of (4.40), whereas F_V reduces to

$$F_V = \exp \frac{-V}{U_T} \quad \text{(weak inversion only).} \tag{4.51}$$

Since F_q contains all other parameters on which Q_i depends, namely $V_G - V_{FB}$, C_{ox}, and N_b, these parameters may change along the channel without affecting the property.

In weak inversion, the fundamental property expressed by (4.43) is thus valid *even for a nonhomogeneous channel*. This is true also for bipolar transistors operated in moderate injection, and can be traced back to the fact that the current is a linear function of the mobile charge density. This linearity exists as long as the charge carriers do not affect the electrostatic potential: indeed, they are dominated respectively by the depletion charge Q_b for MOS transistor in weak inversion, and by majority carriers for the bipolar transistor in moderate injection.

4.5.4 Concept of Pseudo-Resistor

By defining [75, 76] a *pseudo-voltage V^** given by

$$V^* = -K_0 \int_V^\infty F_V(V, V_G) \, dV \tag{4.52}$$

and a *pseudo-resistance* R^* given by

$$R^* = K_0 \int_0^L F_x(x, V_G) \, dx, \tag{4.53}$$

equation (4.43) of the drain current can be written in the form of a *pseudo-Ohm's law*

$$I_D = (V_D^* - V_S^*)/R^*. \tag{4.54}$$

Hence, by similarity with a network of linear resistors, any network obtained by interconnecting the transistors characterized by the same function $F_V(V, V_G)$ (same process) biased (in the general case) at the same gate voltage V_G is *linear with respect to currents*. In other words, at each node of such a network, currents split linearly in the various branches [77].

Thus, any prototype network made of real linear resistors may be converted to a pseudo-resistor network made of transistors only, provided only currents are considered. A ground in the resistor prototype ($V = 0$) corresponds to a *pseudo-ground* in the transistor network ($V^* = 0$) obtained by choosing V large enough to make integral (4.52) negligible. This means that the corresponding side of the transistor is saturated.

The constant K_0 introduced in (4.52) and (4.53) is always positive. Hence, pseudo-voltages for N-channel transistors are always negative. This means that real voltages in the corresponding prototype made of resistors must also be negative (with respect to the ground).

It must be noticed that for P-channel transistors, the "minus" sign in (4.52) must be replaced by a "plus" sign (with the sign conventions defined in Section 2.2). Hence, pseudo-voltages for P-channel transistors are always positive.

The numerical value of K_0 is irrelevant, since it disappears in (4.54), but its dimension can be chosen so as to obtain the dimension of V^* in volts, and that of R^* in ohms.

It should be pointed out that, although a narrow channel does not affect the fundamental property of a transistor, it has an effect on function $F_V(V, V_G)$. This function may differ from that of a wider channel, thereby degrading the linearity of current splitting.

For the *special case of weak inversion*, (4.51) shows that F_V and hence V^* are independent of V_G. The linear pseudo-Ohm's law is thus valid even for transistors having different gate voltages. Now since F_q in (4.50) depends on V_G and is included in F_x, the value of pseudo-resistor R^* given by (4.53) can be controlled separately for each transistor by its gate voltage. Furthermore, linear current splitting is maintained even with narrow channel transistors.

4.6 CHANNEL LENGTH MODULATION

4.6.1 Effective Channel Length

All previous calculations of the drain current [in particular by (4.7) or by the more general expression (4.42)] used the effective value L of the channel length. But this effective length is shorter than the distance L_{SD} separating the source and drain metallurgic junctions.

Let us first consider the flat-band situation, obtained by applying a gate voltage $V_G = V_{FB}$, with flat-band voltage V_{FB} defined by (3.22). In this situation, the electrostatic potential Ψ in the channel remains constant. Indeed, $\Psi = 0$ from deep in the substrate to the surface.

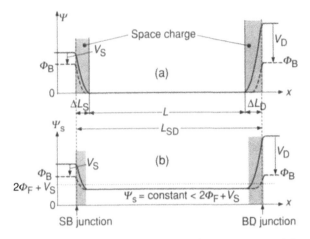

Figure 4.12   Potential along the channel: (a) flat-band; (b) weak inversion

However, at both ends of the channel, a potential barrier is produced by the source and drain junctions. As shown by Figure 4.12(a), this creates space charge regions of length $\Delta L_{S,D}$ that depend on the barrier height $\Phi_B + V_{S,D}$, with the barrier at equilibrium Φ_B given by

$$\Phi_B = U_T \ln \frac{N_{\text{diff}} N_b}{n_i^2},\tag{4.55}$$

where N_{diff} is the doping concentration of the source and drain diffusions. Since $N_b \ll N_{\text{diff}}$, the source and drain space charge regions extend mainly in the P-type channel region. Their lengths can be calculated by double integration of Poisson equation (3.2) along the x-axis with $\rho = -q N_b$. Identification of the result with the barrier height $\Phi_B + V_{S,D}$ then yields

$$\Delta L_{S,D} = \sqrt{\frac{2\epsilon_{si}(\Phi_B + V_{S,D})}{q N_b}}.\tag{4.56}$$

To obtain the effective channel length L, the lengths of the space charge regions must be subtracted from distance L_{SD}; hence,

$$L = L_{SD} - \Delta L_S - \Delta L_D = L_{SD} - \sqrt{\frac{2\epsilon_{si}}{q N_b}} \left(\sqrt{\Phi_B + V_S} + \sqrt{\Phi_B + V_D} \right).\tag{4.57}$$

The effective length is slightly dependent on the value of the source and drain voltages. Differentiation of (4.57) provides

$$\frac{dL}{dV_{S,D}} = -\sqrt{\frac{\epsilon_{si}}{2q N_b(\Phi_B + V_{S,D})}}.\tag{4.58}$$

4.6.2 Weak Inversion

If the gate voltage is increased above the flat-band voltage V_{FB}, the surface potential Ψ_s increases as shown by Figure 3.3. As long as the device remains in weak inversion ($\Psi_s < 2\Phi_F + V_S$), the surface potential is constant along the channel and independent of V_S and V_D, as illustrated in Figure 4.12(b). At both ends of the channel the barrier height is reduced, but the field pattern becomes two-dimensional, since Ψ becomes a function of x and z. Equation (4.57) for the one-dimensional case is no longer exact, but the effective channel length is still modulated by V_D and V_S, resulting in a slight variation of the specific current I_{spec}, and proportional variations of I_F and I_R.

Now, I_F already strongly depends on V_S as expressed by (4.34), so this small additional variation can be neglected. But I_F would be independent of V_D without channel length modulation, so this small variation cannot be neglected. The symmetrical situation exists for the dependency of I_R on V_D and V_S. Hence by (4.58),

$$\frac{dI_{F,R}}{dV_{D,S}} = -\kappa \frac{dI_{F,R}}{dL} \sqrt{\frac{\epsilon_{si}}{2q N_b(\Phi_B + V_{D,S})}} \qquad (4.59)$$

where κ is a correction factor for two-dimensional effects.

4.6.3 Strong Inversion

The variation of surface potential Ψ_s along the channel in strong inversion is depicted in Figure 4.13. It can be assumed to be independent of the gate voltage and to follow the channel voltage according to $\Psi_S = V + \Psi_0$ (see Section 3.6.3). Hence, at equilibrium ($V = V_S = V_D = 0$), $\Psi_s = \Psi_0$ all along the channel. The potential barrier at both ends of the channel is reduced

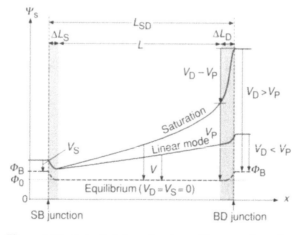

Figure 4.13 Potential along the channel in strong inversion

to $\Phi_B - \Psi_0$. Since $\Psi_0 \cong 2\Phi_F$, combining (3.8) and (4.55) gives

$$\Phi_B - \Psi_0 \cong U_T \ln \frac{N_{\text{diff}}}{N_b}. \tag{4.60}$$

For V_S and $V_D > 0$, this barrier remains constant as long as the transistor is in linear mode ($V_D < V_P$). Hence, effective channel length L remains constant. If the transistor is saturated with $V_D - V_P > 0$, the drain barrier is increased by the same amount; the forward current becomes slightly dependent on the drain voltage. A symmetrical situations exists in reverse saturation. Thus, similar to (4.59) with a different barrier height

$$\frac{dI_{F,R}}{dV_{D,S}} = -\kappa \frac{dI_{F,R}}{dL} \sqrt{\frac{\epsilon_{si}}{2q N_b(\Phi_B - \Psi_0 + V_{D,S} - V_P)}}. \tag{4.61}$$

Experiments show that factor κ correcting for two-dimensional effects tends to decrease when the inversion coefficient IC is increased.

4.6.4 Geometrical Effects

For the usual case of constant width W, I_{spec} given by (4.14) is inversely proportional to L. Hence

$$-\frac{dI_{F,R}}{dL} = \frac{I_{F,R}}{L}, \tag{4.62}$$

and the effect of channel length modulation is identical in forward and reverse saturation.

We have seen in Section 4.5 that the fundamental property of the transistor, which includes symmetrical source–drain characteristics, is independent of the channel geometry. But channel length modulation degrades this property and is itself sensitive to device geometry.

As an example, consider again a concentric circular transistor with a radius R_S of the source and an inner radius $R_D = R_S + L$ of the drain. According to equations (4.43) and (4.46),

$$I_F \propto \frac{1}{\ln(1 + L/R_S)}; \tag{4.63}$$

hence,

$$-\frac{dI_F}{dL} = \frac{I_F}{R_D \ln R_D/R_S} = \frac{I_F}{L} \frac{1 - R_S/R_D}{\ln(R_D/R_S)}, \tag{4.64}$$

which is smaller than result (4.62) for a rectangular channel.

Now if the transistor is in the reverse mode, its saturation current is $-I_R$ and channel length modulation occurs at the source end of the channel. Thus, (4.63) can be rewritten as

$$I_R \propto \frac{1}{\ln \frac{R_D}{R_D - L}}, \tag{4.65}$$

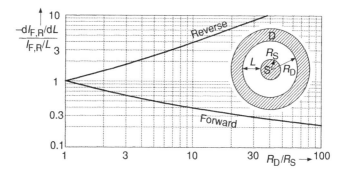

Figure 4.14 Effect of channel length modulation in a circular concentric transistor

giving

$$-\frac{\mathrm{d}I_R}{\mathrm{d}L} = \frac{I_R}{L}\frac{R_D/R_S - 1}{\ln(R_D/R_S)},$$ (4.66)

which is larger than result (4.62). Results (4.64) and (4.66) are plotted in Figure 4.14.

It must be pointed out that in a concentric transistor structure, the inner diffusion is usually the drain, in order to minimize the associated junction capacitance. Forward and reverse modes are then exchanged with respect to Figure 4.14 and the effect of channel length modulation on the forward saturation current is increased. This result can be qualitatively extended to any nonrectangular channel: for a given channel length L, the variation of forward saturation current due to channel length modulation is increased if the source is wider than the drain; it is reduced if the drain is wider than the source.

5 The Small-Signal Model

This Chapter describes the small-signal model of the MOS transistor obtained from the large-signal model after a proper linearization of the large-signal equations at a defined operating point. It starts looking at the dc small-signal model, introducing the source, drain and gate transconductances and higlighting the fundamental relations between them. The transconductances are then expressed in terms of bias covering all modes of inversion. The fundamental transconductance to drain current ratio is then introduced and its use for circuit sizing is illustrated. The small-signal dynamic behavior is introduced directly by first deriving a complete non-quasi-static (NQS) model, introducing the source, drain and gate transadmittances together with the five other admittances and the fundamental relations between them. Their bias dependence over all regions of inversion is presented. The NQS model serves as the basis for deriving the quasi-static (QS) model. The concept of transcapacitances is introduced as a result of a first-order approximation of the transadmittances. In the QS model, the admittances of the NQS model then reduce to the five intrinsic capacitances. The domain of validity of the three different small-signal models is then defined and the use of a NQS model for transient operation is also discussed.

5.1 THE STATIC SMALL-SIGNAL MODEL

5.1.1 Transconductances

5.1.1.1 General expressions

The most important small-signal parameters are without doubt the *transconductances*. The transconductances together with the capacitances determine the transit frequency f_t or the speed of the device, the thermal noise, and indirectly the current consumption. Since the MOS transistor is a four-terminal device, it is controlled by three independent voltages, namely V_G, V_S, and V_D. A transconductance value can therefore be defined for each of these control

Charge-Based MOS Transistor Modeling: The EKV Model for Low-Power and RF IC Design   C. Enz and E. Vittoz
© 2006 John Wiley & Sons, Ltd.

voltages. The total increment of the drain current ΔI_D is given by

$$\Delta I_D = \left.\frac{\partial I_D}{\partial V_S}\right|_{op} \Delta V_S + \left.\frac{\partial I_D}{\partial V_D}\right|_{op} \Delta V_D + \left.\frac{\partial I_D}{\partial V_G}\right|_{op} \Delta V_G$$
$$= -G_{ms}\,\Delta V_S + G_{md}\,\Delta V_D + G_m\,\Delta V_G \qquad (5.1)$$

where G_{ms}, G_{md}, and G_m are the *source, drain* and *gate transconductances* respectively, defined as

$$G_{ms} \triangleq -\left.\frac{\partial I_D}{\partial V_S}\right|_{op}, \qquad (5.2a)$$

$$G_{md} \triangleq \left.\frac{\partial I_D}{\partial V_D}\right|_{op}. \qquad (5.2b)$$

$$G_m \triangleq \left.\frac{\partial I_D}{\partial V_G}\right|_{op}, \qquad (5.2c)$$

where notation *op* stands for the operating point at which the linearization occurs. It can be characterized by the set of the three dc voltages V_G, V_S, V_D. Note that all the small-signal transconductances defined in (5.2) are positive. Now, since according to (4.43) $I_D = I_F(V_G, V_S) - I_R(V_G, V_D)$, (5.1) can be rewritten as

$$\Delta I_D = \underbrace{\frac{\partial I_F}{\partial V_S}}_{=-G_{ms}}\,\Delta V_S + \underbrace{\frac{-\partial I_R}{\partial V_D}}_{=G_{md}}\,\Delta V_D + \underbrace{\left(\frac{\partial I_F}{\partial V_P} - \frac{\partial I_R}{\partial V_P}\right)\frac{\partial V_P}{\partial V_G}}_{=G_m}\,\Delta V_G. \qquad (5.3)$$

Transconductance G_{ms} depends only on the forward current I_F and is therefore independent of the drain voltage V_D, whereas G_{md} depends only on the reverse current I_R and is therefore independent of the source voltage V_S. These tranconductances can be identified on the Q_i versus V plot as illustrated in Figure 5.1.

As can be seen by inspection of this figure, the quantity by which a small variation of source voltage V_S must be multiplied to obtain the corresponding variation of area I_D/β is the

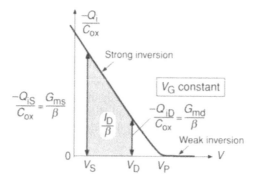

Figure 5.1   Relation of source and drain transconductances with the mobile inverted charge

particular value $-Q_{iS}/C_{ox}$ of $-Q_i/C_{ox}$ taken at the source end of the channel. Hence,

$$G_{ms} = \frac{\beta}{C_{ox}} (-Q_{iS}) = \mu \frac{W}{L} (-Q_{iS}) = G_{spec} \, q_s \qquad (5.4)$$

and, symmetrically,

$$G_{md} = \frac{\beta}{C_{ox}} (-Q_{iD}) = \mu \frac{W}{L} (-Q_{iD}) = G_{spec} \, q_d, \qquad (5.5)$$

where

$$G_{spec} \triangleq I_{spec}/U_T = 2n\beta U_T, \qquad (5.6)$$

and q_s and q_d are the normalized charges at both ends of the channel defined by

$$q_s \triangleq \frac{-Q_{iS}}{Q_{spec}}, \qquad (5.7a)$$

$$q_d \triangleq \frac{-Q_{iD}}{Q_{spec}}. \qquad (5.7b)$$

Note that the two expressions (5.4) and (5.5) are very general, and do not depend on the precise shape of $-Q_i(V)$.

As shown by (4.25), I_F and I_R depend on the differences $V_P - V_S$ and $V_P - V_D$ respectively. A variation of V_S or V_D has the same effect on I_F and I_R as an equal variation of V_P of opposite sign. Therefore we have

$$\frac{\partial I_F}{\partial V_P} = G_{ms} \quad \text{and} \quad \frac{\partial I_R}{\partial V_P} = G_{md}. \qquad (5.8)$$

Furthermore, according to (3.51), $dV_P/dV_G = 1/n$. Hence the expression of G_m in (5.3) becomes

$$G_m = \frac{G_{ms} - G_{md}}{n}. \qquad (5.9)$$

Note that this dependency of the gate transconductance on the source and drain transconductances G_{ms} and G_{md} is independent of the inversion coefficient of the transistor.

When the transistor is saturated, $I_R \ll I_F$ and $G_{md} \ll G_{ms}$ and hence

$$G_m = \frac{G_{ms}}{n} \qquad \text{(saturation).} \qquad (5.10)$$

Using the normalized charges q_s and q_d given in (4.24), expressions (5.4) and (5.5) become

$$g_{ms,d} \triangleq \frac{G_{ms,d}}{G_{spec}} = q_{s,d} = \frac{\sqrt{4i_{f,r}+1}-1}{2} = \frac{2i_{f,r}}{\sqrt{4i_{f,r}+1}+1}, \qquad (5.11)$$

or, by using denormalized variables,

$$G_{ms,d} = n\beta U_T \left(\sqrt{4 I_{F,R}/I_{spec} + 1} - 1 \right). \tag{5.12}$$

The gate transconductance can be obtained by introducing this expression in (5.9), giving

$$G_m = \beta U_T \left(\sqrt{4 I_F/I_{spec} + 1} - \sqrt{4 I_R/I_{spec} + 1} \right). \tag{5.13}$$

However, this result is useful only if the forward and reverse currents are known separately (and not only their difference $I_D = I_F - I_R$).

Since relation (3.48) between voltages and charge cannot be inverted, the transconductances cannot be expressed as functions of voltages in the general case.

5.1.1.2 Approximation in strong inversion

With the strong inversion approximation of the charge discussed in Section 3.6.3, the various transconductances can be found directly on the $V_{TB}(V)$ plot of Figure 3.12(a), as shown in Figure 5.2(a).

This diagram, which also shows the current in function of the bias voltages, will be called the *Jespers–Memelink* diagram [15,77]. It can be used to analyze and synthesize circuits using transistors in strong inversion. Expression (5.9) of the gate transconductance can be verified by simple inspection of this diagram.

The expressions of the source and drain transconductances can be obtained by differentiating the current given by (4.28) or simply by inspection of Figure 5.2.

If β, $V_{S,D}$, and V_P (or V_G) are known, this figure shows that

$$G_{ms,d} = n\beta(V_P - V_{S,D}) = \beta(V_G - V_{T0} - n V_{S,D}). \tag{5.14}$$

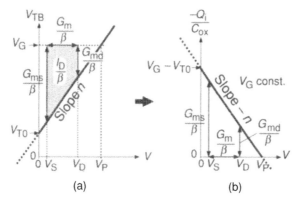

(a)   (b)

Figure 5.2   Drain current and transconductances in strong inversion: (a) Jespers–Memelink diagram; (b) corresponding $Q_i(V)$ plot

If only currents and voltages are known, but not β (the transistor is not yet sized), then remembering that the total area of the triangle is $I_{F,R}/\beta$ (see Figure 4.7), the transconductances can be obtained from

$$G_{ms,d} = \frac{2I_{F,R}}{V_P - V_{S,D}} = \frac{2nI_{F,R}}{V_G - V_{T0} - nV_{S,D}}. \tag{5.15}$$

If the known parameters are β and $I_{F,R}$, then the transconductances are obtained by extracting $(V_P - V_{S,D})$ from the current equation (4.28) and introducing it in (5.14) or in (5.15)

$$G_{ms,d} = \sqrt{2n\beta I_{F,R}}, \tag{5.16}$$

which is the expression used most usually, showing that the transconductance of a transistor in strong inversion is proportional to the square root of the current. It corresponds to the general expression (5.12) for $I_{F,R} \gg I_{spec} = 2n\beta U_T^2$.

Introducing expression (5.14) of the source and drain transconductances in (5.9) gives the simple expression of the gate transconductance in linear mode:

$$G_m = \beta(V_D - V_S) \tag{5.17}$$

that can also be obtained directly by inspection of Figure 5.2.

In *forward saturation*, $I_D = I_F$, $G_m = G_{ms}/n$, $(V_P - V_S) = V_{DSsat}$ according to (4.12), and the inversion coefficient IC is defined by (4.26). The gate transconductance is thus given by one of the following expressions:

$$
\begin{aligned}
G_m &= \beta(V_P - V_S) = \beta V_{DSsat} = \frac{\beta}{n}(V_G - V_{T0} - nV_S) \\
&= \frac{2I_D}{n(V_P - V_S)} = \frac{2I_D}{nV_{DSsat}} = \frac{2I_D}{V_G - V_{T0} - nV_S} \\
&= \sqrt{\frac{2\beta I_D}{n}} = 2\beta U_T\sqrt{IC} = \frac{G_{spec}}{n}\sqrt{IC}.
\end{aligned}
\tag{5.18}
$$

The first line of this equation shows that for a given value of β, V_{DSsat} must be increased to augment the transconductance G_m (with the result of an increase of current). But if the current is given, the second line shows that V_{DSsat} must be decreased to increase G_m (which requires an increase of β).

5.1.1.3 Approximation in weak inversion

The transconductance in weak inversion can be obtained by differentiating approximation (4.34) of the current:

$$G_{ms,d} = -\frac{\partial I_{F,R}}{\partial V_{S,D}} = \frac{I_{F,R}}{U_T}. \tag{5.19}$$

It corresponds to the general expression (5.12) for $I_{F,R} \ll I_{spec} = 2n\beta U_T^2$.

According to (5.9), the gate transconductance is then given by

$$G_m = \frac{G_{ms} - G_{md}}{n} = \frac{I_F - I_R}{nU_T} = \frac{I_D}{nU_T}. \tag{5.20}$$

It is proportional to the total drain current.

5.1.2 Residual Output Conductance in Saturation

According to the fundamental property discussed in Section 4.5, the forward component of the drain current (I_F) does not depend on the drain voltage V_D. As a consequence, the overall drain current in forward saturation (where $I_D = I_F$) should remain constant. However, as analyzed in Section 4.6, the slight variation of channel length caused by drain voltage variations renders I_F slightly dependent on V_D. This corresponds to a parasitic drain transconductance $\partial I_F / \partial V_D$ expressed by (4.59) or (4.61). In forward mode, this transconductance is normally much smaller than the source transconductance G_{ms}. Hence, it can be neglected in linear mode, where it is also smaller than the main drain transconductance G_{md}. This is no longer possible in forward saturation where G_{md} itself becomes negligible.

The symmetrical situation exists in reverse saturation where the parasitic source transconductance dI_R/dV_S can no longer be neglected with respect to the very small main source transconductance G_{ms}.

A convenient way to include these parasitic transconductances due to channel shortening is to replace them by a single *drain-to-source conductance* G_{ds}. Indeed, the total variation of the drain current given by (5.3) then becomes

$$\begin{aligned}
\Delta I_D &= G_m \, \Delta V_G - G_{ms} \, \Delta V_S + G_{md} \, \Delta V_D + G_{ds} \, (\Delta V_D - \Delta V_S) \\
&= G_m \, \Delta V_G - (G_{ms} + G_{ds}) \, \Delta V_S + (G_{md} + G_{ds}) \, \Delta V_D.
\end{aligned} \tag{5.21}$$

According to this equation, G_{ds} has to be accounted for only when G_{md} is very small (forward saturation) or when G_{ms} is very small (reverse saturation).

Since G_{ds} is due to the variation of channel length, its value is proportional to I_F (or I_R in the reverse mode). It can thus be expressed by

$$G_{ds} = I_{F,R}/V_M, \tag{5.22}$$

where V_M is the *channel length modulation voltage* given by

$$V_M = I_{F,R} \frac{dV_{D,S}}{dI_{F,R}} = L \left(-\frac{dV_{D,S}}{dL} \right) \left(-\frac{I_{F,R}}{L} \frac{dL}{dI_{F,R}} \right). \tag{5.23}$$

This fictitious voltage is thus proportional to the channel length L. The first term in parentheses can be obtained from (4.59) or (4.61) and is proportional to the square root of the channel doping concentration N_b.

The second term in parentheses is equal to 1 for constant channel width W (see (4.62)). If W is not constant and increases from source to drain, then this second term is increased in forward mode and decreased in reverse mode, as illustrated in Figure 4.14 for a concentric

Figure 5.3   Convergence of saturation characteristics toward $-V_M$

circular transistor. The value of V_M is thus increased (hence G_{ds} decreased) if the saturated side of the transistor is wider and decreased (hence G_{ds} increased) in the opposite case.

Assuming $V_M \gg V_{D,S}$ (long channel) and constant, it corresponds to the convergence point of saturation currents for various gate voltages, as illustrated in Figure 5.3. In reality, due to two-dimensional effects, V_M tends to increase slowly when the inversion coefficient IC is increased.

5.1.3 Equivalent Circuit

The small-signal equivalent circuit of the transistor, which corresponds to equation (5.21), is shown in Figure 5.4.

The three transconductances discussed in Section 5.1.1 are represented by voltage-controlled current sources (VCCS) of adequate sign connected between the source and drain nodes. The residual conductance in saturation G_{ds}, introduced in Section 5.1.2, is connected in parallel. Note that the small-signal schematic of Figure 5.4 includes only the intrinsic components of the transistor.

The expressions of the intrinsic parameters, derived previously as function of bias voltages and/or currents, are summarized in Table 5.1 for weak and strong inversion.

A new parameter $A_{v\,max}$ defined by

$$A_{v\,max} \triangleq \frac{G_{ms}}{G_{ds}} \tag{5.24}$$

is introduced at the last row of this table. It is the *maximum voltage gain* available from the transistor in common gate configuration (driven from the source with $\Delta V_G = 0$). Indeed, if

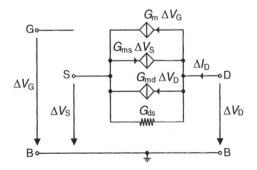

Figure 5.4   Small-signal dc equivalent circuit

Table 5.1 Summary of small-signal parameters

Small-signal parameter	Weak inversion	Strong inversion
$G_{\mathrm{ms,d}}$	$\dfrac{I_{\mathrm{F,R}}}{U_{\mathrm{T}}}$	$n\beta(V_{\mathrm{P}} - V_{\mathrm{S,D}}) = \dfrac{2I_{\mathrm{F,R}}}{V_{\mathrm{P}} - V_{\mathrm{S,D}}} = \sqrt{2n\beta I_{\mathrm{F,R}}}$
G_{m}	$\dfrac{I_{\mathrm{D}}}{nU_{\mathrm{T}}}$	Linear: $\beta(V_{\mathrm{D}} - V_{\mathrm{S}})$
		Saturation: $\beta V_{\mathrm{DSsat}} = \dfrac{2I_{\mathrm{D}}}{nV_{\mathrm{DSsat}}} = \dfrac{I_{\mathrm{D}}}{nU_{\mathrm{T}}\sqrt{IC}} = \sqrt{\dfrac{2\beta I}{n}}$
G_{ds}	$\dfrac{I_{\mathrm{F,R}}}{V_{\mathrm{M}}}$	$\dfrac{I_{\mathrm{F,R}}}{V_{\mathrm{M}}}$
$A_{\mathrm{v\,max}}$	$\dfrac{V_{\mathrm{M}}}{U_{\mathrm{T}}}$	$\dfrac{2V_{\mathrm{M}}}{V_{\mathrm{DSsat}}} = \dfrac{V_{\mathrm{M}}}{U_{\mathrm{T}}\sqrt{IC}}$

the device is forward saturated (hence $G_{\mathrm{md}} = 0$), then the whole current variation $G_{\mathrm{ms}}\,\Delta V_{\mathrm{S}}$ flows through G_{ds}, and the drain voltage variation is

$$\Delta V_{\mathrm{D}} = \left(1 + \frac{G_{\mathrm{ms}}}{G_{\mathrm{ds}}}\right)\Delta V_{\mathrm{S}} = (1 + A_{\mathrm{v\,max}})\,\Delta V_{\mathrm{S}}, \tag{5.25}$$

where the term "1" can be neglected. In common source configuration (driven from the gate with $\Delta V_{\mathrm{S}} = 0$), G_{ms} is replaced by G_{m} and the maximum (negative) voltage gain is $A_{\mathrm{v\,max}}/n$.

5.1.4 The Normalized Transconductance to Drain Current Ratio

The source and drain transconductances can be written as a function of the forward and reverse current respectively by using (5.11), resulting in

$$G_{\mathrm{ms}} = G_{\mathrm{spec}}\, q_{\mathrm{s}} = G_{\mathrm{spec}} \frac{2i_{\mathrm{f}}}{\sqrt{4i_{\mathrm{f}} + 1} + 1} = \frac{G_{\mathrm{spec}}}{2}\left(\sqrt{4i_{\mathrm{f}} + 1} - 1\right), \tag{5.26a}$$

$$G_{\mathrm{md}} = G_{\mathrm{spec}}\, q_{\mathrm{d}} = G_{\mathrm{spec}} \frac{2i_{\mathrm{r}}}{\sqrt{4i_{\mathrm{r}} + 1} + 1} = \frac{G_{\mathrm{spec}}}{2}\left(\sqrt{4i_{\mathrm{r}} + 1} - 1\right). \tag{5.26b}$$

The above equation (5.26a) can be used to derive the important transconductance to current ratio

$$\frac{G_{\mathrm{ms}}\, U_{\mathrm{T}}}{I_{\mathrm{F}}} = \frac{G_{\mathrm{ms}}}{G_{\mathrm{spec}}\, i_{\mathrm{f}}} = \frac{q_{\mathrm{s}}}{q_{\mathrm{s}}^2 + q_{\mathrm{s}}} = \frac{1}{q_{\mathrm{s}} + 1}, \tag{5.27}$$

which can also be written in terms of the forward normalized current

$$\frac{G_{\mathrm{ms}}\, U_{\mathrm{T}}}{I_{\mathrm{F}}} = \frac{2}{\sqrt{4i_{\mathrm{f}} + 1} + 1} = \begin{cases} 1 & \text{for } i_{\mathrm{f}} \ll 1 \\ 1/\sqrt{i_{\mathrm{f}}} & \text{for } i_{\mathrm{f}} \gg 1. \end{cases} \tag{5.28}$$

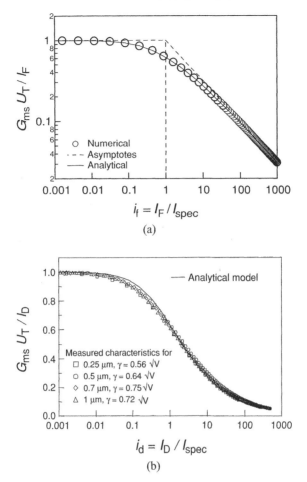

Figure 5.5 $G_{ms} U_T/I_F$ versus forward normalized current: (a) comparison between the asymptotes, a numerical calculation and the analytical expression; (b) comparison between analytical expression and results measured on several long-channel devices in saturation from different technologies

In forward saturation, $I_D = I_F$ and $G_{ms} = n\,G_m$, and hence

$$\frac{G_{ms}\,U_T}{I_D} = \frac{G_m\,n\,U_T}{I_D} = \frac{2}{\sqrt{4i_f + 1} + 1}. \tag{5.29}$$

Equation (5.29) is plotted versus the normalized forward current in Figure (5.5a) together with the unity asymptote valid in weak inversion and the $1/\sqrt{i_f}$ asymptote valid in strong inversion. It perfectly matches the symbols obtained from a numerical simulation with $\Gamma_b = 0.7\sqrt{V}$. Figure (5.5b) shows the same transconductance to drain current ratio characteristic compared to results measured on several technology generations. It is interesting to note that the transconductance to drain current ratio characteristic given by (5.29) is independent of any process parameters (except of course those used for normalizing the forward current).

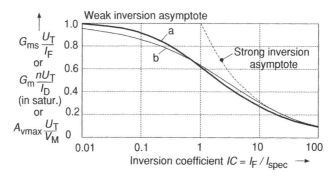

Figure 5.6 Variation of transconductance-to-current ratio and maximum voltage gain with inversion coefficient: (a) from charge model (5.29); (b) from approximation (5.30)

Equation (5.29) is also plotted as curve a in Figure 5.6. It shows that in weak inversion, the transconductance is within 1% of its asymptotic unity value given by (5.20) for $IC \ll 0.01$, and 10% below it at $IC = 0.12$. In strong inversion, the transconductance is 10% below its asymptotic value (5.18) at $IC = 23$, and within 1% of it only for $IC > 2500$. For $IC = 1$, the transconductance is only 62% of the value obtained from either asymptotes. Hence the range of moderate inversion where the full continuous equation (5.29) must be used depends on the required precision.

Curve b in Figure 5.6 is the variation of transconductance obtained from differentiation of the mathematical interpolation of the current (4.39) from weak to strong inversion:

$$G_{ms} \frac{U_T}{I_F} = \underbrace{G_m \frac{nU_T}{I_D}}_{\text{in saturation}} = \frac{1 - \exp(-\sqrt{IC})}{\sqrt{IC}}. \tag{5.30}$$

By introducing expression (5.22) of the residual output conductance in saturation, the maximum voltage gain defined by (5.24) becomes

$$A_{v\,max} = \frac{G_{ms} V_M}{I_F} = G_{ms} \frac{U_T}{I_F} \frac{V_M}{U_T}. \tag{5.31}$$

Figure 5.6 therefore also represents the variation of $A_{v\,max}$ normalized to its maximum value V_M/U_T obtained in weak inversion. This degradation of $A_{v\,max}$ with the increase of inversion coefficient IC is only slightly attenuated by the slow increase of V_M mentioned at the end of Section 5.1.2.

The transconductance-to-drain current ratio is very useful in analog circuit design. It actually shows how much transconductance you get for a given current and can therefore be used as a figure of merit for evaluating the current efficiency of any device including circuits such as operational transconductance amplifiers (OTA). From Figures 5.5 and 5.6, we clearly see that weak inversion offers the highest current efficiency, whereas in strong inversion it decreases as $1/\sqrt{i_f}$. It is important to note that the transconductance-to-drain current characteristic is a ratio; i.e., the transconductance of a given device can always be made larger in strong inversion than in weak inversion by increasing the bias current, but obviously this comes at the price of a lower current efficiency.

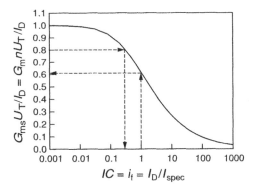

Figure 5.7 The transconductance on drain current ratio characteristic used for sizing a device

As explained by Figure 5.7, the transconductance-to-drain current ratio can be used to size a particular device. Assume you have a certain bias current available and you want to bias a device in saturation in the middle of the moderate inversion with $IC = i_f = 1$. Knowing the bias current and the inversion factor, you can deduce the specific current $I_{\text{spec}} = I_D/IC$ and then the W/L ratio from I_{spec}. You still need another criterion to select independently the W and the L (it could be noise, gain, etc.). You can then calculate the normalized transconductance to be about 0.62 and the actual transconductance to be $G_m = 0.62\, I_D/(n\, U_T)$. Less trivial, if you want a certain transconductance for a given bias current, you can compute the transconductance-to-drain current ratio, say 0.8 in this case and deduce the inversion factor to be about 0.3. Knowing the bias current you can extract the specific current I_{spec} and then the W/L ratio as explained above.

5.2 A GENERAL NQS SMALL-SIGNAL MODEL

The small-signal equivalent circuit described in Section 5.1 is valid only at low frequency (actually at dc to be correct) and hence does not account for the small-signal dynamic behavior. Now each time a terminal voltage is changed, the current changes by a certain amount. In order for this current to change, the inversion charge density within the channel has to change. This means that some incremental charges are either brought to or taken away from the static inversion charge density. This gives rise to transient charge flows (or ac currents) that are obviously not accounted for in the static small-signal model described in Section 5.1. They can be modeled by several capacitors which would result in the QS small-signal model described in Section 5.3. But since the latter is only a first-order approximation of a more general NQS model, we prefer to first derive the general NQS small-signal model and then its first-order QS approximation. In many textbooks, the QS small-signal model is derived first and then extended to the NQS model. But this requires some tedious calculations and difficult explanations that are completely avoided when starting from the general, but more complex, NQS model. Therefore we will start with a description of a very general NQS small-signal model.

The QS and NQS operations are delimited by the QS frequency ω_{qs} corresponding to the intrinsic channel time constant $\tau_{qs} \triangleq 1/\omega_{qs}$. The normalized QS frequency[1] Ω_{qs} is bias

---

[1] Here the actual frequency is denoted by a small cap, whereas its normalized value is denoted with an upper cap. This is an exception to the rule used throughout the book, namely that normalized variables use small caps.

dependent according to [44]:

$$\Omega_{qs} \triangleq \frac{\omega_{qs}}{\omega_{spec}} \triangleq \frac{\tau_{spec}}{\tau_{qs}} = 30 \frac{(q_s + q_d + 1)^3}{4q_s^2 + 4q_d^2 + 12q_s q_d + 10q_s + 10q_d + 5}, \tag{5.32}$$

where

$$\omega_{spec} = \frac{1}{\tau_{spec}} \triangleq \frac{\mu U_T}{L^2}. \tag{5.33}$$

The QS channel cutoff frequency ω_{qs} can be used as a figure of merit since it represents the ultimate frequency that the intrinsic part of the MOS transistor can reach without accounting for the extrinsic components such as the overlap capacitors that further decrease this frequency and are accounted for in the transit frequency f_t discussed in Section 11.3.1. Notice that ω_{spec} scales like $1/L^2$, therefore increasing quadratically when reducing the transistor length. This feature is one of the driving force for CMOS downscaling to achieve faster circuits and also use CMOS for RF circuits.

In saturation ($q_s \gg q_d$), (5.32) reduces to

$$\Omega_{qs} \cong 30 \frac{(q_s + 1)^3}{4q_s^2 + 10q_s + 5} = \begin{cases} 6 & \text{(weak inv.)} \\ \frac{15}{2} q_s = \frac{15}{2} \sqrt{i_f} = \frac{15}{4} \frac{V_P - V_S}{U_T} & \text{(strong inv.).} \end{cases} \tag{5.34}$$

From (5.34), we see that in weak inversion $\omega_{qs} = 6\omega_{spec} = 6\mu U_T/L^2$, which is constant and determined only by the mobility and the transistor length. Note that in weak inversion, $\tau_{qs} = 1/\omega_{qs}$ corresponds to the transit time of the carriers diffusing from source to drain. In strong inversion, Ω_{qs} is proportional to q_s or to $\sqrt{i_f}$ or to the saturation voltage $V_{D\,Ssat} = V_P - V_S$.

For $V_D = V_S$ (or equivalently $q_s = q_d$), (5.32) reduces to

$$\Omega_{qs} \cong 6 (2q_s + 1). \tag{5.35}$$

The normalized QS frequency Ω_{qs} is plotted versus the inversion factor i_f in Figure 5.8(a) which clearly shows the weak and strong inversion asymptotes. Notice that in strong inversion and for $V_D = V_S$, Ω_{qs} is 1.6 times larger than in saturation for the same inversion factor. Ω_{qs} is also plotted versus the normalized pinch-off voltage v_p in Figure 5.8(b) for different values of the normalized drain voltage ranging from $v_d = 0$ to $v_d = 100$, corresponding to saturation. Also shown is the strong inversion asymptote $15/4\ v_p$.

The NQS analysis will not be detailed here, since it is rather lengthy. It can be found in [38,44,78]. The most important result of the latter analysis is that the NQS small-signal behavior can be represented by the equivalent small-signal circuit presented in Figure 5.9. This circuit is made of five admittances, three connected to the gate (Y_{GSi}, Y_{GDi}, and Y_{GBi}) and two connected to the bulk (Y_{BSi}, Y_{BDi}) not accounting for Y_{GBi} and three transadmittances Y_m, Y_{ms}, and Y_{md}

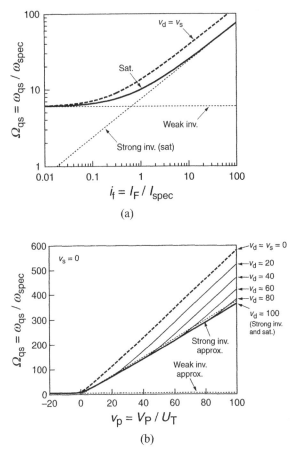

Figure 5.8 Intrinsic QS normalized frequency versus inversion factor (a) and versus pinch-off voltage for different drain voltages (b)

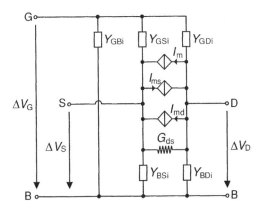

Figure 5.9 The complete intrinsic NQS small-signal equivalent circuit

connected between drain and source and controlled by the gate, source, and drain voltages respectively. These three transadmittances are modeled by three VCCS defined by [44,78]

$$I_m \triangleq Y_m \, \Delta V_G, \tag{5.36a}$$
$$I_{ms} \triangleq Y_{ms} \, \Delta V_S, \tag{5.36b}$$
$$I_{md} \triangleq Y_{md} \, \Delta V_D. \tag{5.36c}$$

Note that these admittances and transadmittances have real and imaginary parts.

In the same way there exists a relation between G_m, G_{ms}, and G_{md} (see equation (5.9)), it can be shown that the same relation also holds for the transadmittances, namely [44,78]

$$Y_m = \frac{Y_{ms} - Y_{md}}{n}, \tag{5.37}$$

which reduces to (5.9) for $\omega \ll \omega_{qs}$.

Note that we need to know only one transadmittance to deduce the other two. Indeed, assuming we know Y_{ms}, we can deduce Y_{md} by using the symmetry property of the device; i.e, Y_{md} can be calculated using the same equation used to calculate Y_{ms} but after permuting V_S and V_D (or i_f and i_r or q_s and q_d). Knowing both Y_{ms} and Y_{md}, we can compute Y_m using (5.37).

The gate-to-bulk admittance is related to the gate-to-source and gate-to-drain admittances by the following fundamental relation [38,44,78]:

$$Y_{GBi} = \frac{n-1}{n} \, (j \, \omega C_{OX} - Y_{GSi} - Y_{GDi}), \tag{5.38}$$

where $C_{OX} \triangleq W \, L \, C_{ox}$. In addition, the bulk-to-source and bulk-to-drain admittances are related to the gate-to-source and gate-to-drain admittances by [38,44,78]

$$Y_{BSi} = (n-1) \, Y_{GSi}, \tag{5.39a}$$
$$Y_{BDi} = (n-1) \, Y_{GDi}. \tag{5.39b}$$

As for the transadmittances, knowing only one admittance is enough to calculate the four other ones. For example, if we know Y_{GSi}, we can calculate Y_{GDi} by symmetry, Y_{BSi} and Y_{BDi} using (5.39) and finally Y_{GBi} by using (5.38).

The NQS intrinsic small-signal model shown in Figure 5.9 is therefore fully characterized for a given operating point when one transadmittance and one admittance are known. All the other components are deduced either from symmetry or by using (5.37), (5.38), and (5.39). In addition, it can be shown that the transadmittances can always be written as the product of the dc transconductance, which accounts for the bias dependence, times a common normalized function $\xi_m[\theta(q_s, q_d)]$, which accounts for the frequency dependence [44,78]

$$Y_{ms} = G_{ms} \, \xi_m[\theta(q_s, \, q_d)], \tag{5.40a}$$
$$Y_{md} = G_{md} \, \xi_m[\theta(q_d, \, q_s)], \tag{5.40b}$$

where $\theta = \omega/\omega_{qs} = \omega\tau_{qs}$. Notice the symmetry property illustrated by the swapping of the position of the q_s and q_d variables as arguments of θ in (5.40).

The exact derivation of the normalized function ξ_m is a bit tedious and will not be described here. The detailed calculation can be found in [38, 44, 78]. Nevertheless, a very good approximation of ξ_m valid in any mode of inversion is given by [44]

$$\xi_m = \frac{\lambda}{\sinh(\lambda)}, \tag{5.41}$$

with $\lambda \triangleq (1 + j) \sqrt{3\theta}$. Note that for $\theta \ll 1$, (5.41) can be approximated by a first-order function [44]

$$\xi_m \cong \frac{1}{1 + j \, \omega\tau_{qs}} \qquad \text{for } \omega\tau_{qs} \ll 1. \tag{5.42}$$

The magnitude and phase of function ξ_m are plotted versus the normalized frequency in Figure 5.10(a) (plain line) and compared to the first-order approximation (5.42) (dashed line) and also the second-order approximation (dashed-dotted line). Figure 5.10(a) shows that the first-order model can be used up to about $\theta \cong 1$ for both magnitude and phase. For $\theta > 1$, the second-order model can be used up to about $\theta \cong 3$ and the full function has to be used above. The dash-double-dot curve corresponds to the QS model and will be discussed later in Section 5.3.

The ξ_m function has been checked against measurements made on several devices and at different biases [44]. An example of such measurements made on a long-channel ($L = 10 \ \mu m$) N-type device in saturation is shown in Figure 5.10(b) for different gate bias ranging from

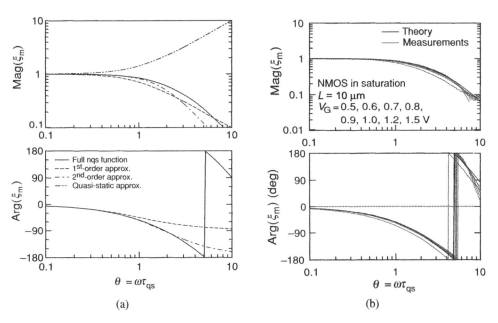

(a)   (b)

Figure 5.10 Normalized transdmittance function ξ_m versus normalized frequency. (a) Evaluated from (5.41). The line corresponds to the approximations given by (5.41), the dashed line corresponds to the first-order approximation given by (5.42), and the dashed-dotted line corresponds to the second-order approximation. (b) Comparison with measurements made on an N-channel device measured at different biases and properly normalized

$V_G = 0.5$ to $1.5\ V$. This plot clearly shows that after proper frequency and bias normalization, all the curves corresponding to the different bias conditions fall on the same normalized ξ_m curve. It also demonstrates that the NQS transadmittances are fully characterized by the single normalized function ξ_m.

Similarly, the Y_{GSi} and Y_{GDi} admittances can be written as a product of a bias-dependent capacitance (representing the QS approximation) times a common normalized function $\xi_c[\Omega(q_s, q_d)]$

$$Y_{GSi} = j\,\omega C_{GSi}(q_s, q_d)\,\xi_c[\theta(q_s, q_d)] = j\,\omega C_{OX}\,c_c(q_s, q_d)\,\xi_c[\theta(q_s, q_d)], \quad (5.43a)$$

$$Y_{GDi} = j\,\omega C_{GSi}(q_d, q_s)\,\xi_c[\theta(q_d, q_s)] = j\,\omega C_{OX}\,c_c(q_d, q_s)\,\xi_c[\theta(q_d, q_s)], \quad (5.43b)$$

where $c_c(q_s, q_d)$ is the normalized gate-to-source intrinsic capacitance

$$c_c(q_s, q_d) \triangleq \frac{C_{GSi}}{C_{OX}} = \frac{q_s}{3}\frac{2q_s + 4q_d + 3}{(q_s + q_d + 1)^2} = \begin{cases} 2/3 & \text{in strong inv. and sat.} \\ \\ q_s & \text{in weak inv. and sat.} \end{cases} \quad (5.44)$$

A good approximation for the normalized function ξ_c used for the admittances is given by [44]

$$\xi_c = \frac{\tilde{\xi}_c(\theta)}{\sqrt[3]{\varrho + (1 - \varrho)\,\tilde{\xi}_c(\theta/2)}}, \quad (5.45)$$

with

$$\tilde{\xi}_c \triangleq 2\,\frac{\cosh(\lambda) - 1}{\lambda\,\sinh(\lambda)} \quad (5.46)$$

and

$$\varrho \triangleq \left[\frac{10r\,(r + 2)^2}{9(r + 1)\,(r^2 + 3r + 1)}\right]^{3/2}, \quad (5.47)$$

where r is defined as

$$r \triangleq \frac{q_s + 1/2}{q_d + 1/2} = \sqrt{\frac{i_f + 1/4}{i_r + 1/4}}. \quad (5.48)$$

Note that r is much smaller than 1 for $q_s \ll 1 \ll q_d$, corresponding to strong inversion and reverse saturation, it is equal to unity when $q_s = q_d$ or when both $q_s \ll 1$ and $q_d \ll 1$ (corresponding to weak inversion), and finally r is much larger than 1 for $q_d \ll 1 \ll q_s$, corresponding to strong inversion and forward saturation. Function ξ_c is plotted in Figure 5.11(a) versus the normalized frequency $\theta \triangleq \omega\,\tau_{qs}$ for different values of r. For $\theta \ll 1$, $\xi_c \cong 1$ and therefore the admittances increase proportionally to θ. For $\theta \gg 1$, $\xi_c \propto 1/\sqrt{\theta}$ meaning that the admittances then only increase proportionally to $\sqrt{\theta}$. The proportionality factor depends on the parameter r. As shown in Figure 5.11(a), there is a small difference between the forward mode and the reverse mode, particularly on the phase, but in forward mode, there is very little difference between weak ($r = 1$) and strong inversion ($r \gg 1$) since ϱ only varies from 1 to 1.17.

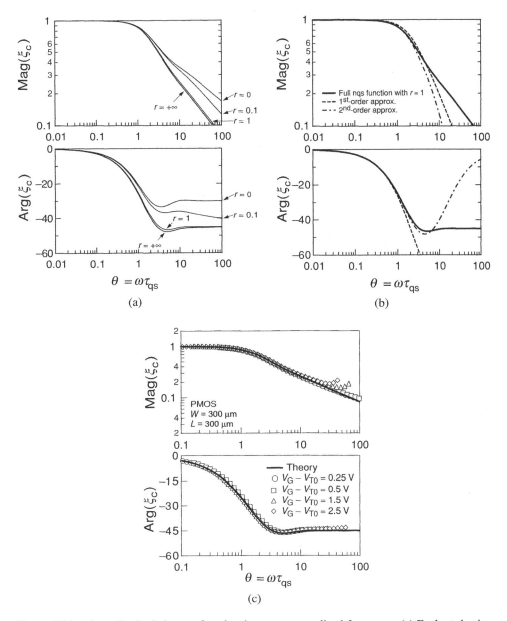

Figure 5.11 Normalized admittance function ξ_c versus normalized frequency. (a) Evaluated using (5.45) to (5.48) for different values of r with $r = 0$ and 0.1 corresponding to strong inversion and reverse saturation ($q_s \ll 1 \ll q_d$), $r = 1$ corresponding to weak inversion and $r = +\infty$ corresponding to strong inversion and forward saturation ($q_d \ll 1 \ll q_s$). (b) Different approximations of ξ_c (for $r = 1$). The line corresponds to the first-order approximation given by (5.49) and (5.45) and the dashed-dotted line corresponds to the second-order approximation. (c) Comparions with measurements made on a P-channel device measured at different biases and properly normalized

Therefore, in forward mode, ξ_c can be approximated by $\tilde{\xi}_c$ given by (5.46), which can be further approximated by the following first-order function:

$$\xi_c \cong \tilde{\xi}_c \cong \frac{1}{1 + j\,\omega\tau_{qs}/2} \qquad \text{for } \omega\tau_{qs} \ll 1. \qquad (5.49)$$

The latter approximation is similar to the transadmittance approximation (5.42) except that the pole is a factor 2 higher than ω_{qs}. Function $\tilde{\xi}_c$ is plotted versus the normalized frequency in Figure 5.11(b) together with the first-order approximation (5.49) (dashed line) and also the second-order approximation (dashed-dotted line). As for the transadmittance approximation, the first-order admittance approximation can be used up to about $\theta \cong 1$ and the second-order approximation up to about $\theta \cong 3$.

As shown in Figure 5.11(c), the normalized admittance function ξ_c has been validated against measurements made on a $300\,\mu m \times 300\,\mu m$ P-channel MOS transistor integrated in a $0.35\,\mu m$ CMOS process and biased at different overdrive voltages. The fact that all the measured points fall onto the same curve illustrates the strength of the bias and frequency normalization processes.

A simpler approximation of both the transadmittances and the admittances will be given in the next section and will lead to the QS model.

5.3 THE QS DYNAMIC SMALL-SIGNAL MODEL

5.3.1 Intrinsic Capacitances

As mentioned in the previous section, for $\theta = \omega\,\tau_{qs} \ll 1$, the NQS function for the admittances ξ_c is about equal to unity and the admittances reduce to the intrinsic capacitances. The normalized gate-to-source and gate-to-drain intrinsic capacitances are derived from (5.43) and (5.44) with $\xi_c = 1$

$$c_{GSi} \triangleq \frac{C_{GSi}}{C_{OX}} = c_c(q_s, q_d) = \frac{q_s}{3}\frac{2q_s + 4q_d + 3}{(q_s + q_d + 1)^2}, \qquad (5.50a)$$

$$c_{GDi} \triangleq \frac{C_{GDi}}{C_{OX}} = c_c(q_d, q_s) = \frac{q_d}{3}\frac{2q_d + 4q_s + 3}{(q_s + q_d + 1)^2}, \qquad (5.50b)$$

whereas the gate-to-bulk, source-to-bulk, and drain-to-bulk intrinsic capacitances are derived from (5.38) and (5.39) respectively

$$c_{GBi} \triangleq \frac{C_{GBi}}{C_{OX}} = \frac{n-1}{n}(1 - c_{GSi} - c_{GDi}), \qquad (5.51a)$$

$$c_{BSi} \triangleq \frac{C_{BSi}}{C_{OX}} = (n-1)\,c_{GSi}, \qquad (5.51b)$$

$$c_{BDi} \triangleq \frac{C_{BDi}}{C_{OX}} = (n-1)\,c_{GDi}. \qquad (5.51c)$$

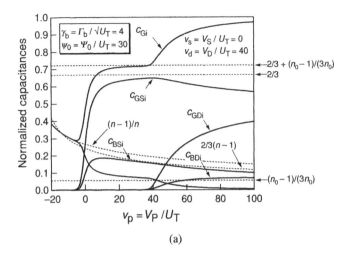

(a)

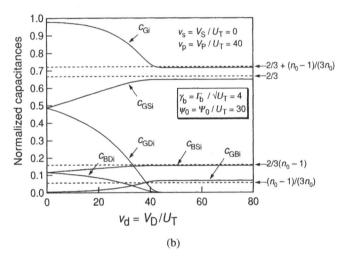

(b)

Figure 5.12 Normalized intrinsic capacitances versus the pinch-off voltage (a) and versus the drain voltage (b)

They are plotted in Figure 5.12 for $\Gamma_b = 4\sqrt{U_T}$ and $\Psi_0 = 30\,U_T$, together with the total gate capacitance

$$c_{Gi} \triangleq c_{GSi} + c_{GDi} + c_{GBi} = \frac{1}{n}\,(n-1+c_{GSi}+c_{GDi}), \qquad (5.52)$$

where (5.51a) has been used.

Figure 5.12 also shows the approximate values of the intrinsic normalized capacitances in strong inversion and saturation

$$c_{GSi} \cong \frac{2}{3}, \qquad (5.53\text{a})$$

$$c_{GDi} \cong 0, \qquad (5.53\text{b})$$

$$c_{GBi} \cong \frac{n-1}{3n}, \tag{5.53c}$$

$$c_{BSi} \cong (n-1)\frac{2}{3}, \tag{5.53d}$$

$$c_{BDi} \cong 0. \tag{5.53e}$$

Figure 5.12(b) also shows that for $v_d = v_s$, $c_{GSi} = c_{GDi}$ and $c_{BSi} = c_{BDi}$. It might be surprising that c_{GSi} and c_{GDi} are not exactly equal to 1/2 for $v_d = v_s = 0$ in Figure 5.12(b), but c_{GSi} and c_{GDi} are actually only equal to 1/2 asymptotically in very strong inversion and a little below 1/2 in strong inversion.

In weak inversion, the normalized intrinsic capacitances are given by

$$c_{GSi} \cong q_s, \tag{5.54a}$$

$$c_{GDi} \cong q_d, \tag{5.54b}$$

$$c_{GBi} \cong \frac{n-1}{n}, \tag{5.54c}$$

$$c_{BSi} \cong (n-1)\, q_s, \tag{5.54d}$$

$$c_{BDi} \cong (n-1)\, q_d. \tag{5.54e}$$

Since in weak inversion $q_s \ll 1$ and $q_d \ll 1$, the intrinsic capacitances are dominated by c_{GBi}.

5.3.2 Transcapacitances

In the QS regime, $\theta = \omega\,\tau_{qs} \ll 1$ and the first-order transadmittance function (5.42) can be further approximated as

$$\xi_m = \frac{Y_{ms}}{G_{ms}} = \frac{Y_{md}}{G_{md}} \cong 1 - j\,\omega\tau_{qs} \qquad \text{for } \omega\tau_{qs} \ll 1. \tag{5.55}$$

The source transadmittance can then be written as

$$Y_{ms} \cong G_{ms}\,(1 - j\,\omega\tau_{qs}) = G_{ms} - j\omega\,G_{ms}\,\tau_{qs} = G_{ms} - j\,\omega\,C_{ms}, \tag{5.56}$$

where $C_{ms} \triangleq G_{ms}\,\tau_{qs}$ is defined as the *source transcapacitance*. A transcapacitance is similar to a transconductance except that the output current is proportional to the derivative of the control voltage instead of the control voltage itself. Similarly, the drain transadmittance can be written as

$$Y_{md} \cong G_{md}\,(1 - j\,\omega\tau_{qs}) = G_{md} - j\,\omega\,C_{md}, \tag{5.57}$$

with $C_{md} \triangleq G_{md}\,\tau_{qs}$ being the *drain transcapacitance*. Using (5.37), the gate transadmittance is then given by

$$Y_m \cong G_m\,(1 - j\,\omega\tau_{qs}) = G_m - j\,\omega\,C_m, \tag{5.58}$$

with $C_m \triangleq G_m \tau_{qs}$ being the *gate transcapacitance* which is related to the source and drain transcapacitances by

$$C_m = \frac{C_{ms} - C_{md}}{n}. \tag{5.59}$$

The QS approximation of ξ_m given by (5.55) is also plotted versus θ in Figure 5.10. The phase is identical to the first-order approximation (5.42), but as shown in Figure 5.10, the magnitude increases proportionally to θ instead of decreasing. This is a clear limitation of the QS approximation. It is important to note that any charge-based model implemented in a circuit simulator such as Spice without specific NQS model will show this behavior. It is therefore important to remember that the QS model is valid only to about a fraction of the QS frequency ω_{qs} (typically $\omega_{qs}/3$).

By definition, the QS time constant is related to the transconductances and transcapacitances by

$$\tau_{qs} = \frac{C_{ms}}{G_{ms}} = \frac{C_{md}}{G_{md}} = \frac{C_m}{G_m}. \tag{5.60}$$

From (5.32) and (5.4), we can derive the expression for the source and drain transcapacitances

$$c_{ms} \triangleq \frac{C_{ms}}{C_{OX}} = n \frac{q_s}{15} \frac{4q_s^2 + 4q_d^2 + 12q_sq_d + 10q_s + 10q_d + 5}{(q_s + q_d + 1)^3}, \tag{5.61a}$$

$$c_{md} \triangleq \frac{C_{md}}{C_{OX}} = n \frac{q_d}{15} \frac{4q_s^2 + 4q_d^2 + 12q_sq_d + 10q_s + 10q_d + 5}{(q_s + q_d + 1)^3}, \tag{5.61b}$$

$$c_m \triangleq \frac{C_m}{C_{OX}} = \frac{q_s - q_d}{15} \frac{4q_s^2 + 4q_d^2 + 12q_sq_d + 10q_s + 10q_d + 5}{(q_s + q_d + 1)^3}. \tag{5.61c}$$

They are plotted versus v_p and v_d in Figure 5.13 for $\Gamma_b = 4 \sqrt{U_T}$ and $\Psi_0 = 30 \ U_T$. As shown in Figure 5.13, the approximate values in strong inversion and saturation are given by

$$c_{ms} \cong n \frac{4}{15}, \tag{5.62a}$$

$$c_{md} \cong 0, \tag{5.62b}$$

$$c_m \cong \frac{4}{15}. \tag{5.62c}$$

It is important to note that these transcapacitances are of the same order of magnitude than the intrinsic capacitances and therefore cannot be neglected. We will see in Section 12.2 that neglecting, for example, C_m can lead to a large phase error on Y_m at RF (see Figure 12.3).

5.3.3 Complete QS Circuit

The complete QS intrinsic small-signal schematic is shown in Figure 5.14, where the transadmitances are modeled by three VCCS defined by (5.36) with Y_m, Y_{ms}, and Y_{md} given by (5.58), (5.56), and (5.57) respectively.

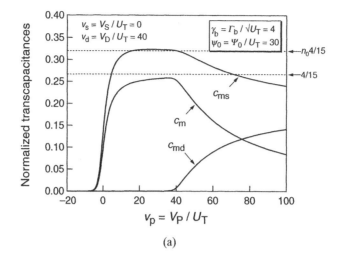

(a)

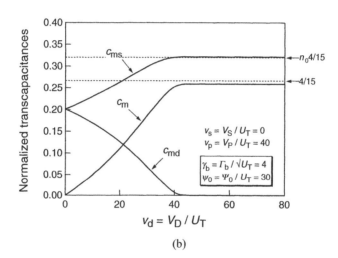

(b)

Figure 5.13   Normalized transcapacitances versus the pinch-off voltage (a) and versus the drain voltage (b)

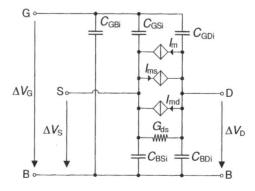

Figure 5.14   The complete intrinsic QS small-signal equivalent circuit

5.3.4 Domains of Validity of the Different Models

It is important to clearly understand the domains of validity of the three small-signal models shown in Figures 5.4, 5.9, and 5.14. The dc small-signal model of Figure 5.4 is strictly valid only at $\omega = 0$. Now it can be used for many low-frequency circuit analyses to evaluate for example the dc gain. It can also be used when extrinsic capacitors, including interconnects, dominate, as is often the case.

It is important to notice that very often the frequency limitation is not given by the intrinsic QS frequency but by the parasitic extrinsic elements or by some other component such as a load capacitor. Actually it is mainly when working at RF that the intrinsic frequency limitations become important and should be accounted for. But even at RF, the NQS regime should be avoided and therefore the NQS small-signal model of Figure 5.9 is seldom used. There are nevertheless some cases where the device might operate in NQS mode. For example, in some circuits like G_m-C filters where nonminimum length P-channel transistors are used in the current mirror of the transconductor, the combination of nonminimum length and lower mobility of P-channel device might drive the device close to the NQS regime. The additional phase shift introduced by the NQS effect in the current mirror might then change the frequency behavior of the filter. It is therefore important to check whether any device operates close to the NQS regime. Also, increasing the QS frequency can simply be done by increasing the bias current at the cost of a higher power dissipation and a possible increase in the minimum required supply voltage in order to adapt for the increase of V_{DSsat}. It is then useful to evaluate the QS frequency and set it at a frequency just high enough for avoiding any NQS effects but without increasing the bias current prohibitively in order to maintain the power consumption as low as possible.

Another situation where NQS effects might appear is when very fast clock signals are used with very steep slopes. At the start of the step signal, the transistor might momentarily be driven into NQS regime. Although this is usually a large-signal transient problem, it might be useful to discuss it here.

Although it is possible to derive an exact small-signal solution, it is much more difficult (maybe even impossible) to derive an analytical large-signal NQS model. Only first- or second-order approximations have been derived up to now [79]. On the other hand, large-signal transient simulations can be done quite easily by simply cutting the transistor channel into several slices [80]. This can easily be done by replacing in a circuit simulator a single transistor by a cascade of N fictitious transistors in series each having a length L/N, where L is the length of the original transistor [80]. This is illustrated in Figure 5.15(a), where 10 transistors connected in series have been used. Of course, the middle transistors should model only the intrinsic part of the device and hence all the extrinsic components such as the overlap capacitors and the access resistors have to be sized accordingly. In the following example, we simulated only the intrinsic behavior and therefore all the transistors include only the intrinsic part. The transistor is biased in strong inversion with a 1-V gate and drain voltage, corresponding to $V_G - V_{T0} \cong 0.4$ V or $IC \cong 28$. The source voltage is biased at 0 V. To illustrate the NQS behavior, a negative step voltage of -100 mV is applied at the source as illustrated in Figure 5.16. Figure 5.15(b) shows how this step voltage propagates along the channel from the source to the drain. It clearly shows that the effect on the channel voltage close to the drain is not instantaneous and it takes some time for the step to reach the drain. The source and drain currents I_S and I_D are plotted versus time in Figure 5.15(c). It shows that the source and drain currents

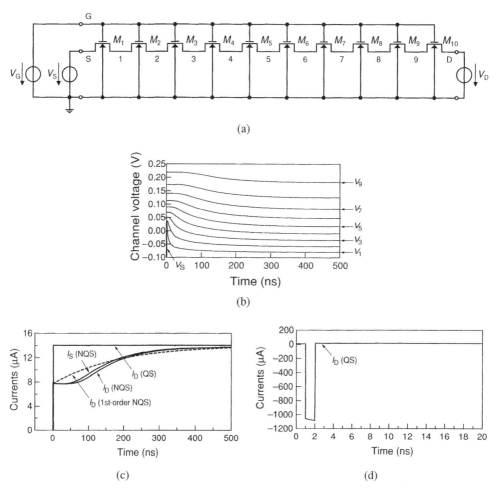

(a)

(b)

(c)   (d)

Figure 5.15 (a) NQS transient simulations by slicing the channel of a single transistor of length L into several intrinsic transistors in series having a length equal to L/N. (b) Voltages along the channel due to a step voltage at the source. (c) Drain and source currents due to the step voltage at the source. (d) Zoom on the QS drain current showing the negative Dirac impulse occurring at the origin

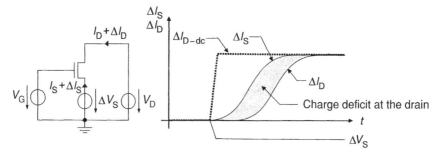

Figure 5.16 NQS transient simulations and drain and source currents showing the charge deficit at the drain

are not equal and that the drain current lags the source current. This situation is reproduced in a slightly larger scale in Figure 5.16. Note that the area between the source current (top curve) and the drain current (lower curve) represents charges that actually never reach the drain and therefore constitutes a charge deficit at the drain.

The result obtained from the QS model of a single transistor with the same total length and bias is also plotted in Figure 5.15(c). As expected from the expression of the source transadmittance (5.56), the response to this step voltage is a large negative peak (actually a Dirac impulse) corresponding to the derivative of the input voltage due to the transcapacitance part of the source transadmittance, followed by an instantaneous change of the drain current due to the transconductance part of the source transadmittance. Since the QS model ensures charge conservation, the negative spike represents the charge deficit at the drain observed in the NQS simulations but these charges are actually taken away from the drain current immediately instead of not reaching the drain as is the case in the NQS simulation. This is illustrated in Figure 5.15(d), where the QS drain current $I_D(QS)$ shown in Figure 5.15(c) has been zoomed out to show the negative impulse. Notice the large value of the negative impulse amplitude which in some cases may give rise to simulation (convergence) problems.

Also plotted in Figure 5.15(c) is the result obtained from the time domain equivalent of the first-order NQS approximation given by (5.42). As expected it differs from the NQS simulation at small time values but then follows quite nicely the NQS simulation after about 150 ns. The major difference with the QS simulation is that the spike has now disappeared, avoiding any simulation problems. Note that the second-order NQS approximation would of course give an even better result.

6 The Noise Model

This chapter is dedicated to the modeling of the noise in the MOS transistor focusing mainly on the intrinsic part and assuming that the device has a long channel and hence does not account for short-channel effects which are described in Section 9.4. There are mainly two kinds of noises coming from the channel region, namely the thermal noise and the flicker or $1/f$ noise. Both can be described by a local noise source which depends on the position along the channel. Section 6.1 shortly presents the different methods to calculate the power spectral density (PSD) of the drain current fluctuations, by integration of the local noise source along the channel. The method will then be applied for both the thermal noise and the flicker noise. The low-frequency thermal noise is then derived in Section 6.2. The thermal noise parameter is defined as the ratio of the drain thermal noise conductance to the output conductance at $V_D = V_S$ (or equivalently the source transconductance), and the thermal excess noise factor is the ratio of the drain thermal noise conductance and the gate transconductance. The use of these two parameters is illustrated by several circuit examples. Section 6.3 is devoted to flicker noise, which arises from several sources: the fluctuation of the number of mobile carriers in the channel, discussed in Section 6.3.1, the fluctuation of the mobility, presented in Section 6.3.2, and finally the flicker noise coming from the access resistances, described in Section 6.3.3. The bias dependence of all the three noise sources is particularly emphasized. All three contributions are compared in Section 6.3.4 using measured data from the literature. Finally, the scaling properties of the flicker noise are discussed in Section 6.3.5.

6.1 NOISE CALCULATION METHODS

6.1.1 General Expression

The origin of noise is related to local random fluctuations of the carrier velocity or the carrier density. These local fluctuations can be modeled by adding a random current to the local dc current as shown in Figure 6.1(a). They then propagate to the terminals resulting in fluctuations of the voltages or currents around the dc operating point. There are clearly two different parts in the noise analysis: the microscopic part which consists in deriving the statistics of the stochastic

Charge-Based MOS Transistor Modeling: The EKV Model for Low-Power and RF IC Design   C. Enz and E. Vittoz
© 2006 John Wiley & Sons, Ltd.

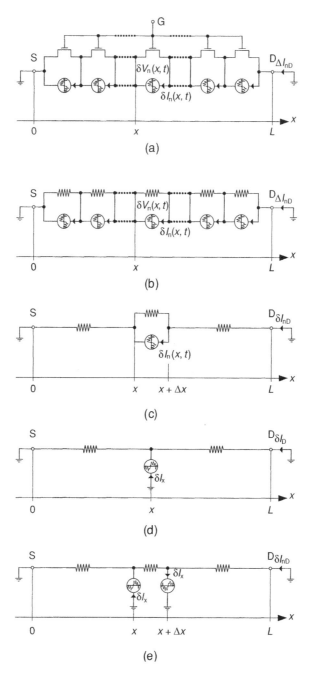

Figure 6.1 Equivalent circuits of the MOS transistor noisy channel illustrating different noise analysis approaches. (a) The Langevin method used by Klassen and Prins [81], where a distributed noise source is added to the differential equation giving the drain current. (b) Circuit obtained from (a) after linearization of the differential equation and keeping only the first-order terms. (c) Equivalent small-signal circuit approach: since the circuit of (b) is linear, the effect of each local noise source on the drain current fluctuation can be calculated separately and summed at the drain. (d) The impedance field method, where the current response δI_D at the drain due to a current δI_x injected at position x along channel and the corresponding current $A_i(x) \triangleq \delta I_D/\delta I_x$ are calculated. The noise is then evaluated from the derivative of $A_i(x)$ with respect to x

process such as the velocity fluctuation and translating it into a variation of the local current, and the macroscopic part which consists in calculating the response of the terminal currents (or voltages) to these fluctuations or in other words how these local fluctuations propagate to the terminals and produce variations of the terminal voltages or currents. In this part we will mainly focus on the macroscopic part, namely finding a common expression for the PSD of the drain current fluctuations due to all the microscopic local noise sources located in the channel. At this stage, it is important to point out that these local fluctuations are always small and consequently, the analysis of the propagation of the noise sources to the terminal voltages or currents reduces to a linear analysis. Hence, the principle of superposition can be applied for adding the effects of all the noise sources along the channel. In principle, since these noise sources are random process, they might be spatially correlated, which should be accounted for when summing their effects. However, in all of the cases discussed below, it will be assumed that the local sources are spatially uncorrelated and therefore their PSD can be summed.

Several methods have been used for the calculation of noise in the MOS transistor. A detailed description of all these different methods goes beyond the objective of this book, but they will be briefly illustrated below. The first approach is based on the Langevin method and was used by Klaassen and Prins [81] initially for deriving the thermal noise and was then extended by Klaassen to also account for the flicker noise [82] and by Langevelde to include also the induced gate noise [83]. It starts with the differential equation giving the drain current (4.5), to which a Langevin noise source $\delta I_n(x, t)$ is added as shown in Figure 6.1(a). This distributed noise source depends on the position along the channel and induce channel voltage fluctuations $\delta V_n(x, t)$ that should normally be accounted for if the $\delta I_n(x, t)$ would be large-signal fluctuations. Since these voltage fluctuations are much smaller than the thermodynamic voltage, the current differential equation can be linearized, resulting in the equivalent small-signal circuit shown in Figure 6.1(b). Since this equivalent circuit is now linear, the superposition principle can be used. The problem can then be reduced to isolating a single noise source in the channel and calculate its effect δI_{nD} on the drain terminal current as shown in Figure 6.1(c). The total effect is then obtained by summing the contributions of all the noise sources along the channel assuming they are spatially uncorrelated. This latter approach will be called the *equivalent small-signal circuit approach*.[1] Note that a similar transmission line approach has also been used [84]. A third approach is the *impedance field method* (IFM) initially introduced by Shockley [85] and modified afterward by van Vliet [86,87]. As illustrated in Figure 6.1(d), the channel is excited by a current δI_x at position x along the channel, and the current response δI_D at the drain terminal and the corresponding current gain $A_i(x) \Delta \delta I_D / \delta I_x$ are calculated. The noise due to the local noise source at position x is then evaluated from the so-called *impedance field*[2] corresponding to the derivative of the current gain with respect to x at each position. Taking the derivative with respect to position corresponds actually to taking the difference between the current gain at positions $x + \Delta x$ and x, which is illustrated in Figure 6.1(e). If δI_x is set equal to the local noise source $\delta I_n(x)$, then the equivalent circuit of Figure 6.1(e) reduces to the circuit of Figure 6.1(c). The IFM is therefore equivalent to the small-signal circuit approach. A more rigorous demonstration shows that actually the three approaches mentioned above are equivalent [156].

---

[1] It is sometimes also called the *two-transistor approach*.
[2] The name *impedance field* comes from the fact that the terminal voltage fluctuations were initially evaluated resulting in a gain between the excitation current and the voltage fluctuation that has the dimension of an impedance (it is actually a transimpedance).

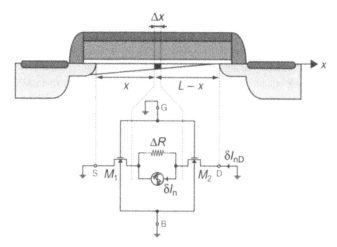

Figure 6.2   MOS transistor cross-section with an infinitesimal noisy piece of channel between points x and $x + \Delta x$ and split into two noiseless transistors M_1 and M_2

In the following, we will derive a general analysis based on the more intuitive equivalent small-signal circuit approach, offering a common framework for the treatment of both thermal and flicker noise.

To start, let us assume that the channel is noiseless except for a slice of the channel comprised between positions x and $x + \Delta x$ which is noisy and has a resistance ΔR as shown in Figure 6.2. The microscopic noise due to the channel slice is modeled by a current source δI_n having a PSD $S_{\delta I_n^2}$ and connected between x and $x + \Delta x$ in parallel with the resistance ΔR of the slice. Note that a current source (Norton source) is used because the physical origin of noise is a random fluctuation of the carrier velocity and/or charge density, resulting in fluctuations of the local current which is then represented by a noise current source added to the dc current. The transistor can then be split into two noiseless transistors on each side of the noise current source, namely transistor M_1 of length x on the source side of point x and transistor M_2 of length $L - x$ on the drain side. As mentioned above, it can be assumed that the noise voltage δV_n across resistance ΔR is much smaller than U_T and therefore a small-signal approach can be used. Both transistors M_1 and M_2 can then be replaced by their low-frequency small-signal equivalent circuits. For frequencies much below the channel cutoff frequency ω_{qs} (see equation (5.32) for definition), the capacitive coupling can be neglected. As illustrated in Figure 6.3, the equivalent circuits of transistors M_1 and M_2 reduce to two simple conductances, of values $G_s \triangleq G_{md1}$ on the source side and $G_d \triangleq G_{ms2}$ on the drain side. Since ΔR can obviously be neglected compared to the series connection of G_s and G_d, the drain current fluctuation δI_{nD} is then given by

$$\delta I_{nD} = G_{ch}\Delta R\delta I_n,\tag{6.1}$$

where conductance G_{ch} corresponds simply to the series connection of G_s and G_d

$$\frac{1}{G_{ch}(x)} \triangleq \frac{1}{G_s(x)} + \frac{1}{G_d(x)}.\tag{6.2}$$

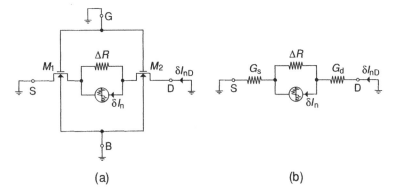

Figure 6.3   Two-transistor equivalent circuit used for low-frequency noise calculation

Note that both G_s and G_d and hence also G_{ch} depend on the position x of the local noise source. The noise source δI_n also depends on the position x along the channel and might also depend on frequency ω (for example when considering $1/f$ noise).

The PSD of the drain current fluctuations due to noise current source δI_n is then given by

$$S_{\delta I_{nD}^2}(\omega, x) = G_{ch}^2(x)\Delta R^2(x)S_{\delta I_n^2}(\omega, x). \tag{6.3}$$

The PSD of the total noise current fluctuation at the drain $S_{\Delta I_{nD}^2}$ due to all the different sections along the channel is obtained by summing their elementary contributions $S_{\delta I_{nD}^2}$ assuming that the contribution of each slice at different positions along the channel remain uncorrelated. This translates into integrating the elementary contributions over the channel from source to drain, resulting in[3]

$$S_{\Delta I_{nD}^2}(\omega) = \int_0^L G_{ch}^2(x)\Delta R^2(x)\frac{S_{\delta I_n^2}(\omega, x)}{\Delta x}\,dx. \tag{6.4}$$

Note that $S_{\delta I_n^2}$ has to be divided by Δx in (6.4) to represent the contribution of the noise current source by unit length.

As explained above, we have chosen to represent the noise source by a current source (Norton source) in parallel with the elementary section because it is closer to the physical origin of noise. As shown in Figure 6.4, we could of course also use an equivalent noise voltage source (Thévenin equivalent) defined by

$$\delta V_n = \Delta R \delta I_n, \tag{6.5}$$

or in terms of PSD,

$$S_{\delta V_n^2}(\omega, x) = \Delta R^2 S_{\delta I_n^2}(\omega, x). \tag{6.6}$$

---

[3] Note that the contribution of one elementary slice to the drain current is written as $S_{\delta I_{nD}^2}$, whereas the total contribution of all the sections is written as $S_{\Delta I_{nD}^2}$.

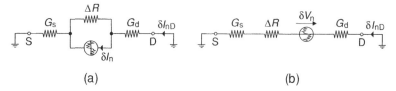

(a)   (b)

Figure 6.4   Two-transistor equivalent circuits: (a) using the original noise current source of Figure 6.3; (b) circuit with an equivalent noise voltage source

The drain current fluctuation due to δV_n is then given as

$$S_{\delta I_{\mathrm{nD}}^2}(\omega,\, x) = G_{\mathrm{ch}}^2(x) S_{\delta V_{\mathrm{n}}^2}(\omega,\, x).\tag{6.7}$$

If the noise current sources δI_n are spatially uncorrelated, so are the noise voltage sources δV_n and (6.4) becomes

$$S_{\Delta I_{\mathrm{nD}}^2}(\omega) = \int_0^L G_{\mathrm{ch}}^2(x)\frac{S_{\delta V_{\mathrm{n}}^2}(\omega,\, x)}{\Delta x}\,\mathrm{d}x.\tag{6.8}$$

Equation (6.4) or (6.8) will be used below to derive the low-frequency thermal noise and flicker noise PSD in Sections 6.2 and 6.3 respectively.

6.1.2 Long-Channel Simplification

Conductance G_{ch} is actually nothing else than the channel conductance G_{ch} at point x. If the mobility is assumed to be constant, it is then given by

$$G_{\mathrm{ch}} = \frac{\mathrm{d}I_\mathrm{D}}{\mathrm{d}V} = \mu(-Q_i)\frac{W}{L} = G_{\mathrm{spec}}\, q_\mathrm{i},\tag{6.9}$$

where $G_{\mathrm{spec}} \triangleq I_{\mathrm{spec}}/U_\mathrm{T} = 2n\beta U_\mathrm{T}$. The resistance ΔR of a section, again assuming a constant mobility, is given by

$$\Delta R = \frac{\Delta V}{I_\mathrm{D}} = \frac{\Delta x}{W\mu(-Q_i)}.\tag{6.10}$$

Note that the derivation in the case velocity saturation and mobility reduction due to the vertical field have to be accounted for is much more tedious and is presented in Section 9.4.
   Combining (6.9) and (6.10) into (6.3) results in

$$S_{\delta I_{\mathrm{nD}}^2}(\omega,\, x) = \left(\frac{\Delta x}{L}\right)^2 S_{\delta I_{\mathrm{n}}^2}(\omega,\, x).\tag{6.11}$$

The PSD of the total drain current fluctuation is then given by

$$S_{\Delta I_{\mathrm{nD}}^2}(\omega) = \int_0^L \left(\frac{\Delta x}{L}\right)^2 \frac{S_{\delta I_{\mathrm{n}}^2}(\omega,\, x)}{\Delta x}\,\mathrm{d}x = \frac{1}{L^2}\int_0^L \Delta x\, S_{\delta I_{\mathrm{n}}^2}(\omega,\, x)\,\mathrm{d}x.\tag{6.12}$$

A more formal derivation of (6.12) is given in Appendix A2 at the end of this chapter, following the approach of Klaassen and Prins [81]. Note that the above model ignored the capacitive coupling of the noise generated in the channel to the gate. The latter will be analyzed in the high-frequency noise model described in Chapter 13.

6.2 LOW-FREQUENCY CHANNEL THERMAL NOISE

6.2.1 Drain Current Thermal Noise PSD

The PSD of the drain current fluctuations due to thermal noise in the channel can be evaluated using (6.4) or (6.8). The PSD of the noise current source of one section is then simply given from (6.10):

$$S_{\delta I_n^2} = \frac{4kT}{\Delta R} = 4kT \frac{W\mu(-Q_i)}{\Delta x}, \tag{6.13}$$

where k is the Boltzmann constant and T is the absolute temperature.

The PSD of the total drain current fluctuation $S_{\Delta I_{nD}^2}$ is obtained from (6.12) as

$$S_{\Delta I_{nD}^2} \triangleq 4kT \cdot G_{nD} = 4kT \frac{1}{L^2} \int_0^L W\mu[-Q_i(x)] \, dx$$

$$= 4kT\mu \frac{W}{L^2} \int_0^L [-Q_i(x)] \, dx, \tag{6.14}$$

where it has been assumed that the mobility μ is constant. The *thermal noise conductance at the drain* G_{nD} is then defined as

$$G_{nD} \triangleq \mu \frac{W}{L^2} \int_0^L [-Q_i(x)] \, dx = \frac{\mu}{L^2} |Q_I|, \tag{6.15}$$

where Q_I is the total inversion charge in the channel given by

$$Q_I \triangleq W \int_0^L Q_i(x) \, dx. \tag{6.16}$$

Hence, for a mobility that is independent of the electric field, the noise conductance is proportional to the total inversion charge "stored" in the channel. Equation (6.15) can be written in normalized form as

$$g_{nD} \triangleq \frac{G_{nD}}{G_{spec}} = \int_0^1 q_i(\xi) \cdot d\xi = q_I \triangleq \frac{Q_I}{Q_{spec}}. \tag{6.17}$$

The total normalized inversion charge q_I can be evaluated using the drain current expression (4.21), repeated here for convenience

$$i_d = -(2q_i + 1)\frac{dq_i}{d\xi},$$

and making a change of variable by expressing $d\xi$ as

$$d\xi = -\frac{2q_i + 1}{i_d}dq_i. \tag{6.18}$$

Replacing $d\xi$ in (6.17) by (6.18) results in

$$g_{nD} = q_I = \int_0^1 q_i(\xi)\,d\xi = -\frac{1}{i_d}\int_{q_s}^{q_d} q_i(2q_i + 1)\,dq_i =$$
$$= \frac{1}{6}\frac{4q_s^2 + 3q_s + 4q_sq_d + 3q_d + 4q_d^2}{q_s + q_d + 1}, \tag{6.19}$$

where the expression of the normalized drain current (4.22)

$$i_d = (q_s^2 + q_s) - (q_d^2 + q_d)$$

has been used.

For $V_D = V_S$ (or $q_s = q_d$) and in saturation (i.e., for $q_s \gg q_d$), (6.19) can be simplified as

$$g_{nD} = q_I = \begin{cases} q_s & \text{for } V_D = V_S(q_s = q_d) \\ q_s\frac{\frac{2}{3}q_s+\frac{1}{2}}{q_s+1} & \text{in saturation } (q_s \gg q_d). \end{cases} \tag{6.20}$$

In SI and saturation, $q_s \gg 1$ and (6.20) reduces to

$$g_{nD} = q_I = \frac{2}{3}q_s, \tag{6.21}$$

In WI, $q_s \ll 1$ and $q_d \ll 1$, and (6.19) becomes

$$g_{nD} = q_I = \frac{1}{2}(q_s + q_d) = \frac{1}{2}(i_f + i_r). \tag{6.22}$$

The thermal noise conductance is then obtained as

$$G_{nD} = G_{spec}\,g_{nD} = \frac{I_{spec}}{U_T}\frac{i_f + i_r}{2} = \frac{I_F + I_R}{2U_T}. \tag{6.23}$$

The PSD is then given by

$$S_{\Delta I_{nD}^2} = 4kT\,G_{nD} = 4kT\frac{I_F + I_R}{2U_T} = 2q\,(I_F + I_R), \tag{6.24}$$

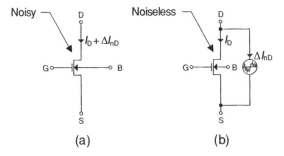

Figure 6.5 Thermal noise equivalent circuit with noiseless transistor

which corresponds to full shot noise of both the forward and the reverse components of the drain current [88]. This result might be surprising since the above derivation used the expression of the local thermal noise and integrated its effect on the drain current over the channel. In weak inversion the current is dominated by diffusion and, similar to a bipolar transistor, the noise can be interpreted as shot noise related to the potential barriers at the source and the drain. Nevertheless, it can be shown that expression (6.15) (or its normalized form (6.17)), which was derived with the assumptions of thermal noise due to the channel conductance, is valid in all regions of inversion [89, 90].

Assuming that the channel thermal noise is the only source of noise within the transistor, for frequencies much below the channel cutoff frequency ω_{qs}, the noisy transistor can then be modeled as a noiseless device to which a noisy current source ΔI_{nD} is connected between drain and source as shown in Figure 6.5. This noisy current source has a PSD given by (6.14), where the noise conductance G_{nD} is given by

$$G_{nD} = G_{spec} \, q_I, \tag{6.25}$$

where q_I is given by (6.19).

6.2.2 Thermal Noise Excess Factor Definitions

Several thermal noise excess factors can be defined according to the definitions introduced initially by van der Ziel [91]. The *thermal noise parameter* related to the drain terminal δ_{nD} is defined as[4]

$$\delta_{nD} \triangleq \frac{G_{nD}}{G_{ds0}}, \tag{6.26}$$

where G_{ds0} is the drain-to-source conductance at $V_{DS} = 0$,

$$G_{ds0} = G_{ms} = G_{spec} \, q_s. \tag{6.27}$$

[4] Van der Ziel initially used γ for the thermal noise parameter defined by (6.26) and α for the noise excess factor defined by (6.30). The most important noise excess factor from a circuit design point of view is the one given by (6.30), which has been called γ in many papers instead of α as it was defined initially by Van der Ziel [91]. We will keep the circuit design definition of γ and rename the Van der Ziel's γ as δ.

The δ_{nD} parameter shows how much the thermal noise of the active device deviates from the value it takes when it operates as a passive resistor of conductance G_{ds0}. Since for $V_{DS} = 0$ the noise conductance G_{nD} is equal to the channel conductance G_{ds0}, the noise parameter δ_{nD} is then equal to unity.

Assuming constant mobility, (6.25) can be used together with (6.27) in (6.26), allowing δ_{nD} to be written in terms of the normalized charges as

$$\delta_{nD} = \frac{q_1}{q_s}. \tag{6.28}$$

In saturation, from (6.20), δ_{nD} is equal to

$$\delta_{nD} = \frac{2}{3}\frac{q_s + 3/4}{q_s + 1} = \begin{cases} 1/2 & \text{WI and saturation } (q_s \ll 1) \\ 2/3 & \text{SI and saturation } (q_s \gg 1). \end{cases} \tag{6.29}$$

Note that the δ_{nD} thermal noise parameter compares the thermal noise conductance evaluated at a given operating point that is not necessarily the same as the one used to define the output conductance G_{ds0} (i.e., $V_{DS} = 0$). It is therefore not very useful for circuit design and is used more for modeling purposes.

For circuit design, it is more useful to define another figure of merit, γ_{nD}, named the *thermal noise excess factor* related to the drain and defined as

$$\gamma_{nD} \overset{\Delta}{=} \frac{G_{nD}}{G_m} = \frac{g_{nD}}{g_m} = \frac{nq_1}{q_s - q_d}. \tag{6.30}$$

γ_{nD} represents how much noise is generated at the drain of a transistor for a given gate transconductance. Contrary to the δ_{nD} thermal noise parameter, the noise conductance and the gate transconductance used in the definition (6.30) are evaluated *at the same operating point*. γ_{nD} has a direct impact on the noise performance of circuits.

The smaller γ_{nD}, the better the noise performance of the device. Note that γ_{nD} can become quite large in the linear region when V_D tends to V_S. Indeed, in this region, the gate transconductance gets smaller as the drain-to-source voltage decreases, but the thermal noise conductance does not decrease, resulting in a degradation of the γ_{nD} noise excess factor. At the limit when V_D becomes equal to V_S, the gate transconductance becomes zero and γ_{nD} tends to infinity.

Note that γ_{nD} is also a figure of merit that can be used for any transconductor (even for circuit transconductors) to evaluate how much thermal noise is generated for a given transconductance. The smaller γ_{nD}, the better the transconductor.

The δ_{nD} thermal noise parameter and the γ_{nD} noise excess factor are related by

$$\gamma_{nD} = \frac{G_{nD}}{G_{ds0}}\frac{G_{ds0}}{G_m} = \delta_{nD}\frac{G_{ds0}}{G_m} = \delta_{nD}\,n\frac{G_{ds0}}{G_{ms} - G_{md}} = \delta_{nD}\,n\frac{q_s}{q_s - q_d}. \tag{6.31}$$

In saturation, $G_{md} = 0$ and $q_d = 0$, resulting in

$$\gamma_{nD} = \delta_{nD}\,n\frac{G_{ds0}}{G_{ms}} = \delta_{nD}\,n = \begin{cases} \dfrac{n}{2} & \text{WI and saturation} \\ n\dfrac{2}{3} & \text{SI and saturation,} \end{cases} \tag{6.32}$$

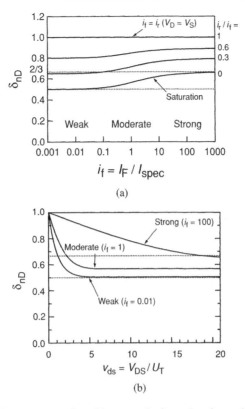

Figure 6.6 Thermal noise parameter δ_{nD}: (a) versus the inversion factor for a given i_r/i_f ratio; (b) versus v_{ds} for a given i_f

since $G_{ds0} = G_{ms}$. For $n = 1.5$, the thermal noise factor in SI and in saturation is approximately equal to unity.

The thermal noise parameter δ_{nD} is plotted versus the inversion factor for different values of the i_r/i_f ratios in Figure 6.6(a) and versus the normalized $v_{ds} \triangleq V_{DS}/U_T$ voltage for a given inversion factor in Figure 6.6(b).

It is important to notice that the above results have been obtained with the assumption of constant mobility along the channel. The latter assumption is valid for long-channel devices, where the lateral electric field E_x remains much smaller than some critical field E_c defined in Section 9.1. As soon as E_x approaches E_c, the carrier velocity starts to saturate and the mobility can no longer be considered as constant along the channel. Also, the carrier temperature starts to rise, increasing the thermal noise. These effects affect the thermal noise excess factor, which can become larger than its long-channel value. These effects will be discussed in detail in Section 9.4.

6.2.3 Circuit Examples

The effect of thermal noise and more particularly the use of the noise parameters and noise excess factor are illustrated with three examples. The first is shown in Figure 6.7 and corresponds

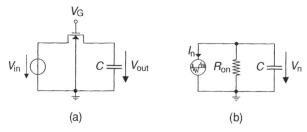

(a)   (b)

Figure 6.7   Elementary sample-and-hold circuit used to illustrate the use of the thermal noise parameter to evaluate the rms noise voltage on capacitor C

to a simple sample-and-hold function implemented with a MOS transistor operating as a switch
and a hold capacitor on which the input voltage is sampled and held. To close the switch, the
gate voltage is set to a sufficiently high value such that the on resistance R_{on} and the corresponding $R_{on} C$ time constant are low enough to allow the voltage on the capacitor to settle at
a value equal to the input voltage. After the output voltage has settled, the switch is opened
and the voltage is sampled on capacitor C. Due to the thermal noise of the switch, in addition
to the input voltage, there is also a noise voltage that is sampled. To evaluate the rms value
of this noise voltage, the PSD of the noise voltage on capacitor C has to be evaluated first.
As shown in Figure 6.7(b), this is done by setting the input voltage to zero and replacing the
transistor by its equivalent small-signal circuit including the noise current between source and
drain accounting for the thermal noise voltage of the channel. The equivalent circuit simplifies
to a simple parallel RC network as shown in Figure 6.7(b). The output noise voltage is then
simply equal to the low-pass filtered current noise according to

$$V_n = \frac{-R_{on}}{1 + j\omega/\omega_c} I_n,$$ (6.33)

where $\omega_c \triangleq 1/(R_{on}C)$ is the cutoff frequency. The PSD of the output noise voltage is then
given by

$$S_{V_n^2} = \left| \frac{R_{on}}{1 + j\omega/\omega_c} \right|^2 S_{I_n^2} = \frac{R_{on}^2}{1 + (\omega/\omega_c)^2} S_{I_n^2},$$ (6.34)

with $S_{I_n^2}$ given by

$$S_{I_n^2} = 4kT G_{nD}.$$ (6.35)

In this particular example, after the voltage has settled, the V_{DS} voltage is equal to zero and
therefore it is better to use the definition of the noise parameter for G_{nD}

$$G_{nD} = \delta_{nD} G_{ms} = G_{ms} = G_{spec} q_s,$$ (6.36)

since for $V_D = V_S$, $\delta_{nD} = 1$. Note that $R_{on} = 1/G_{ms}$. Even though the transistor is usually an
active device, in this particular case it operates as a simple (nonlinear) resistor and the variance
of the output voltage can be found directly by applying the Bode theorem (see Appendix 6.4),

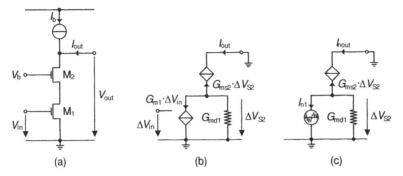

Figure 6.8   Linearized MOS transconductor using transistor M_1 biased in the linear region, illustrating the use of the thermal noise excess factor

resulting in

$$V_n^2 = \frac{kT}{C}. \tag{6.37}$$

Another example with a MOS transistor biased in the linear region is the linearized transconductor shown in Figure 6.8. It is similar to a cascode stage except that in this case the driver transistor M_1 is biased in the linear region instead of saturation to take advantage of the linear transconductance. Indeed, in the linear region, the gate transconductance is given by

$$G_m = \beta V_{DS} \tag{6.38}$$

and is therefore set by the drain-to-source voltage, which has to be chosen smaller than the pinch-off voltage in order to bias the transistor in the linear region. The overall transconductance G_m is obtained from the analysis of the small-signal equivalent circuit of Figure 6.8(b), resulting in

$$G_{m\,eq} \triangleq \frac{-I_{out}}{\Delta V_{in}} = \frac{G_{m1}}{1 + G_{md1}/G_{ms2}} \cong G_{m1} \quad \text{for } G_{ms2} \gg G_{md1}. \tag{6.39}$$

This means that for $G_{ms2} \gg G_{md1}$, the overall transconductance $G_{m\,eq}$ is equal to the transconductance G_{m1} of M_1. The latter condition suggests to bias M_2 in weak inversion in order to get the largest source transconductance for the imposed bias current.

A first-order noise analysis can be carried out by setting the ΔV_{in} to zero and assuming that the cascode transistor M_2 and the bias current source can be made noiseless, the only noise contributor being transistor M_1. With the help of the small-signal circuit shown in Figure 6.8(c), the output noise current is given by

$$I_{n\,out} = -\frac{I_{n1}}{1 + G_{md1}/G_{ms2}} \cong I_{n1} \quad \text{for } G_{ms2} \gg G_{md1}, \tag{6.40}$$

which means that all the current noise generated by transistor M_1 is conveyed to the output. The output noise current PSD is then equal to the PSD of transistor M_1:

$$S_{I^2_{n\,out}} \triangleq 4kT\,G_{n\,out} \cong S_{I^2_{n1}} = 4kT\,G_{nD1}. \tag{6.41}$$

As mentioned above, this linearized transconductor can be evaluated in terms of noise performance by evaluating its noise excess factor

$$\gamma_{n\,eq} \triangleq \frac{G_{n\,out}}{G_{m\,eq}} \cong \frac{G_{nD1}}{G_{m1}} = \gamma_{nD1}, \tag{6.42}$$

which is simply equal to the thermal noise excess factor γ_{nD1} of transistor M_1. Since M_1 is biased in strong inversion and in the linear region, (6.30) simplifies to

$$q_{11} \cong \frac{2}{3} \frac{q_{s1}^2 + q_{s1}q_{d1} + q_{d1}^2}{q_{s1} + q_{d1}}, \tag{6.43}$$

and the thermal noise excess factor becomes

$$\gamma_{nD1} = \frac{n_1 q_{11}}{q_{s1} - q_{d1}} = n_1 \frac{2}{3} \frac{q_{s1}^2 + q_{s1}q_{d1} + q_{d1}^2}{q_{s1}^2 - q_{d1}^2} = n_1 \frac{2}{3} \frac{1 + \alpha + \alpha^2}{1 - \alpha^2}, \tag{6.44}$$

where

$$\alpha \triangleq \frac{q_{d1}}{q_{s1}}, \tag{6.45}$$

which in strong inversion is equal to

$$\alpha = \frac{v_{p1} - v_{d1}}{v_{p1}} = 1 - \varepsilon, \tag{6.46}$$

with $\varepsilon \triangleq v_{d1}/v_{p1}$. Since the drain voltage of M_1 has to be much smaller than its pinch-off voltage $\varepsilon \ll 1$ and hence (6.44) can be approximated by

$$\gamma_{nD1} \cong \frac{n_1}{\varepsilon} = n_1 \frac{v_{p1}}{v_{d1}} \cong \frac{V_{G1} - V_{T0n}}{V_{DS}}. \tag{6.47}$$

For biasing M_1 in the linear region, the overdrive voltage is necessarily larger than the V_{DS} voltage, resulting in a thermal noise excess factor larger than the value obtained in saturation (about equal to $n\,2/3 \cong 1$).

The last example is shown in Figure 6.9 and corresponds to a diode-connected MOS transistor. The noise PSD of the voltage fluctuation across capacitor C can be evaluated from the small-signal circuit given in Figure 6.9(b), where the conductance G_m corresponds to the small-signal conductance of the diode-connected transistor M and the noise current source to its thermal noise. The noise voltage fluctuation V_n is then given by

$$V_n = -\frac{1}{G_m} \frac{1}{1 + j\omega/\omega_c} I_n, \tag{6.48}$$

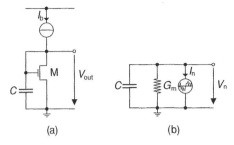

Figure 6.9 Diode-connected MOS transistor used to illustrate the use of the thermal noise excess factor in circuits

where $\omega_c \triangleq G_m/C$ is the cutoff frequency of the low-pass filtered white noise. The corresponding PSD is then given by

$$S_{V_n^2} = \frac{1}{G_m^2} \frac{1}{1 + (\omega/\omega_c)^2} S_{I_n^2}, \tag{6.49}$$

with $S_{I_n^2}$ given by (6.35) with G_{nD} equal to

$$G_{nD} = \gamma_{nD}\, G_m. \tag{6.50}$$

Rewriting $S_{V_n^2} = 4kT R_n(\omega)$, where $R_n(\omega)$ corresponds to the gate input-referred equivalent noise resistance given by

$$R_n(\omega) = \frac{\gamma_{nD}}{G_m} \frac{1}{1 + (\omega/\omega_c)^2}. \tag{6.51}$$

Note that $R_n(\omega)$ is frequency dependent. The noise voltage variance is then obtained by integrating the PSD over the frequency from 0 to $+\infty$. Since the cutoff frequency and the noise resistance are respectively proportional and inversely proportional to G_m, the integral does not depend on G_m. This is similar to the simple RC network discussed in Appendix 6.4 except for the additional γ_{nD} parameter. The variance is therefore given by

$$\overline{V_n^2} = \gamma_{nD} \frac{kT}{C}. \tag{6.52}$$

The thermal noise voltage variance is therefore γ_{nD} times that of a passive RC circuit. Since for long-channel devices, γ_{nD} can be smaller than 1 (but is always larger than 1/2), the noise generated on capacitor C by an active transistor connected like a diode can be smaller than that obtained from a passive RC circuit. For a long-channel device, in strong inversion and saturation, $\gamma_{nD} \cong 1$ and there is almost no difference. As will be shown in Section 9.4, for short-channel devices (actually at high lateral electric field), the noise excess factor γ_{nD} can become significantly larger than 1, degrading the performance of analog circuits.

6.3 FLICKER NOISE

In addition to the thermal noise of the channel described above, the MOS transistor also exhibits flicker or $1/f$ noise. As its name suggests, flicker noise is characterized by a PSD that is inversely proportional to frequency. It therefore mainly dominates at low frequency, below the so-called corner frequency f_k defined as the frequency at which $1/f$ noise contributes equally than the channel thermal noise to the total noise PSD (referred indifferently at the drain or at the gate). Because the $1/f$ noise scales inversely proportional to the gate area, it is becoming a major issue for analog IC design in deep and ultradeep submicron devices. Corner frequencies of several tens of megahertz are now typical, and hence low-frequency analog circuits are usually totally dominated by $1/f$ noise. Techniques exist to reduce or even eliminate this low-frequency noise. The most obvious one is to size the gate area in order to bring down the corner frequency to an acceptable value. This is done at the expense of higher capacitances, which require higher transconductance and hence higher current for the same characteristic frequency. Other circuit techniques such as chopper stabilization and correlated double sampling can be used to eliminate the $1/f$ noise [92, 93].

There are basically two main causes to this $1/f$ noise. The first results from carrier fluctuations of the inversion charge due to trapping in traps located in the oxide close to the Si–SiO$_2$ interface, whereas the second originates from fluctuations of the carrier mobility. Each of these causes will be presented below.

6.3.1 Carrier Number Fluctuations (Mc Worther Model)

The flicker noise due to carrier fluctuations originates from the fluctuations of the inversion charge close to the Si–SiO$_2$ interface due to variations of the interfacial oxide charge resulting from dynamic trapping/detrapping of mobile carriers from the channel into slow border traps [91, 94–96].

Consider again a section of the channel comprised between x and $x + \Delta x$. The current at position x is obtained from (4.5)

$$I_D = WqN(x)\mu \frac{dV}{dx},$$

where $N(x) = -Q_i(x)/q$ is the number of carriers per unit area. If a number of carriers get trapped at a position x, the relative current fluctuation is then given by

$$\frac{\delta I_D(x)}{I_D} = \frac{\delta N}{N} + \frac{\delta \mu}{\mu}, \tag{6.53}$$

where the mobility fluctuation term is induced by the influence of the trapping on the scattering mechanism. The mobility being affected by the trapping mechanism hence depends on the number of trapped charges per unit area N_t according to [97, 98]

$$\frac{1}{\mu} = \frac{1}{\mu_0} + \tilde{\alpha}_c N_t = \frac{1}{\mu_0} + \alpha_c |Q_t|, \tag{6.54}$$

where $Q_t \triangleq -q N_t$ is the trapped charge density and $\alpha_c \triangleq \tilde{\alpha}_c/q$ is the Coulomb scattering coefficient which is about 10^4 Vs/C for electrons and 10^5 Vs/C for holes in silicon [95, 97, 98]. Accordingly, $\tilde{\alpha}_c$ is about 1.6×10^{-15} Vs for electrons and 1.6×10^{-14} Vs for holes in silicon. Accounting for this scattering mechanism, (6.53) can be rewritten as

$$\frac{\delta I_D(x)}{I_D} = \left(\frac{1}{N} \frac{dN}{dN_t} + \frac{1}{\mu} \frac{d\mu}{dN_t} \right) \delta N_t = \left(\frac{1}{N} \frac{dN}{dN_t} - \tilde{\alpha}_c \, \mu \right) \delta N_t. \qquad (6.55)$$

We can relate δN and δN_t considering that the fluctuation δQ_t of the trapped charge density causes a variation $\delta \Psi_s$ of the surface potential which produces a change of all the charges that depend directly on Ψ_s, namely the inversion charge, the depletion charge, and the gate charge. These other charges vary according to the charge conservation principle, resulting in [99]

$$\delta Q_g + \delta Q_b + \delta Q_i = -\delta Q_t, \qquad (6.56)$$

where δQ_g, δQ_b and δQ_i are the induced fluctuations of the gate, depletion, and inversion charge densities respectively.[5] They can be related to the fluctuation of the surface potential $\delta \Psi_s$ according to [99]

$$\delta Q_g = -C_{ox} \, \delta \Psi_s, \qquad (6.57a)$$
$$\delta Q_b = -C_b \, \delta \Psi_s, \qquad (6.57b)$$
$$\delta Q_i = -C_i \, \delta \Psi_s. \qquad (6.57c)$$

It follows that [99]

$$R \triangleq \frac{\delta N}{\delta N_t} = \left| \frac{\delta Q_i}{\delta Q_t} \right| = \frac{C_i}{C_i + C_{ox} + C_d}. \qquad (6.58)$$

It can be shown from (4.3) and assuming $V = $ const., that $C_i \cong -Q_i/U_T$ and therefore [98]

$$R \cong \frac{Q_i}{Q_i + Q^*} = \frac{N}{N + N^*}, \qquad (6.59)$$

where $Q_i = -q N$ and

$$Q^* = -q N^* = -U_T C_{ox} \left(1 + \frac{C_d}{C_{ox}} \right). \qquad (6.60)$$

From the definition (3.70), the term $1 + C_d/C_{ox}$ in (6.60) is actually the slope factor in weak inversion n_w which is approximately equal to the slope factor n. Equation (6.60) then reduces

---

[5] Note that the additional variation of the interface traps δQ_{it}, originally included in the analysis presented in [98,99], has been neglected considering that they are much smaller than the variations of the other charges.

to

$$Q^* \cong -nU_T C_{ox} = \frac{Q_{spec}}{2}, \tag{6.61}$$

with Q_{spec} given by (3.42). Equation (6.59) then becomes

$$R = \frac{\delta N}{\delta N_t} \cong \frac{Q_i}{Q_i + Q_{spec}/2} = \frac{q_i}{q_i + 1/2}. \tag{6.62}$$

Using (6.62), the relative local current fluctuation (6.55) can be written as

$$\frac{\delta I_D(x)}{I_D} = \left(\frac{1}{q_i + 1/2} + \alpha\mu \right) \frac{\delta N_t}{N_{spec}}, \tag{6.63}$$

where $N_{spec} \triangleq -Q_{spec}/q = 2kTnC_{ox}/q^2$ and $\alpha \triangleq \alpha_c(-Q_{spec}) = \tilde{\alpha}_c \cdot N_{spec}$ is a coefficient related to the Coulomb scattering coefficient.

The corresponding PSD of the local noise current source δI_n normalized to the square of the dc current is then given by

$$\left. \frac{S_{\delta I_n^2}}{I_D^2} \right|_{\Delta N} = \left(\frac{1}{q_i + 1/2} + \alpha\mu \right)^2 \frac{S_{\delta N_t^2}}{N_{spec}^2}. \tag{6.64}$$

The PSD of the trap charge density fluctuation $S_{\delta N_t^2}$ depends essentially on the trapping mechanisms into the oxide. For tunneling process, the trapping probability decreases exponentially with the distance from the Si–SiO$_2$ interface into the oxide. The trapping charge density fluctuation PSD is then defined by [91, 94–96]

$$S_{\delta N_t^2} = \frac{kT\lambda N_T}{W\Delta x f}, \tag{6.65}$$

where f is the frequency, λ is the tunneling attenuation distance (≈ 0.1 nm) [95], and N_T the oxide volumetric trap density per unit energy in eV$^{-1} \cdot$ m^{-3} evaluated close to the Fermi energy level. Note that N_T is obtained from measurements and typically ranges from 10^{-17} to 10^{-16} eV$^{-1} \cdot$ cm^{-3}.

The fluctuation of the drain current due to an elementary section is then given by (6.3)

$$\left. \frac{S_{\delta I_{nD}^2}}{I_D^2} \right|_{\Delta N} = G_{ch}^2 \Delta R^2 \left. \frac{S_{\delta I_n^2}}{I_D^2} \right|_{\Delta N} = \left(\frac{\Delta x}{L} \right)^2 \left. \frac{S_{\delta I_n^2}}{I_D^2} \right|_{\Delta N} \tag{6.66}$$

$$= \left(\frac{\Delta x}{L} \right)^2 \left(\frac{1}{q_i + 1/2} + \alpha\mu \right)^2 \frac{S_{\delta N_t^2}}{N_{spec}^2},$$

where a constant mobility is assumed.[6] Finally, the relative PSD of the total fluctuation of the

---

[6] This is in contradiction with the scattering mechanism described above which would imply that the mobility is dependent on the inversion charge and hence on bias. Nevertheless, this approximation allows to derive a simple first-order approximation of the bias dependence of the flicker noise.

drain current is obtained by integration according to (6.12)

$$\frac{S_{\Delta I_{nD}^2}}{I_D^2}\bigg|_{\Delta N} = \frac{1}{L^2} \int_0^L \Delta x \; \frac{S_{\delta I_n^2}}{I_D^2}\bigg|_{\Delta N} dx = S_D|_{\Delta N} \; K_D(q_s, q_d)|_{\Delta N}, \tag{6.67}$$

with

$$S_D|_{\Delta N} \triangleq \frac{q^4 \lambda N_T}{kT \, W \, L \, n^2 \, C_{ox}^2 \, f}. \tag{6.68}$$

If we assume that $S_D|_{\Delta N}$ is only weakly bias dependent, most of the bias dependence is accounted for by the unitless factor $K_D(q_s, q_d)|_{\Delta N}$ defined by

$$
\begin{aligned}
K_D(q_s, q_d)|_{\Delta N} &\triangleq \frac{1}{4} \int_0^1 \left(\frac{1}{q_i + 1/2} + \alpha\mu \right)^2 d\xi \\
&= \frac{1}{4i_d} \int_{q_d}^{q_s} \left(\frac{1}{q_i + 1/2} + \alpha\mu \right)^2 (2q_i + 1) \, dq_i \\
&= \frac{1}{2i_d} \ln\left(\frac{1 + 2q_s}{1 + 2q_d} \right) + \frac{\alpha\mu}{1 + q_s + q_d} + \left(\frac{\alpha\mu}{2} \right)^2,
\end{aligned} \tag{6.69}
$$

with $\xi \triangleq x/L$ and $i_d = q_s^2 - q_d^2 + q_s - q_d = (q_s - q_d)(1 + q_s + q_d)$.

The bias-dependent factor $K_D(q_s, q_d)|_{\Delta N}$ is plotted versus the inversion factor in saturation in Figure 6.10(a) for two values of the $\alpha\mu$ product. The value $\alpha\mu = 0.4$ used in Figure 6.10 has been taken from Table 6.1 [100].

In very strong inversion $K_D(q_s, q_d)|_{\Delta N}$ tends to

$$K_D(q_s, q_d)|_{\Delta N} \cong \left(\frac{\alpha\mu}{2} \right)^2 \quad \text{for } q_s, q_d \gg 1, \tag{6.70}$$

whereas in weak inversion $K_D(q_s, q_d)|_{\Delta N}$ reduces to

$$K_D(q_s, q_d)|_{\Delta N} \cong \left(1 + \frac{\alpha\mu}{2} \right)^2 \quad \text{for } q_s, q_d \ll 1, \tag{6.71}$$

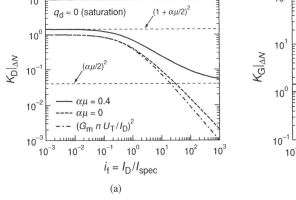
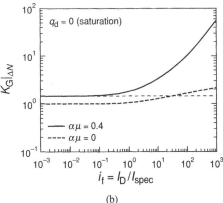

(a)   (b)

Figure 6.10 Bias-dependent factors $K_D(q_s, q_d)|_{\Delta N}$ and $K_G(q_s, q_d)|_{\Delta N}$ versus the inversion factor i_f in saturation and for $\eta = 1/2$

which is equal to unity when the mobility fluctuations induced by the carrier trapping mechanism can be ignored (i.e., for $\alpha = 0$).

As shown in Figure 6.10(a), the behavior of $K_D(q_s, q_d)|_{AN}$ very much depends on the value of the $\alpha\mu$ product. For values of $\alpha\mu$ that get close to unity, $K_D(q_s, q_d)|_{AN}$ starts to saturate to $(\alpha\mu/2)^2$ in very strong inversion. On the other hand, for small values of $\alpha\mu$, Figure 6.10(a) shows that $K_D(q_s, q_d)|_{AN}$ is very close to the square of normalized G_m/I_D curve.

For circuit design, it is more useful to refer the flicker noise PSD at the gate by dividing $S_{\Delta I_{nD}^2}\big|_{AN}$ by G_m^2, resulting in

$$S_{\Delta V_G^2}\Big|_{AN} \triangleq \frac{S_{\Delta I_{nD}^2}\big|_{AN}}{G_m^2} = S_D|_{AN}\ K_D(q_s, q_d)|_{AN}\ \left(\frac{I_D}{G_m}\right)^2 \tag{6.72}$$
$$= S_G|_{AN}\ K_G(q_s, q_d)|_{AN}\,,$$

where

$$S_G|_{AN} \triangleq \frac{q^2\ kT\ \lambda\ N_T}{W\ L\ C_{ox}^2\ f}. \tag{6.73}$$

Assuming again that $S_G|_{AN}$ is only weakly bias dependent, most of the bias dependence is captured by the factor $K_G(q_s, q_d)|_{AN}$ defined by

$$K_G(q_s, q_d)|_{AN} \triangleq (1 + q_s + q_d)^2\ K_D(q_s, q_d)|_{AN}. \tag{6.74}$$

The bias-dependent term $K_G(q_s, q_d)|_{AN}$ is plotted versus the inversion factor in saturation in Figure 6.10(b) for the same values of $\alpha\mu$ used in Figure 6.10(a). As explained above, when the correlation term $\alpha\mu$ is much smaller than 1, $K_D(q_s, q_d)|_{AN}$ is approximately equal to $(G_m n U_T/I_D)^2$ in saturation and hence $K_G(q_s, q_d)|_{AN}$ is only weakly bias dependent. As shown in Figure 6.10(b), for $\alpha = 0$, it approximately changes only by a factor 2 over 6 decades of current. This is no more the case when $\alpha\mu$ gets closer to 1. In this case the gate-referred flicker noise starts to increase significantly in strong inversion. Figure 6.10(b) also indicates that the flicker noise referred to the gate in saturation and due to number fluctuation is minimum in weak inversion.

The source of flicker noise coming from the fluctuation of the mobility will be discussed in the next section.

6.3.2 Mobility Fluctuations (Hooge Model)

In the Hooge model [101], the drain current noise results from the fluctuations of the carrier mobility. The PSD of the local noise current source of an elementary section is given by [91]

$$\frac{S_{\delta I_n^2}}{I_D^2}\bigg|_{\Delta\mu} = \frac{\alpha_H\ q}{W\ \Delta x\ (-Q_i)\ f}, \tag{6.75}$$

where α_H is the Hooge parameter which is unitless and ranges from about 10^{-4} to 10^{-6}. Assuming a constant average mobility, the fluctuation of the drain current due to an elementary channel section is then given by (6.11)

$$\left.\frac{S_{\delta I_{nD}^2}}{I_D^2}\right|_{\Delta\mu} = \left(\frac{\Delta x}{L}\right)^2 \left.\frac{S_{\delta I_n^2}}{I_D^2}\right|_{\Delta\mu} = \frac{\Delta x\,\alpha_H\,q}{W\,L^2\,(-Q_i)\,f}. \tag{6.76}$$

The PSD of the total fluctuation of the drain current is then given by

$$\left.\frac{S_{\Delta I_{nD}^2}}{I_D^2}\right|_{\Delta\mu} = S_D|_{\Delta\mu}\,K_D(q_s,\,q_d)|_{\Delta\mu}, \tag{6.77}$$

with

$$S_D|_{\Delta\mu} \triangleq \frac{\alpha_H\,q^2}{kT\,W\,L\,nC_{ox}\,f}, \tag{6.78}$$

and where $K_D(q_s,\,q_d)|_{\Delta\mu}$ accounts for the bias dependence and is defined by

$$\begin{aligned} K_D(q_s,\,q_d)|_{\Delta\mu} &\triangleq \int_0^1 \frac{d\xi}{2q_i(\xi)} = \frac{1}{i_d}\int_{q_d}^{q_s}\left(1+\frac{1}{2q_i}\right)dq_i = \\ &= \frac{1}{i_d}\left[q_s - q_d + \frac{1}{2}\ln\left(\frac{q_s}{q_d}\right)\right] \\ &= \frac{1}{1+q_s+q_d}\left[1+\frac{\ln(q_s/q_d)}{2(q_s-q_d)}\right]. \end{aligned} \tag{6.79}$$

The bias-dependent factor $K_D(q_s,\,q_d)|_{\Delta\mu}$ is plotted versus the inversion factor in saturation (i.e., assuming a constant ratio $q_s/q_d = 100$) in Figure 6.11(a). In weak inversion, $K_D(q_s,\,q_d)|_{\Delta\mu}$ is approximately given by

$$K_D(q_s,\,q_d)|_{\Delta\mu} \cong \frac{\ln(q_s/q_d)}{2(q_s-q_d)} = \frac{\ln(i_f/i_r)}{2i_d} = \frac{V_{DS}/U_T}{2i_d}, \tag{6.80}$$

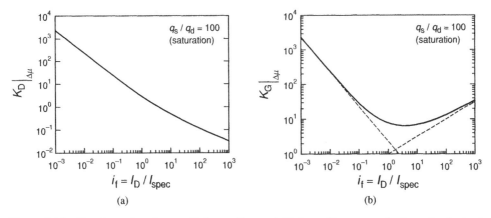

(a)   (b)

Figure 6.11 Bias-dependent factors $K_D(q_s,\,q_d)|_{\Delta\mu}$ and $K_G(q_s,\,q_d)|_{\Delta\mu}$ versus the inversion factor i_f in saturation (i.e., for $q_s/q_d = 100$)

whereas in strong inversion, it is given by

$$K_D(q_s, q_d)|_{\Delta\mu} \cong \frac{1}{q_s + q_d} = \frac{1}{\sqrt{i_f} + \sqrt{i_r}}. \tag{6.81}$$

The gate-referred noise PSD of the drain current fluctuations PSD is given by

$$S_{\Delta V_G^2}\Big|_{\Delta\mu} \triangleq \frac{S_{\Delta I_{nD}^2}\Big|_{\Delta\mu}}{G_m^2} = S_D|_{\Delta\mu} \, K_D(q_s, q_d)|_{\Delta\mu} \left(\frac{I_D}{G_m}\right)^2 \tag{6.82}$$

$$= S_G|_{\Delta\mu} \, K_G(q_s, q_d)|_{\Delta\mu},$$

where

$$S_G|_{\Delta\mu} \triangleq \frac{kT \, n \, \alpha_H}{W \, L \, C_{ox} \, f}, \tag{6.83}$$

and

$$K_G(q_s, q_d)|_{\Delta\mu} \triangleq (1 + q_s + q_d)^2 \, K_D(q_s, q_d)|_{\Delta\mu}$$

$$= (1 + q_s + q_d) \left[1 + \frac{\ln(q_s/q_d)}{2(q_s - q_d)}\right]. \tag{6.84}$$

Bias-dependent factor $K_G(q_s, q_d)|_{\Delta\mu}$ is plotted versus the inversion factor i_f in saturation in Figure 6.11(b). It shows that for a given q_s/q_d ratio (or equivalently a given V_{DS} voltage), $K_G(q_s, q_d)|_{\Delta\mu}$ is decreasing like $1/i_d$ in weak inversion

$$K_G(q_s, q_d)|_{\Delta\mu} \cong \frac{\ln(q_s/q_d)}{2(q_s - q_d)} = \frac{\ln(i_f/i_r)}{2i_d} = \frac{V_{DS}/U_T}{2i_d}. \tag{6.85}$$

In strong inversion, $K_G(q_s, q_d)|_{\Delta\mu}$ is approximately given by

$$K_G(q_s, q_d)|_{\Delta\mu} \cong q_s + q_d = \sqrt{i_f} + \sqrt{i_r}, \tag{6.86}$$

and hence increases like $\sqrt{i_f}$ in very strong inversion and in saturation.

Unlike $S_{\Delta V_G^2}\Big|_{\Delta N}$, which is minimum in weak inversion, Figure 6.11(b) shows that $S_{\Delta V_G^2}\Big|_{\Delta\mu}$ is minimum in moderate inversion (in saturation as well as in the linear region).

6.3.3 Additional Contributions Due to the Source and Drain Access Resistances

An additional contribution arises from the $1/f$ noise generated in the source and drain access resistances [102, 103]. The latter is modeled by two voltage sources in series with the source and drain resistances R_S and R_D respectively. Assuming that $R_S = R_D = R_a/2$, the PSD of the resulting drain current fluctuations is given by

$$S_{\Delta I_{nD}^2}\Big|_{\Delta R} = \frac{G_{ms}^2 + G_{md}^2}{[1 + (G_{ms} + G_{md})\, R_a/2]^2}\, S_{\Delta V_R^2}, \tag{6.87}$$

where $S_{\Delta V_R^2}$ is the PSD of the $1/f$ noise voltage sources in series with the access resistances. Assuming that $(G_{ms} + G_{md})\, R_a/2 \ll 1$, the PSD normalized to the square of the drain current is then given by

$$\frac{S_{\Delta I_{nD}^2}}{I_D^2}\Big|_{\Delta R} \cong (G_{ms}^2 + G_{md}^2)\, \frac{S_{\Delta V_R^2}}{I_D^2}. \tag{6.88}$$

Now, the PSD of the voltage source $S_{\Delta V_R^2}$ is related to the PSD of the resistance fluctuation $S_{\Delta R^2}$ by

$$S_{\Delta V_R^2} = I_D^2\, S_{\Delta R^2}. \tag{6.89}$$

Equation (6.88) then reduces to

$$\frac{S_{\Delta I_{nD}^2}}{I_D^2}\Big|_{\Delta R} = (G_{ms}^2 + G_{md}^2)\, S_{\Delta R^2} = (q_s^2 + q_d^2)\, G_{spec}^2\, S_{\Delta R^2}. \tag{6.90}$$

In strong inversion and in saturation, (6.90) becomes

$$\frac{S_{\Delta I_{nD}^2}}{I_D^2}\Big|_{\Delta R} = G_{ms}^2\, S_{\Delta R^2} = q_s^2\, G_{spec}^2\, S_{\Delta R^2} = 2n\beta I_D\, S_{\Delta R^2}. \tag{6.91}$$

Assuming that $S_{\Delta R^2}$ is only weakly bias dependent, (6.91) shows that the contribution of the access resistances to the flicker noise at the drain in strong inversion sharply increases with the drain current (actually proportionally to I_D^3). Hence the access resistances will mostly contribute at high current level and should be negligible in moderate and weak inversion.

6.3.4 Total 1/f Noise at the Drain

The total $1/f$ noise at the drain is given by the sum of the different contributions described above [95, 100]

$$\frac{S_{\Delta I_{nD}^2}}{I_D^2} = \frac{S_{\Delta I_{nD}^2}}{I_D^2}\Big|_{\Delta N} + \frac{S_{\Delta I_{nD}^2}}{I_D^2}\Big|_{\Delta \mu} + \frac{S_{\Delta I_{nD}^2}}{I_D^2}\Big|_{\Delta R}. \tag{6.92}$$

$S_{\Delta I_{nD}^2}/I_D^2$ and the different contributions have been computed using the parameters given in Table 6.1 which have been taken from reference [100]. The results obtained for $V_{DS} = 50$ mV[7] are plotted versus the inversion factor in Figure 6.12. It can be seen that the contribution coming

---

[7] The $V_{DS} = 50$ mV bias voltage is not very representative for analog circuit design since most of the time the transistors are biased in saturation, except for switches which have a zero V_{DS}. It has been chosen mainly to be able to compare the results to the experimental measurements presented in [100], which are the only recent measurements found where the PSD is plotted versus the current over a wide bias range.

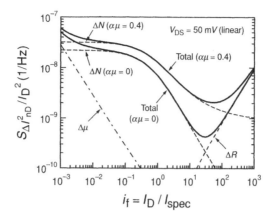

Figure 6.12 Total $1/f$ noise PSD at the drain in the linear region normalized to the square of the drain current and comparison of the different contributions versus the inversion factor. The parameters used for the calculation are given in Table 6.1 and are taken from [100]

from the number fluctuation dominates over a wide bias range (i.e., for $10^{-2} < i_f < 10^2$), whereas $S_{\Delta I_{nD}^2}/I_D^2$ is dominated by the mobility fluctuation at very weak inversion (i.e., for $i_f < 10^{-2}$) and by the contribution coming from the access resistances in very strong inversion (i.e., for $i_f > 10^2$). Note that the case with $\alpha\mu = 0$ looks similar to the result obtained in [100], where the term due to the mobility correlation is used as a fitting parameter and was found to be negligible (corresponding to the curve with $\alpha\mu = 0$ in Figure 6.12).

From a circuit design point of view, it is more interesting to look at the total flicker noise referred at the gate. The latter is plotted in Figure 6.13 versus the inversion factor using the same parameters than in Figure 6.12. As mentioned above, the number fluctuation contribution is dominating $S_{\Delta V_G^2}$ for i_f ranging from 10^{-2} to a bit less than 10^2. In this range and for $\alpha\mu = 0$, $S_{\Delta V_G^2}$ stays almost constant. It starts to increase drastically in very strong inversion and tends to increase also in very weak inversion.

The data plotted in Figures 6.12 and 6.13 correspond to a transistor biased in the linear region. It can be shown that the curves do not change drastically when moving into saturation, as long as the drain-to-source voltage is not kept constant and made too large in order to ensure saturation also in very strong inversion when sweeping the inversion factor from weak to strong inversion

Table 6.1 Typical values of parameters taken from [100] and used in Figures 6.12 and 6.13 for the evaluation of (6.92)

T	W	L	t_{ox}	μ	n	I_{spec}
[K]	[μm]	[μm]	[nm]	$\left[\frac{cm^2}{V \cdot s}\right]$	–	[μA]
300	10	0.18	3.5	560	1.4	57.5
V_{DS}	λ	N_T	α_c	f	α_H	$S_{\Delta R^2}$
[mV]	[nm]	$\left[\frac{1}{eV \cdot cm^3}\right]$	$\left[\frac{V \cdot s}{C}\right]$	[Hz]	–	$\left[\frac{\Omega^2}{Hz}\right]$
50	0.1	7.7×10^{17}	10^4	10	10^{-6}	10^{-6}

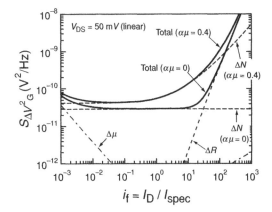

Figure 6.13 Total $1/f$ noise PSD referred at the gate in the linear region and comparison of the different contributions versus the inversion factor. The parameters used for the calculation are given in Table 6.1 and are taken from [100]

(remember that the required drain-to-source voltage for biasing the transistor in saturation in weak inversion is only a few U_T, which might not be enough for ensuring saturation in very strong inversion). If a constant drain-to-source voltage large enough for biasing the transistor in saturation also in the very strong inversion region is maintained, the gate-referred noise increases in weak inversion due to the increase of the mobility fluctuation term. This can be understood from (6.85) which indicates that $S_G|_{\Delta\mu}$ is actually proportional to V_{DS}.

Of course, the results shown in Figures 6.12 and 6.13 have to be taken with some precaution since they strongly depend on process parameters which can change significantly from one technology to another. Also, the model presented above as well as the computation assumed a constant average mobility and ignored all the short-channel effects. Nevertheless, this gives a first idea of the bias dependence of the flicker noise and allows to state that the flicker noise should be minimum in the moderate inversion region.

6.3.5 Scaling Properties

An important property of the $1/f$ noise is that it scales inversely proportional to the gate area. The product of the $1/f$ noise PSD referred at the gate times the gate area can be defined as a figure of merit for the $1/f$ noise. This figure of merit is strongly related to the oxide thickness and decreases when reducing the oxide thickness. Indeed, from the number fluctuation PSD referred to the gate (6.72), it can be seen that the figure of merit $W\,L\,S_{\Delta V_G^2}\big|_{\Delta N}$ is proportional to the product $N_T\,t_{ox}^2$. In case the oxide trap density remains about constant when scaling down the oxide, the noise should diminish significantly with scaled technologies.

The PSD referred to the gate using the mobility fluctuation model as given by (6.83) is proportional to the product $\alpha_H\,t_{ox}$. If the Hooge parameter α_H is assumed to be constant, the figure of merit for mobility fluctuation therefore decreases also when scaling the oxide thickness.

Therefore, at constant gate area, the $1/f$ noise improves with scaled technologies, at a rate proportional to t_{ox}^n where n is comprised between 1 and 2 depending on the bias region. But, when one really wants to take advantage of the scaling, then the ratio $t_{ox}^n/(WL)$ should be considered. If all the scaled dimensions are proportional to the scaling factor $\kappa < 1$, then the $1/f$ noise increases like $1/\kappa$ in the worst case (when $1/f$ noise is dominated by mobility fluctuations) or in the best case remains constant (when the $1/f$ noise is dominated by the number fluctuations).

Further discussion of the impact of downscaling technology on $1/f$ noise is presented in [96], including the $1/f$ noise also present in the gate leakage current and its correlation to the channel flicker noise.

6.4 APPENDICES

Appendix A1: The Nyquist and Bode Theorems

It is probably useful to recall here some fundamental properties of thermal noise in passive RLC networks such as the one shown in Figure A.1. The first is called the Nyquist theorem and states that the PSD of the voltage fluctuation between two terminals of a passive RLC network is simply given by the real part of the impedance Z obtained when looking into the port

$$S_{V_n^2}(f) = 4kT \, \Re\{Z(j2\pi f)\} \,. \tag{A.1}$$

The variance of the voltage V_n is then given by

$$\overline{V_n^2} = \int_0^{+\infty} S_{V_n^2}(f) \cdot \mathrm{d}f = 4kT \cdot \int_0^{+\infty} \Re\{Z(j2\pi f)\} \cdot \mathrm{d}f. \tag{A.2}$$

Instead of calculating the integral given by (A.2), a more powerful mean to evaluate $\overline{V_n^2}$ is to use the Bode theorem (Figure A.2), which states that

$$\overline{V_n^2} = kT \left(\frac{1}{C_\infty} - \frac{1}{C_0} \right), \tag{A.3}$$

where C_∞ is defined by

$$\frac{1}{C_\infty} \triangleq \lim_{s \to +\infty} s \, Z(s), \tag{A.4}$$

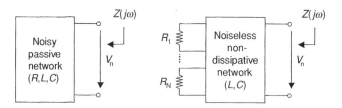

Figure A.1   The Nyquist theorem for thermal noise in passive RLC circuits

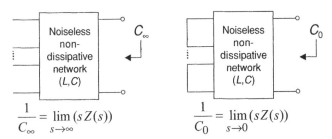

$$\frac{1}{C_\infty} = \lim_{s\to\infty} (s\,Z(s)) \qquad\qquad \frac{1}{C_0} = \lim_{s\to 0} (sZ(s))$$

Figure A.2   The Bode theorem for thermal noise in passive RLC circuits

which corresponds to the capacitance obtained when looking into the port after having removed all resistances from the circuit (or set them to infinity) as shown in Figure A.2.

C_0 is defined as

$$\frac{1}{C_0} \triangleq \lim_{s\to 0} s\,Z(s), \tag{A.5}$$

which corresponds to the capacitance obtained when looking into the port after having replaced every resistance by a short circuit (or set them to zero) as shown in Figure A.2.

Usually, C_∞ and C_0 can be obtained by inspection of the corresponding circuit by applying simple circuit transformation rules.

Both theorems can be illustrated on the simple first-order RC circuit shown in Figure A.3. The impedance $Z(j\,\omega)$ of the RC circuit of Figure A.3 is given by

$$Z(j\,\omega) = \frac{R}{1 + j\,\omega RC} = \frac{R}{1 + (\omega RC)^2} - j\,\frac{\omega R^2 C}{1 + (\omega RC)^2}. \tag{A.6}$$

The PSD of the thermal voltage fluctuations is then obtained from the Nyquist theorem as

$$S_{V_n^2}(f) = 4kT \cdot \Re\{Z(j2\pi f)\} = \frac{4kT \cdot R}{1 + (\omega RC)^2}. \tag{A.7}$$

The variance can then be calculated by evaluating the integral

$$\overline{V_n^2} = 4kTR \int_0^{+\infty} \frac{df}{1 + (2\pi f RC)^2} = \frac{kT}{C}. \tag{A.8}$$

The later result is obtained directly without computing any integral by using the Bode theorem

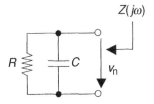

Figure A.3   Applying the Nyquist and Bode theorems to a first-order RC circuit

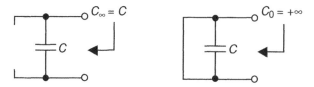

Figure A.4   Applying the Nyquist and Bode theorems to a first-order RC circuit

as illustrated in Figure A.4. C_∞ is obtained directly from circuit inspection as $C_\infty = C$, whereas

$$\frac{1}{C_0} = \lim_{s \to 0} \frac{sR}{1 + sRC} = 0. \tag{A.9}$$

The noise variance is then obtained from (A.3) as

$$\overline{V_n^2} = \frac{kT}{C}. \tag{A.10}$$

Appendix A2: General Noise Expression

The following derivation is taken from [81]. The expression of the drain current valid in all regions of inversion is given by (4.5), which is repeated here for convenience

$$I_D = \mu W(-Q_i) \frac{dV}{dx}.$$

The drain current is decomposed into a dc value I_{D0} and a fluctuation $\delta I_{nD}(x, t)$ resulting from the local current fluctuation $\delta I_n(x, t)$ at point x along the channel

$$I_{D0} + \delta I_{nD}(x, t) = \mu W(-Q_i) \frac{dV}{dx} + \delta I_n(x, t), \tag{A.11}$$

where $V = V_0 + \delta V(x, t)$ with $\delta V(x, t)$ being the fluctuation of the channel voltage at point x. Expanding the current in a series and neglecting the second-order terms lead to

$$\delta I_{nD}(x, t) = \frac{d}{dx} [\mu W(-Q_{i0}) \delta V(x, t)] + \delta I_n(x, t), \tag{A.12}$$

where $Q_{i0} \triangleq Q_i(V_0)$.
  Integrating from $x = 0$ to $x = L$,

$$\Delta I_{nD}(x, t) L = \int_0^L \frac{d}{dx} [\mu W(-Q_{i0}) \delta V(x, t)] dx + \int_0^L \delta I_n(x, t) dx \tag{A.13}$$

$$= \int_0^L \delta I_n(x, t) dx,$$

since in short-circuit condition, $\delta V = 0$ at $x = 0$ and $x = L$. The autocorrelation function of ΔI_{nD} is then given by

$$R_{\Delta I_{nD}^2}(\tau) = \overline{\delta I_{nD}(t)\, \delta I_{nD}(t+\tau)} \tag{A.14}$$

$$= \frac{1}{L^2} \int_0^L \int_0^L \overline{\delta I_n(x,\, t)\, \delta I_n(x',\, t+\tau)}\, dx\, dx'.$$

The PSD of ΔI_{nD} is then given by

$$S_{\Delta I_{nD}^2}(\omega) = \frac{1}{L^2} \int_0^L \int_0^L S_{\delta I_n \delta I_n'}(x,\, x',\, \omega)\, dx\, dx', \tag{A.15}$$

where $S_{\delta I_n \delta I_n'}(x,\, x',\, \omega)$ is the cross-power spectral density (CPSD) between noise at points x and x'. If there is no spatial correlation, then $S_{\delta I_n \delta I_n'}(x,\, x',\, \omega)$ is a Dirac impulse located at point $x' - x$

$$S_{\delta I_n \delta I_n'}(x,\, x',\, \omega) = F(x,\, \omega)\, \delta(x' - x). \tag{A.16}$$

Equation (A.15) then reduces to

$$S_{\Delta I_{nD}^2}(\omega) = \frac{1}{L^2} \int_0^L F(x,\, \omega)\, dx. \tag{A.17}$$

Applying (A.17) to only one noisy section between x and $x + \Delta x$ leads to

$$S_{\delta I_{nD}^2}(\omega) = \frac{F(x,\, \omega)}{\Delta x}, \tag{A.18}$$

which relates $F(x,\, \omega)$ to the PSD of the noise of a single section $S_{\delta I_n^2}(x,\, \omega)$

$$F(x,\, \omega) = \Delta x\, S_{\delta I_n^2}(x,\, \omega). \tag{A.19}$$

Finally, (A.17) can be written as

$$S_{\Delta I_{nD}^2}(\omega) = \frac{1}{L^2} \int_0^L \Delta x\, S_{\delta I_n^2}(x,\, \omega)\, dx. \tag{A.20}$$

7 Temperature Effects and Matching

This chapter models variations of the transistor characteristics, which are very important issues in circuit design. Section 7.2 considers variations with the temperature, which constitute the main external perturbation on transistor characteristics. Section 7.3 is dedicated to the problem of mismatch between the characteristics of supposedly identical transistors, which is a very important limitation to the performance of most analog circuits. These two kinds of variations will be characterized as variations of the basic model parameters established so far, and will be traced back to variations of physical parameters.

7.1 INTRODUCTION

The purpose of a transistor model is to describe its electrical characteristics. These are essentially the static and dynamic relationships between the voltages applied at its various terminals and the currents flowing through them.

The model introduced so far describes how these electrical characteristics depend on physical parameters. These physical parameters have been lumped into a reduced set of model parameters. As long as the physical parameters remain constant, the characteristics and the model parameters remain constant. However, some physical parameters may change, thereby modifying the transistor characteristics.

Aging is the consequence of parameters changing with *time*. It will not be considered here.

Many parameters are changing with the *temperature*, which can be considered the main external perturbation on the transistor.

The characteristics of two or more transistors designed to be identical do not match perfectly. This *mismatch* is the consequence of parameters changing in *space*.

A list of all basic parameters that influence the characteristics of the transistor is given in Table 7.1. It indicates which of them depend on the temperature and which of them may change spatially, resulting in mismatch.

Charge-Based MOS Transistor Modeling: The EKV Model for Low-Power and RF IC Design   C. Enz and E. Vittoz
© 2006 John Wiley & Sons, Ltd.

Table 7.1 Basic physical parameters: dependence on temperature and effect on mismatch

	U_T	n_i	N_b	$\epsilon_{ox}, \epsilon_{si}$	Q_{fc}	Φ_{ms}	t_{ox}	L, W	μ
Temperature	Y	Y	N	Ne	N	Y	N	N	Y
Mismatch	N	N	Y	N	Y	Y	Y	Y	Y

Y = yes, N = no, Ne = negligible

Table 7.2 Intermediate parameters and model parameters: dependence on temperature and effect on mismatch

	Φ_F	C_{ox}	V_{FB}	Γ_b	Ψ_0	$n\vert_V$	V_{T0}	β	Q_{spec}	I_{spec}
Defin.	3.8	3.20	3.22	3.30	3.66	3.34	3.58	4.8	3.42	4.14
Temp.	Y	Ne	Y	Ne	Y	Y	Y	Y	Y	Y
Match	Y	Y	Y	Y	Y	Y	Y	Y	Y	Y

Y = yes, N = no, Ne = negligible

The temperature is assumed here to be constant in space. As can be seen, most of the physical parameters are independent of either space or temperature. The temperature coefficient of dielectric constants ϵ_{ox} and ϵ_{si} is small enough to be neglected.

Table 7.2 is a list of most of the additional parameters that have been derived so far by combining some of the basic parameters, including the most important device parameters n, V_{T0}, and β. Most of them depend on temperature and all of them are subject to mismatch.

7.2 TEMPERATURE EFFECTS

7.2.1 Variation of Basic Physical Parameters

As indicated in Table 7.1, only four of the basic physical parameters that control the transistor characteristics have a nonnegligible dependence on temperature T.

The dependence of $U_T = kT/q$ is obvious. This fundamental specific voltage, which has been used to normalize other voltages, is proportional to the absolute temperature. Its temperature coefficient is $k/q = 86\mu V/°K$.

The intrinsic carrier concentration of silicon has the value $n_i = 1.45 \times 10^{10} cm^{-3}$ at $300°K$, but it is strongly dependent on T since [67]

$$n_i \propto T^{3/2} \exp \frac{-V_{Gap}}{2U_T}, \tag{7.1}$$

where V_{Gap} is the voltage corresponding to the energy band gap of silicon. This band gap voltage itself depends slightly on the temperature, as shown by Figure 7.1. For the usual range of ambient temperature, this variation can be linearized as shown in the same figure. Then

$$n_i \propto T^{3/2} \exp \frac{-(V_{G0} - aT)}{2U_T} \propto T^{3/2} \exp \frac{-V_{G0}}{2U_T}, \tag{7.2}$$

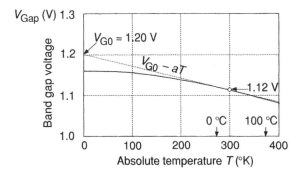

Figure 7.1 Temperature dependence of the silicon band gap [67]

where $V_{G0} \cong 1.20\,\text{V}$ is called the *extrapolated band gap voltage*. This result is plotted in Figure 7.2. Most of the temperature variation is due to the exponential term; therefore, (7.2) can be approximated by keeping the term $T^{3/2}$ constant at a value $T_0^{3/2}$:

$$n_i = n_{i\infty} \exp \frac{-V_{G0}}{2U_T},$$ (7.3)

where $n_{i\infty} = 1.73 \times 10^{20}\,\text{cm}^{-3}$ for $V_{G0} = 1.20\,\text{V}$ and $T_0 = 300°\text{K}$. This approximation is also plotted in Figure 7.2. It departs from (7.2) by about 40% at the limits of the displayed temperature range. If more precision is needed, the difference may be reduced to less than 5% (within the line thickness in the figure) by artificially increasing the value of V_{G0} to 1.28 V (and adapting the value of $n_{i\infty}$ to $8.11 \times 10^{20}\,\text{cm}^{-3}$ so as to maintain $n_i = 1.45 \times 10^{10}\,\text{cm}^{-3}$ at 300 °K).

The Fermi potential Φ_F given by (3.8) depends on the temperature through U_T and n_i. By using approximation (7.3) of n_i, it can be expressed as

$$\Phi_F = \frac{V_{G0}}{2} - U_T \ln \frac{n_{i\infty}}{N_b}.$$ (7.4)

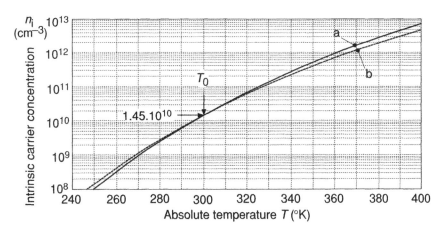

Figure 7.2 Temperature variation of intrinsic concentration n_i. (a) given by (7.2); (b) approximation given by (7.3)

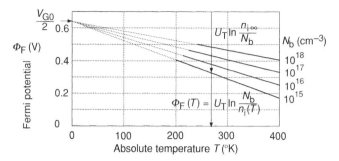

Figure 7.3   Temperature variation of Fermi potential

This variation with temperature is plotted in Figure 7.3 for several values of doping concentration N_b. The value of V_{G0} has been increased by 6.6% to partially compensate for the $T^{3/2}$ term in n_i, as mentioned above.

The temperature coefficient of Φ_F is obtained by differentiation of (7.4) or more simply by inspection of Figure 7.3:

$$\frac{d\Phi_F}{dT} = -\frac{k}{q} \ln \frac{n_{i\infty}}{N_b} = -\frac{1}{T}\left(\frac{V_{G0}}{2} - \Phi_F\right). \tag{7.5}$$

It is independent of temperature and its absolute values decreases when N_b is increased. This value ranges between -1.1 and -1.6 mV/$^\circ$K.

Voltage Φ_{ms}, the difference between the extraction potentials, depends on the gate and substate materials. If both are silicon as in most standard technologies, then Φ_{ms} depends on their isolated Fermi potentials. By applying (7.4) to express the gate Fermi potential Φ_{FG}, with N_b replaced by the gate doping concentration N_g, we obtain for a P-doped gate (same type as that of the local substrate):

$$\Phi_{ms} = \Phi_{FG} - \Phi_F = U_T \ln \frac{N_g}{N_b} > 0. \tag{7.6}$$

For an N-doped gate (same type as that of source and drain, opposite to that of the local substrate), the sign of Φ_{FG} is changed, giving

$$\Phi_{ms} = -\Phi_{FG} - \Phi_F = -V_{G0} + U_T \ln \frac{n_{i\infty}^2}{N_g N_b} < 0. \tag{7.7}$$

It is worth noticing that changing the gate doping from P-type to N-type changes Φ_{ms} by

$$\Phi_{ms}(\text{N-gate}) - \Phi_{ms}(\text{P-gate}) = -\left(V_{G0} - 2U_T \ln \frac{n_{i\infty}}{N_g}\right). \tag{7.8}$$

Indeed, if the gate is highly doped (N_g approaches $n_{i\infty}$), the Fermi level moves from being close to the valence band to being close to the conduction band, thereby crossing the entire band gap. Since Φ_{ms} is part of the flat-band voltage V_{FB} according to (3.22), the threshold voltage is then reduced by the same amount.

From (7.6), the temperature coefficient of Φ_{ms} for an P-doped gate is

$$\frac{d\Phi_{ms}}{dT} = \frac{k}{q} \ln \frac{N_g}{N_b} = \frac{\Phi_{ms}}{T} > 0. \tag{7.9}$$

From (7.7), the temperature coefficient of Φ_{ms} for an N-doped gate is

$$\frac{d\Phi_{ms}}{dT} = \frac{k}{q} \ln \frac{n_{i\infty}^2}{N_g N_b} = \frac{1}{T}(V_{G0} + \Phi_{ms}) > 0. \tag{7.10}$$

The temperature coefficient is positive in both cases, since $N_g > N_b$.

If the gate is very strongly doped, then $N_g \cong n_{i\infty}$ and the temperature coefficient for both types of gate doping becomes

$$\frac{d\Phi_{ms}}{dT} \cong \frac{k}{q} \ln \frac{n_{i\infty}}{N_b} = -\frac{d\Phi_F}{dT} > 0. \tag{7.11}$$

Indeed, for a very high doping, Figure 7.3 shows that the temperature coefficient of Φ_{FG} tends to zero and

$$\Phi_{ms} \cong \text{constant} - \Phi_F. \tag{7.12}$$

This relation is exact for a metal gate, since its extraction potential is independent of temperature.

The last basic parameter of Table 7.1 to be considered is mobility μ. The mobility of electrons and holes in silicon is affected by several scattering mechanisms [67]. Scattering due to acoustic phonons increases with temperature, thereby reducing the mobility. Scattering due to ionized impurities has an opposite effect, since it increases at low temperatures; it also increases with impurity concentration N_b. Within the range of ambient temperatures, their combined effect can be approximated by

$$\mu \propto T^{-\alpha}. \tag{7.13}$$

In nondoped silicon, α is approximately 2.5 for electrons and 2.7 for holes. It decreases to about 1 for $N_b = 10^{18} \text{ cm}^{-3}$ [67].

From 7.13, the relative temperature coefficient of the mobility can be expressed as

$$\frac{d\mu/\mu}{dT} = -\frac{\alpha}{T} \tag{7.14}$$

corresponding to a range of -0.3 to -1% per degree at ambient temperatures.

All possible effects of the temperature on the behavior of the transistor as described by the model derived in Chapters 3 and 4 will be accounted for by simply including the variation with temperature of the basic physical parameters discussed so far.

However, for a better understanding of the temperature behavior of circuits, it is useful to express more explicitly the effect of temperature variations on the model.

7.2.2 Variation of the Voltage–Charge Characteristics

Since neither Q_{fc} nor C_{ox} depends significantly on temperature, the dependence of the flat-band voltage V_{FB} is that of Φ_{ms} given by (7.9) or (7.10). If the gate is strongly doped, it can be reasonably approximated by (7.11):

$$\frac{dV_{FB}}{dT} = \frac{d\Phi_{ms}}{dT} \cong -\frac{d\Phi_F}{dT}, \tag{7.15}$$

with $d\Phi_F/dT$ given by (7.5).

Since the substrate modulation factor Γ_b given by (3.30) does not depend on temperature, the variation with ΔT of the threshold function V_{TB} (3.33) is only ΔV_{FB}, as illustrated in Figure 7.4. The slope $n(\Psi_s)$ remains unchanged and $\Delta(-Q_i/C_{ox})$ in strong inversion at constant surface potential would be $-\Delta V_{FB}$.

However, what is imposed is not the surface potential, but the value of channel voltage $V = \Psi_s - \Psi_0$ at the source and at the drain, and Ψ_0 depends on temperature according to (3.66) repeated here for convenience:

$$\Psi_0 = 2\Phi_F + v_{sh}U_T.$$

The normalized voltage shift v_{sh} is approximately independent of temperature and smaller than 4 according to Figure 3.10. The temperature coefficient of Ψ_0 is thus close to the double of that of Φ_F given by (7.5):

$$\frac{d\Psi_0}{dT} = 2\frac{d\Phi_F}{dT} + \frac{V_{sh}}{T} = -\frac{V_{G0} - 2\Phi_F - V_{sh}}{T} = -\frac{V_{G0} - \Psi_0}{T}. \tag{7.16}$$

As represented in Figure 7.4, the variation of threshold voltage, ΔV_{T0}, is the combined effect of ΔV_{FB} and $\Delta \Psi_0$:

$$\Delta V_{T0} = \Delta V_{FB} + n\Delta\Psi_0, \tag{7.17}$$

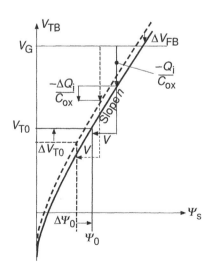

Figure 7.4 Effect of an increase of temperature on voltages and charge

where slope n should be evaluated at $V = 0$: $n = n_0$ given by (3.73). Hence

$$\frac{dV_{T0}}{dT} = \frac{dV_{FB}}{dT} + n_0 \frac{d\Psi_0}{dT} \tag{7.18}$$

as can be obtained directly by differentiation of expression (3.58) of the threshold V_{T0}.

Now, for a highly doped gate, introducing approximation (7.15) with (7.16) and (7.5) yields an explicit expression for the temperature coefficient of V_{T0}:

$$\frac{dV_{T0}}{dT} = (2n_0 - 1)\frac{d\Phi_F}{dT} + n_0 \frac{V_{sh}}{T} = \frac{(n_0 - 1/2)(2\Phi_F - V_{G0}) + n_0 V_{sh}}{T}, \tag{7.19}$$

which is always negative. Using the first expression of $d\Phi_F/dT$ in (7.5), this temperature coefficient can also be written as

$$\frac{dV_{T0}}{dT} = \frac{k}{q}\left[(1 - 2n_0)\ln\frac{n_{i\infty}}{N_b} + n_0 v_{sh}\right], \tag{7.20}$$

showing that, except for the slight variation of n_0 to be discussed further, the temperature coefficient of V_{T0} is independent of the temperature. Practical values are ranging from -2.5 to -1 mV/$^\circ$K.

Although the slope factor n at constant Ψ_s is independent of temperature, at constant $V = \Psi_s - \Psi_0$ it varies through the variation of Ψ_0. By using (3.34), and (7.16),

$$\frac{dn}{dT} = \frac{d}{d\Psi_0}\left(1 + \frac{\Gamma_b}{2\sqrt{\Psi_0 + V}}\right)\frac{d\Psi_0}{dT} = \frac{n - 1}{2T}\frac{V_{G0} - \Psi_0}{V + \Psi_0}, \tag{7.21}$$

which is always positive. Practical values are never larger than 0.1%/$^\circ$K; therefore, this variation can be *neglected* in most situations.

Neglecting the variation of n, the specific charge density Q_{spec} defined by (3.42) and used to normalize the density of inverted charge is simply proportional to U_T; hence,

$$\frac{dQ_{spec}/Q_{spec}}{dT} = \frac{1}{T}. \tag{7.22}$$

Since n can be considered constant, the variation of the pinch-off voltage at constant gate voltage obtained from expression (3.63) becomes

$$\frac{dV_P}{dT} = -\frac{1}{n}\frac{dV_{T0}}{dT}. \tag{7.23}$$

The temperature coefficient of $-Q_i$ for fixed voltages depends on the mode of operation. For strong inversion, differentiating approximation (3.64) and introducing (7.23) yield

$$\frac{d(-Q_i/C_{ox})}{dT} = -\frac{dV_{T0}}{dT}. \tag{7.24}$$

This relation is also illustrated in Figure 7.4.

The behavior is more complicated in weak inversion. Approximation (3.49) can be rewritten:

$$-Q_i = 2nC_{ox}U_T \exp \frac{V_P - V}{U_T},$$

(7.25)

where U_T and V_P depend on temperature. Differentiation gives

$$\frac{dQ_i/Q_i}{dT} = \frac{V + U_T - V_P}{TU_T} - \frac{1}{nU_T}\frac{dV_{T0}}{dT}.$$

(7.26)

Since in weak inversion $V - V_P$ is usually much smaller than $V_{G0} - 2\Phi_F$, an inspection of expression (7.19) shows that the last term dominates. As for strong inversion above, the charge increases with temperature. But the relative increase is maximum in weak inversion.

In summary, among the three device parameters controlling the charge–voltage characteristics of the transistor, only V_{T0} is significantly dependent on the temperature. The variation of C_{ox} is negligible and that of n can usually be neglected.

7.2.3 Variation of the Voltage–Current Characteristics

The only additional parameter needed to obtain the current from the mobile charge density is the transfer parameter β defined by (4.8). It is proportional to the mobility; hence, from (7.14),

$$\frac{d\beta/\beta}{dT} = -\frac{\alpha}{T}.$$

(7.27)

As seen previously, β can be replaced as the third parameter controlling the current by the specific current I_{spec} that is used to normalize all currents. Differentiation of its expression (4.14) gives

$$\frac{dI_{spec}/I_{spec}}{dT} = \frac{2 - \alpha}{T}.$$

(7.28)

We have seen that $\alpha > 2$ for nondoped or lightly doped silicon, resulting in a negative temperature coefficient of I_{spec}. For a doping concentration $N_b = 10^{16}$ cm^{-3}, $\alpha \cong 2$, the variation of U_T^2 compensates that of β, and I_{spec} is approximately independent of the temperature. This compensation disappears and the coefficient becomes positive when α is decreased below 2 by further increasing N_b.

The variation of current with temperature for constant bias voltages depends on the mode of operation.

For strong inversion, the differentiation of approximation (4.28) gives

$$\frac{dI_{F,R}/I_{F,R}}{dT} = \frac{d\beta/\beta}{dT} - \frac{2}{n(V_P - V_{S,D})}\frac{dV_{T0}}{dT}.$$

(7.29)

Since the temperature coefficients of V_{T0} and β are both negative, compensation occurs for a particular value of $n(V_P - V_{S,D}) = V_G - V_{T0} - nV_{S,D}$. This value is ranging between 300 and 600 mV. Above this limit value, the variation of mobility dominates and the current decreases with increasing temperature. Below this limit, the variation of threshold dominates and the current increases with the temperature.

For weak inversion, the differentiation of approximation (4.34) gives, after introduction of (7.28)

$$\frac{dI_{F,R}/I_{F,R}}{dT} = \frac{V_{S,D} - V_P + (2 - \alpha)U_T}{TU_T} - \frac{1}{nU_T}\frac{V_{T0}}{dT}, \tag{7.30}$$

which is very close to expression (7.26) for the charge. Here again the last term due to the variation of threshold dominates. The current for constant bias voltages increases with temperature.

This dependence is *very large*, typically 5% per degree, due to the strong effect of threshold variations on the drain current (large transconductance to current ratio). For this reason, a transistor in weak inversion should always be biased at constant current, and *never at constant gate and source voltages*.

The specific conductance used to normalize transconductances is given by (5.6). Neglecting again the variation of n, we obtain by differentiation

$$\frac{dG_{spec}/G_{spec}}{dT} = \frac{1 - \alpha}{T} < 0. \tag{7.31}$$

Since $\alpha > 1$, this coefficient is always negative.

The transconductance for a given value of β may be imposed by the voltages or by the current.

For strong inversion and *constant voltages*, the differentiation of (5.14) yields

$$\frac{dG_{ms,d}/G_{ms,d}}{dT} = \frac{d\beta/\beta}{dT} - \frac{1}{n(V_P - V_{S,D})}\frac{dV_{T0}}{dT}. \tag{7.32}$$

By comparing with (7.29), we can see that the sensitivity to threshold variations is half of that for the current. The compensation of the variation of β therefore occurs for half the value of $n(V_P - V_{S,D})$, in the range 150–300 mV.

For strong inversion and *constant current*, the differentiation of (5.16) yields

$$\frac{dG_{ms,d}/G_{ms,d}}{dT} = \frac{1}{2}\frac{d\beta/\beta}{dT} = -\frac{\alpha}{2T}. \tag{7.33}$$

The relative temperature coefficient of the transconductance is always negative and independent of the current.

As mentioned above, a transistor in weak inversion should always be biased at *constant current*, the differentiation of (5.19) yields

$$\frac{dG_{ms,d}/G_{ms,d}}{dT} = -\frac{1}{T}. \tag{7.34}$$

Results (7.33) and (7.34) are comparable. The temperature coefficient of the transconductance at constant current is always negative. It is about $-0.3\%/°K$ at ambient temperature.

7.2.4 Variation of the Current–Charge Characteristics

Relationship (4.19) between the normalized current and charge can be rewritten with nonnormalized variables:

$$\frac{\beta}{2n}\left(\frac{-Q_{\text{iS,D}}}{C_{\text{ox}}}\right)^2 + \beta U_{\text{T}}\frac{-Q_{\text{iS,D}}}{C_{\text{ox}}} = I_{\text{F,R}}, \tag{7.35}$$

where the first term may be neglected in weak inversion and the second term may be neglected in strong inversion. Differentiating separately these two components of the inverted charge gives for strong inversion

$$\frac{\mathrm{d}Q_{\text{iS,D}}/Q_{\text{iS,D}}}{\mathrm{d}T} = -\frac{1}{2}\frac{\mathrm{d}\beta/\beta}{\mathrm{d}T} = \frac{\alpha}{2T}. \tag{7.36}$$

and for weak inversion

$$\frac{\mathrm{d}Q_{\text{iS,D}}/Q_{\text{iS,D}}}{\mathrm{d}T} = -\frac{\mathrm{d}\beta/\beta}{\mathrm{d}T} - \frac{1}{T} = \frac{\alpha - 1}{T}. \tag{7.37}$$

These results are comparable. The temperature coefficient of the charge at *constant current* is always positive. It is about $+0.3\%/^\circ$K at ambient temperature.

7.3 MATCHING

7.3.1 Introduction

Matching of the characteristics of two or several transistors is a very important consideration for analog circuits. Even if transistors are exactly identical in their structure and layout, their electrical behaviors are not exactly identical. This is due to spacial fluctuations of the physical parameters that control these behaviors, as listed in Table 7.1.

As summarized in Section 4.4.2, the static voltage–current characteristics of long-channel transistors require only three device parameters, namely V_{T0}, β, and n. Hence, the mismatch of their characteristics is completely characterized by the mismatch of these three parameters.

The sensitivity of these parameters to temperature has been discussed in Section 7.2. If there is any difference in the temperature of two or several transistors, due to a gradient of temperature on the chip, it produces a proportional difference in the parameters. This gradient may be stationary (due to a source of heat on the chip), or it can be transient in time (due to a change of ambient temperature that is too fast with respect to the chip thermal time constant).

Now, even if the temperature does not vary throughout the chip, some variations of the three device parameters remain, which can be traced back to those of the physical parameters, as illustrated in Figure 7.5.

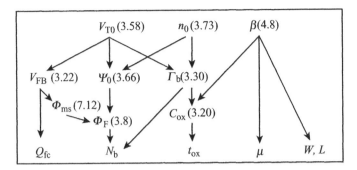

Figure 7.5 Dependence of V_{T0}, $n_0 = n(V = 0)$, and β on basic physical parameters

7.3.2 Deterministic Mismatch

Let us assume that the average value of each physical parameter is slightly different between devices. This might be due to a spatial gradient, or to some local difference due to different environments. The resulting small differences of device parameters can be calculated by differentiation of the various equations summarized in Figure 7.5.

We obtain for the difference ΔV_{T0} of threshold voltages,

$$\Delta V_{T0} = [(2n_0 - 1)U_T + (n_0 - 1)\Psi_0]\frac{\Delta N_b}{N_b} + \left(\frac{Q_{fc}}{C_{ox}} - 2(n_0 - 1)\Psi_0\right)\frac{\Delta C_{ox}}{C_{ox}} - \frac{Q_{fc}}{C_{ox}}\frac{\Delta Q_{fc}}{Q_{fc}},$$
(7.38)

for the difference Δn of slope factors (evaluated at $n = n_0$ where it is the most sensitive to variations of N_b):

$$\frac{\Delta n}{n} = \frac{n_0 - 1}{n_0}\left[\left(\frac{1}{2} - \frac{U_T}{\Psi_0}\right)\frac{\Delta N_b}{N_b} - \frac{\Delta C_{ox}}{C_{ox}}\right],$$
(7.39)

and for the difference $\Delta\beta$ of transfer parameters,

$$\frac{\Delta\beta}{\beta} = \frac{\Delta\mu}{\mu} + \frac{\Delta C_{ox}}{C_{ox}} + \frac{\Delta W}{W} - \frac{\Delta L}{L}.$$
(7.40)

To illustrate the importance of each term, let us take a realistic example with $\Psi_0 = 0.8$ V, $n = 1.4$, and $Q_{fc}/C_{ox} = -200$ mV. The effects of a 1% variation of each physical parameters are summarized in Table 7.3.

Table 7.3 Effect of 1% variation of the physical parameters on the three device parameters: for example $\Psi_0 = 0.8$ V, $n = 1.4$ and $Q_{fc}/C_{ox} = -200$ mV

		Q_{fc}	N_b	C_{ox}	μ	W	L
(7.38)	ΔV_{T0} (mV)	2.00	3.67	−8.40	0	0	0
(7.39)	$\Delta n/n$ (%)	0	0.13	−0.29	0	0	0
(7.40)	$\Delta\beta/\beta$(%)	0	0	1.00	1.00	1.00	−1.00

As can be seen, n is the less sensitive parameter. Its sensitivity to C_{ox} is only $(n_0 - 1)/n_0$ times that of β, and that to N_b about half of this. Hence, the contribution of Δn can often be neglected in a first approximation.

Some correlation exists between the variations of the three parameters, since they all depend on C_{ox}. They would be fully correlated if the only varying parameter was C_{ox}.

Moreover, variations of V_{T0} and n are further correlated by their common dependence on N_b. They would be fully correlated if the only varying parameter was N_b.

To evaluate the importance of these possible correlations, let us calculate the sensitivity of the saturated drain current (forward current I_F) to variations of the three device parameters. As obtained in Chapter 4, this current is a function of $V_P - V_S$ according to

$$I_F = I_{spec} F_I(v_p - v_s) = 2n\beta U_T^2 F_I \left(\frac{V_G - V_{T0}}{nU_T} - \frac{V_S}{U_T} \right), \tag{7.41}$$

where F_I may be the inverse of the function defined by (4.25). Hence, by differentiation with constant V_S,

$$\Delta I_F = \left(\frac{\Delta\beta}{\beta} + \frac{\Delta n}{n} \right) I_F + \underbrace{\frac{I_{spec}}{n} \frac{dF_I}{dV_P}}_{G_m} \left[\Delta V_G - \Delta V_{T0} - (V_G - V_{T0}) \frac{\Delta n}{n} \right]. \tag{7.42}$$

The gate transconductance G_m can be identified as the factor of ΔV_G (independently of the exact form of F_I). We obtain finally

$$\frac{\Delta I_F}{I_F} = \frac{\Delta\beta}{\beta} + \frac{\Delta n}{n} + \frac{G_m}{I_F} \left[\Delta V_G - \Delta V_{T0} - (V_G - V_{T0}) \frac{\Delta n}{n} \right]. \tag{7.43}$$

For $\Delta V_G = 0$, this equation gives the mismatch of saturation currents for transistors having the same gate and source voltages:

$$\frac{\Delta I_F}{I_F} = \frac{\Delta\beta}{\beta} - \frac{G_m}{I_F} \Delta V_{T0} + \left[1 - \frac{G_m}{I_F}(V_G - V_{T0}) \right] \frac{\Delta n}{n}. \tag{7.44}$$

In this situation (for example in a current mirror), $\Delta\beta/\beta$ contributes directly to $\Delta I_F/I_F$, whereas ΔV_{T0} is weighted by G_m/I_F, the transconductance to current ratio that is only a function of the inversion coefficient IC, as shown by Figure 5.6. When IC is increased, G_m/I_F decreases, thereby reducing the effect of ΔV_{T0}.

Expressing ΔV_G for $\Delta I_F = 0$ in (7.43) gives the mismatch of gate voltages for transistors having the same saturation current and the same source voltage:

$$\Delta V_G = \Delta V_{T0} - \frac{I_F}{G_m} \frac{\Delta\beta}{\beta} + \left[(V_G - V_{T0}) - \frac{I_F}{G_m} \right] \frac{\Delta n}{n}. \tag{7.45}$$

In this situation (for example a differential pair), ΔV_{T0} contributes directly to ΔV_G, whereas $\Delta\beta/\beta$ is weighted by I_F/G_m. When IC is increased, I_F/G_m increases, thereby increasing the effect of $\Delta\beta/\beta$.

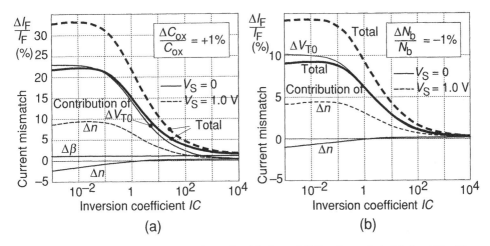

Figure 7.6 Contributions to current mismatch with $\Psi_0 = 0.8$ V and $n = 1.4$: (a) for $\Delta C_{ox}/C_{ox} = +1\%$ alone; (b) for $\Delta\beta/\beta = -1\%$ alone

In both cases, the factor weighting $\Delta n/n$ is a function of the inversion coefficient, since $V_G - V_{T0}$ can be expressed from (4.25) as

$$V_G - V_{T0} = nU_T \left[\sqrt{1 + 4IC} + \ln\left(\sqrt{1 + 4IC} - 1\right) - (1 + \ln 2)\right] + nV_S, \qquad (7.46)$$

whereas from (5.29),

$$\frac{G_m}{I_F} = \frac{2}{nU_T\left(\sqrt{1 + 4IC} + 1\right)}. \qquad (7.47)$$

The mismatch of currents calculated by introducing (7.46) and (7.47) into (7.44) separately for differences in C_{ox} and N_b of 1% is plotted in Figure 7.6, using values given in Table 7.3.

As can be seen, the effect of ΔV_{T0} increases drastically in weak inversion ($IC < 1$), whereas that of $\Delta\beta$ dominates in strong inversion ($IC \gg 1$) for variations of C_{ox}. Their (additive) correlation is therefore relevant only in moderate inversion, when both contributions are comparable.

For $V_S = 0$, Δn causes a slight reduction of mismatch in very weak inversion, which can again be neglected.

For $V_S > 0$, the current mismatch is increased by the contribution of Δn, especially in weak inversion. This can be explained by the fact that the effective gate voltage threshold $V_{T0} + nV_S$ becomes dependent on n.

In strong inversion, some terms are negligible in (7.46) and (7.47), hence (7.44) is reduced to

$$\frac{\Delta I_F}{I_F} = \frac{\Delta\beta}{\beta} - \frac{G_m}{I_F}\Delta V_{T0} - \left(1 + \frac{G_m}{I_F}nV_S\right)\frac{\Delta n}{n}, \qquad (7.48)$$

whereas in weak inversion it becomes

$$\frac{\Delta I_F}{I_F} = \frac{\Delta\beta}{\beta} - \frac{\Delta V_{T0}}{nU_T} + \left(1 - \ln IC - \frac{V_S}{U_T}\right)\frac{\Delta n}{n}. \qquad (7.49)$$

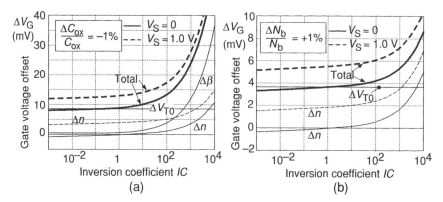

Figure 7.7 Separate contributions to gate voltage offset with $\Psi_0 = 0.8\,\text{V}$ and $n = 1.4$: (a) for $\Delta C_{ox}/C_{ox} = -1\%$ alone; (b) for $\Delta N_b/N_b = +1\%$ alone

The mismatch of gate voltages (gate voltage offset) calculated by introducing (7.46) and (7.47) into (7.45) separately for differences in C_{ox} and N_b of 1% is plotted in Figure 7.7, using values given in Table 7.3.

As can be seen, the contributions of $\Delta\beta$ and Δn increase in strong inversion ($IC \gg 1$), whereas that of ΔV_{T0} dominates in weak inversion for $V_S = 0$. Correlations through ΔC_{ox} and ΔN_b are additive, but relevant only in moderate inversion, when the contributions are comparable.

For $V_S = 0$, Δn causes a slight reduction of offset in very weak inversion, which can be neglected.

For $V_S > 0$, the offset is increased by an amount equal to $V_S\Delta n$, which can again be explained by the fact that the effective threshold becomes dependent on n.

In strong inversion, (7.45) can be approximated by

$$\Delta V_G = \Delta V_{T0} - \frac{I_F}{G_m}\frac{\Delta\beta}{\beta} + \left(\frac{I_F}{G_m} + nV_S\right)\frac{\Delta n}{n}. \tag{7.50}$$

This expression shows that the correlations through ΔC_{ox} and ΔN_b are indeed additive, since the signs of ΔV_{T0}, Δn, and $-\Delta\beta$ are the same, as we can see in Table 7.3.

In weak inversion, (7.45) can be reduced to

$$\Delta V_G = \Delta V_{T0} - nU_T\frac{\Delta\beta}{\beta} + [(\ln IC - 1)U_T + V_S]\Delta n, \tag{7.51}$$

showing that the correlated Δn slightly reduces the offset for $V_S/U_T < 1 - \ln IC$, but augments it above this limit.

Appropriate layout techniques can be used to virtually eliminate the systematic mismatch discussed in this section [104].

Transistors to be matched should be implemented as close as possible, in order to minimize the effect of gradient and/or other variations of low spacial frequency.

Their environments should be fully identical, in order to avoid local variations of critical physical parameters.

In order to eliminate the effect of a constant gradient, each device can be split into two half-width devices located each side of an axis of symmetry and connected in parallel. Best is the implementation of pairs in quad configurations, since it also ensures the same environment for the two transistors.

We have considered differences between the average values of parameters across the devices. However, except for $V_{DS} = 0$, Q_i is not constant along the channel. It is larger close to the source (in forward mode) and therefore the current is more affected by a difference of physical parameters close to the source. Hence, transistors to be matched should have the same source–drain orientation to avoid an additional effect of gradients.

The mismatch that remains after elimination of these systematic variations is due to statistical fluctuations of the physical parameters across the area of the transistor channel.

7.3.3 Random Mismatch

Let us assume that the random fluctuations of a parameter P are not spatially correlated. It can be shown [105, 106] that the standard deviation of the difference ΔP of average values of P across two separate regions of area WL is given by

$$\sigma(\Delta P) = \frac{A_P}{\sqrt{WL}},\qquad(7.52)$$

where A_P is the *area proportionality constant* for parameter P.

This relation can be understood by considering the particular example of C_{ox} illustrated in Figure 7.8.

If for an element of surface of *unit area*,

$$\frac{\sigma_u(\Delta C_{ox})}{C_{ox}} = A_{C_{ox}},\qquad(7.53)$$

then for the total area, the standard deviation of the difference of total capacitances is increased by the square root of the area (the variance increases linearly with the area), whereas the capacitance increases with the area. Hence,

$$\frac{\sigma(\Delta C_{ox})}{C_{ox}} = \frac{A_{C_{ox}}\sqrt{WL}}{WL} = \frac{A_{C_{ox}}}{\sqrt{WL}}.\qquad(7.54)$$

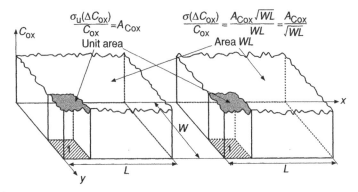

Figure 7.8   Dependence of $\sigma(\Delta C_{ox}/C_{ox})$ on channel area WL

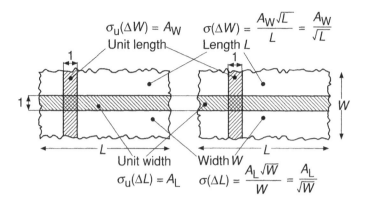

Figure 7.9 Dependence of $\sigma(\Delta W)$ and $\sigma(\Delta L)$ on L and W

The same averaging across area takes place for the fluctuations of parameters Q_{fc}, N_{b}, and μ.

The situation is somewhat different for the fluctuations of W and L. First, they are best characterized as *absolute values* and not as a percentages. Furthermore, they are not averaged across an area, but along a distance. Indeed, as illustrated by Figure 7.9, noncorrelated fluctuations of width are averaged along the length, and noncorrelated fluctuations of length are averaged along the width. If for an element of surface of *unit length*,

$$\sigma_{\text{u}}(\Delta W) = A_{\text{W}}, \tag{7.55}$$

then for the total length L, it becomes

$$\sigma(\Delta W) = \frac{A_{\text{W}}\sqrt{L}}{L} = \frac{A_{\text{W}}}{\sqrt{L}}, \tag{7.56}$$

and symmetrically,

$$\sigma(\Delta L) = \frac{A_{\text{L}}\sqrt{W}}{W} = \frac{A_{\text{L}}}{\sqrt{W}}. \tag{7.57}$$

Proportionality constants A_{W} and A_{L} could be different if the etching process defining the gate is anisotropic. We will assume that they are identical, of value $A_{\text{W,L}} = A_{\text{W}} = A_{\text{L}}$.

Now, the device parameter affected by variations of W and L is $\beta \propto W/L$. If it is small, the relative difference of aspect ratios can be expressed as

$$\frac{\Delta(W/L)}{W/L} = \frac{\Delta W}{W} - \frac{\Delta L}{L}, \tag{7.58}$$

the variance of which is

$$\sigma^2\left(\frac{\Delta(W/L)}{W/L}\right) = \sigma^2\left(\frac{\Delta W}{W}\right) + \sigma^2\left(\frac{\Delta L}{L}\right). \tag{7.59}$$

Introducing (7.56) and (7.57) with $A_W = A_L = A_{W,L}$ yields

$$\sigma^2 \left(\frac{\Delta(W/L)}{W/L} \right) = \frac{A_{W,L}^2}{WL} \left(\frac{1}{L} + \frac{1}{W} \right) = \frac{A_{W,L}^2}{(WL)^{3/2}} \left(\sqrt{\frac{W}{L}} + \sqrt{\frac{L}{W}} \right). \qquad (7.60)$$

The mismatch of W/L decreases faster than that of other relevant parameters when the channel area is increased while keeping W/L constant. Its contribution to the mismatch of β can therefore be made negligible by increasing WL. Moreover, its variance is minimum for a square channel and increases with approximately the square root of the aspect ratio.

It should be emphasized that the proportionality constant A_P was defined by comparing the mismatch of *two identical transistors* (same nominal values of W and L). If one transistor is infinitely large, then the mismatch is reduced by $\sqrt{2}$. Hence, for two transistors of different WL, it becomes

$$\sigma(\Delta P) = \frac{A_P}{\sqrt{2}} \sqrt{\frac{1}{W_1 L_1} + \frac{1}{W_2 L_2}}. \qquad (7.61)$$

Now, except for very large dimensions, the mismatch of two transistors of different sizes is increased by side effects. Hence, a ratio M/N of $\beta \propto W/L$ is best obtained by a series and/or parallel combination of M and N identical elementary transistors of same W and L. The resulting mismatch is then given by

$$\sigma(\Delta P) = \frac{A_P}{\sqrt{2WL}} \sqrt{\frac{1}{M} + \frac{1}{N}}. \qquad (7.62)$$

Using the variances defined above for the (uncorrelated) mismatches of the basic parameters, the variances of the three device parameters are obtained directly from (7.38), (7.39), and (7.40):

$$\sigma^2(\Delta V_{T0}) = [(2n_0 - 1)U_T + (n_0 - 1)\Psi_0]^2 \, \sigma^2 \left(\frac{\Delta N_b}{N_b} \right)$$

$$+ \left[\frac{Q_{fc}}{C_{ox}} - 2(n_0 - 1)\Psi_0 \right]^2 \sigma^2 \left(\frac{\Delta C_{ox}}{C_{ox}} \right)$$

$$+ \left(\frac{Q_{fc}}{C_{ox}} \right)^2 \sigma^2 \left(\frac{\Delta Q_{fc}}{Q_{fc}} \right), \qquad (7.63)$$

$$\sigma^2(\Delta n) = (n_0 - 1)^2 \left[\left(\frac{1}{2} - \frac{U_T}{\Psi_0} \right)^2 \sigma^2 \left(\frac{\Delta N_b}{N_b} \right) + \sigma^2 \left(\frac{\Delta C_{ox}}{C_{ox}} \right) \right] \qquad (7.64)$$

$$\sigma^2 \left(\frac{\Delta \beta}{\beta} \right) = \sigma^2 \left(\frac{\Delta \mu}{\mu} \right) + \sigma^2 \left(\frac{\Delta C_{ox}}{C_{ox}} \right) + \sigma^2 \left[\frac{\Delta(WL)}{W/L} \right]. \qquad (7.65)$$

According to (7.52), the same equations are applicable if each $\sigma(P)$ is replaced by the corresponding area proportionality constant A_P.

Now, the mismatch of these three parameters is possibly correlated through ΔC_{ox} and ΔN_b. In order to evaluate the practical importance of this correlation, an example is illustrated in Figure 7.10. The mismatch of drain currents and gate voltages is calculated for three cases, with the same realistic values of $\sigma(\Delta V_{T0})$ and $\sigma(\frac{\Delta \beta}{\beta})$.

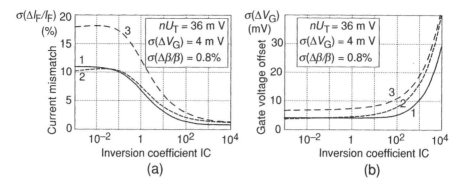

Figure 7.10 Mismatch of (a) drain currents and (b) gate voltages of saturated transistors. Case 1: uncorrelated ΔV_{T0}, $\Delta\beta$, and Δn (hence $\Delta n = 0$). Case 2: Fully correlated ΔV_{T0}, $\Delta\beta$, and Δn, with $\sigma_n = 0.2\%$ and $V_S = 0$. Case 3: same as case 2, but with $V_S = 1$V

In the first case, we assume *no correlation* between the three device parameters, which is possible only if ΔC_{ox} and ΔN_b are negligible. Then Δn is also negligible and V_S has no effect on matching. We obtain from (7.44) and (7.45) [104]

$$\sigma\left(\frac{\Delta I_F}{I_F}\right) = \sqrt{\sigma^2\left(\frac{\Delta\beta}{\beta}\right) + \left(\frac{G_m}{I_F}\right)^2 \sigma^2(\Delta V_{T0})}, \qquad (7.66)$$

$$\sigma(\Delta V_G) = \sqrt{\sigma^2(\Delta V_{T0}) + \left(\frac{I_F}{G_m}\right)^2 \sigma^2\left(\frac{\Delta\beta}{\beta}\right)}. \qquad (7.67)$$

In the second case, we assume *perfect correlation* between the three parameters, and $V_S = 0$. This is possible only if the sole source of mismatch is ΔC_{ox}. Then, according to Table 7.3, the sign of $\Delta\beta$ is opposite to that of Δn and ΔV_{T0}. Mismatch is then calculated by replacing each variation by its variance in equations (7.44) and (7.45), taking into account this sign difference.

The third case is the same, but with $V_S = 1$ V.

Cases 1 and 2 are the two extreme cases for $V_S = 0$. Real cases will be in between, with only partial correlation between device parameters. As can be seen, the maximum ratio between these two extremes is about $\sqrt{2}$ when the effects of ΔV_{T0} and $\Delta\beta$ are equal in moderate inversion. So the correlation can be compensated for by doubling the value of WL. The mismatch of slopes, Δn, has a negligible effect.

As already pointed out, for $V_S > 0$, the effective threshold $V_{T0} + nV_S$ becomes dependent on n, and the mismatch in weak inversion is increased by Δn, as illustrated in Case 3. If ΔV_{T0} and Δn are dominated by ΔC_{ox}, then $\sigma(\Delta V_{T0})$ is simply increased by $\sigma(\Delta n)V_S$.

If ΔN_b dominates, which is usually the case in deep submicron processes, then ΔV_{T0} and Δn are fully correlated, with no correlation with $\Delta\beta$.

The random variation of ΔN_b is essentially due to the limited total number N_{tot} of ionized impurities in the depletion layer. Assuming a Poisson distribution, then

$$\sigma^2\left(\frac{\Delta N_b}{N_b}\right) = \frac{2}{N_{tot}} = \frac{2}{W L t_d N_b}, \qquad (7.68)$$

where the factor 2 comes from the fact that we compare two transistors. The thickness t_d of the depletion layer given by 3.26 can be evaluated at $\Psi_s = \Psi_0$:

$$t_d = \sqrt{\frac{2\Psi_0 \epsilon_{si}}{q N_b}}, \tag{7.69}$$

resulting in

$$\sigma\left(\frac{\Delta N_b}{N_b}\right) = \frac{1}{\sqrt{WL}} \cdot \sqrt[4]{\frac{2q}{\Psi_0 \epsilon_{si} N_b}}. \tag{7.70}$$

Now, if ΔN_b dominates and if $\Psi_0 \gg U_T$, then 7.63 can be approximated by

$$\sigma(\Delta V_{T0}) = (n_0 - 1)\Psi_0 \sigma\left(\frac{\Delta N_b}{N_b}\right) = \frac{\Gamma_b}{2}\sqrt{\Psi_0} \cdot \sigma\left(\frac{\Delta N_b}{N_b}\right) \tag{7.71}$$

where n_0 has been replaced by its expression (3.73). By introducing $\sigma(\frac{\Delta N_b}{N_b})$ given by (7.70) and Γ_b given by (3.30), we obtain the area proportionality constant for threshold mismatch

$$A_{VT0} \triangleq \sqrt{WL}\sigma(\Delta V_{T0}) = \frac{1}{C_{ox}}\sqrt[4]{\frac{q^3 \epsilon_{si} N_b \Psi_0}{2}}. \tag{7.72}$$

Now, the comparison of (7.64) with (7.63) for $\Psi_0 \gg U_T$ shows that if mismatch is dominated by ΔN_b, then

$$A_n \triangleq \sqrt{WL}\sigma(\Delta n) = \frac{A_{VT0}}{2\Psi_0} = \frac{1}{2C_{ox}}\sqrt[4]{\frac{q^3 \epsilon_{si} N_b}{2\Psi_0^3}}. \tag{7.73}$$

It should be pointed out that (7.72) and (7.73) are slightly pessimistic, since t_d in (7.69) is evaluated at $\Psi_s = \Psi_0$, which is its minimum value. For $V_S > 0$, $\Psi_s \geq \Psi_0 + V_S$ and t_d is increased, thereby decreasing $\Delta N_b/N_b$ according to (7.68) [107]. However, this small effect can be neglected as compared to the global effect of increasing V_S. Indeed, since ΔV_{T0} and Δn are fully correlated by their common origin ΔN_b, the mismatch of effective gate threshold voltages $V_{T0} + n V_S$ is

$$\sigma(V_{T0} + n V_S) = \sigma(V_{T0})\left(1 + \frac{V_S}{2\Psi_0}\right), \tag{7.74}$$

This expression describes explicitly, for the particular case of ΔV_{T0} dominated by ΔN_b, the effect of V_S illustrated by curve 3 in Figure 7.10. An important consequence of this increase of mismatch due to $V_S > 0$ appears in a differential pair: its input offset voltage is increased by the nonzero source voltage if the two paired transistors are not put in a separate well connected to their sources.

Part II

The Extended Charge-Based Model

This second part models several nonideal effects that should be added to the core model developed in Part I, in order to best describe modern MOS transistors, in particular those realized in deep submicron processes. Chapter 8 focuses on effects that already affect long-channel devices, whereas Chapter 9 specifically adresses the short-channel effects. Finally, the passive devices that must be added to account for the extrinsic part of the transistor are modeled in Chapter 10.

8 Nonideal Effects Related to the Vertical Dimension

All the nonideal effects discussed in this chapter remain compatible with expression (4.40) of the drain current. Hence, they do not affect the fundamental property of symmetry discussed in Section (4.5). Section 8.2 shows the impact of the mobility reduction resulting from a large gate voltage. Current and tranconductances are reduced, but their ratio is not much affected. Section 8.3 investigates the case of nonuniform vertical doping, with a detailed analysis of two particular profiles. The following sections examine the consequences of the very high substrate doping and very thin gate oxide introduced in aggressively scaled-down processes. Polysilicon gate depletion discussed in Section 8.4 can be accounted for by a reduction of specific charge and current, by an increase of the slope factor for voltages, and by an increased threshold voltage. Band gap widening due to quantum effects is examined in Section 8.5. It has the same qualitative effect on charge, current, and threshold voltage as polydepletion, with which it is usually combined. Gate leakage current due to tunneling through a very thin oxide is analyzed in Section 8.6. Negligible if the oxide is thicker than 3 nm, this current increases by more than a factor 10 for each 0.2 nm of thickness reduction.

8.1 INTRODUCTION

In this chapter we still consider a long and wide channel that is homogeneous in the x direction (along the channel). As demonstrated in Section 4.5, the fundamental property of symmetry is not affected; hence, the drain current can still be decomposed in a forward component I_F and a reverse component I_R. We will separately discuss the main nonideal effects that are related to the vertical dimension, and show how they can be accounted for by modifying the basic model presented in Part I.

8.2 MOBILITY REDUCTION DUE TO THE VERTICAL FIELD

In a MOS transistor, the current flows very close to the silicon surface. As a consequence, the mobility of current carriers is lower than deep inside the substrate (typically two to three times

Charge-Based MOS Transistor Modeling: The EKV Model for Low-Power and RF IC Design   C. Enz and E. Vittoz
© 2006 John Wiley & Sons, Ltd.

lower), due to various scattering mechanisms [67]. This mobility is further reduced if the vertical field E_z becomes too large. This field dependent mobility can be approximated by [1]

$$\mu_z = \frac{\mu_0}{1 + E_{\text{eff}}/E_0}, \tag{8.1}$$

where μ_0 is the low-field surface mobility, $E_0 \cong 4 \times 10^7$ V/m the electric field at which the mobility starts to decrease significantly, and E_{eff} the average field in the inversion layer, approximated by

$$E_{\text{eff}} = \frac{1}{2} [E_{zs} + E_{zb}], \tag{8.2}$$

where E_{zs} and E_{zb} are the values of the vertical electric field at the surface and just below the inversion layer respectively.

According to the Gauss law illustrated by Figure 3.2,

$$E_{zs} = -(Q_i + Q_b)/\epsilon_{si} \quad \text{and} \quad E_{zb} = -Q_b/\epsilon_{si}. \tag{8.3}$$

Hence

$$E_{\text{eff}} = -\frac{1}{\epsilon_{si}} \left(Q_b + \frac{Q_i}{2} \right). \tag{8.4}$$

Introducing (8.4) in expression (8.1) of the field-dependent mobility yields

$$\mu_z = \frac{\mu_0}{1 - \frac{Q_b + Q_i/2}{\epsilon_{si} E_0}} = \frac{\mu_0}{1 + \theta(q_b + q_i/2)}. \tag{8.5}$$

In the second form, the charge is normalized to Q_{spec} defined by (3.42) and θ depends on E_0 according to

$$\theta = \frac{Q_{\text{spec}}}{\epsilon_{si} E_0} = \frac{2nC_{\text{ox}}U_T}{\epsilon_{si} E_0}. \tag{8.6}$$

The mobility reduction is a function of the inverted charge density Q_i and of the bulk depletion charge density Q_b. The latter is given by (3.55) as a linearized function of Q_i, where the slope factor n can be replaced by its value n_w given by (3.68) resulting in

$$q_b = \frac{\psi_p - q_i}{1 + \frac{2}{\gamma_b}\sqrt{\psi_p}}. \tag{8.7}$$

Introducing this result in (8.5) provides the variation of mobility with mobile charge q_i:

$$\frac{\mu_z}{\mu_0} = \frac{1}{k_1 q_i + k_2}, \tag{8.8}$$

where

$$k_1 = \theta \left(\frac{1}{2} - \frac{1}{1 + \frac{2}{\gamma_b}\sqrt{\psi_p}} \right) \quad \text{and} \quad k_2 = 1 + \frac{\theta \psi_p}{1 + \frac{2}{\gamma_b}\sqrt{\psi_p}}. \tag{8.9}$$

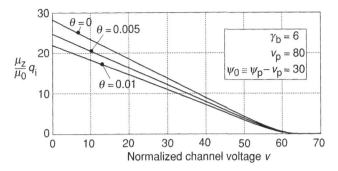

Figure 8.1 Effect of mobility reduction due to the vertical electric field

Alternatively, ψ_p may be replaced by the pinch-off voltage v_p by means of their relationship (3.47), their difference being very close to ψ_0, according to (3.66).

The correction for mobility reduction can be introduced by multiplying the $q_i(v)$ characteristics of Figure 3.11 by μ_z/μ_0. Using q_i as a parameter, $q_i \frac{\mu_z}{\mu_0}$ can be calculated by (8.8), whereas the channel voltage v is obtained from (3.48). Results for several realistic values of parameter θ are plotted in Figure 8.1.

For $\theta = 0$, there is no mobility reduction and the function is identical to that of Figure 3.11. For $\theta > 0$, the function is reduced. As can be seen from (8.5), this reduction is more important in strong inversion where $q_i \gg 1$. However, due to the effect of depletion charge q_b given by (8.7), it is also present in weak inversion, especially if ψ_p is large, corresponding to a large value of V_G.

If the transfer parameter β (4.8) is *calculated for* $\mu = \mu_0$, then the forward and reverse components of the drain current are obtained after replacing q_i by $\mu_z q_i/\mu_0$ in (4.16). Hence (4.18) becomes

$$i_{f,r} = \int_0^{q_{s,d}} \frac{2q_i + 1}{k_1 q_i + k_2}\, dq_i = \frac{1}{k_1}\left[2q_{s,d} + \left(1 - 2\frac{k_2}{k_1}\right)\ln\left(1 + \frac{k_1}{k_2}q_{s,d}\right)\right]. \qquad (8.10)$$

Using $q_{s,d}$ as a parameter and equation (3.48) to obtain $v_{s,d}$, this result is plotted as curve a in Figure 8.2 for a large value of θ. If θ tends to zero, the coefficient k_1 also tends to zero

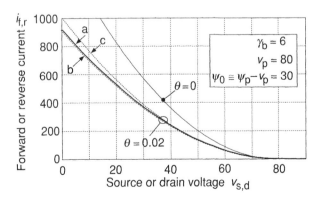

Figure 8.2 Effect of mobility reduction on current: (a) exact result (8.10); (b) third-order series expansion (8.11); (c) second-order series expansion (8.12)

and expression (8.10) diverges numerically. It can be approximated by its third-order series expansion in which the term $2q_{s,d}$ is canceled:

$$i_{f,r} = \frac{1}{k_2}\left[q_{s,d} + \left(1 - \frac{k_1}{2k_2}\right)q_{s,d}^2 - \frac{2k_1}{3k_2}q_{s,d}^3\right], \tag{8.11}$$

which is represented as curve b in Figure 8.2.

Further simplification is possible by limiting the expansion to the second-order term and neglecting $\frac{k_1}{2k_2} \ll 1$:

$$i_{f,r} = \frac{1}{k_2}(q_{s,d} + q_{s,d}^2). \tag{8.12}$$

Hence, in this approximation, the current is simply reduced by the factor k_2. As shown by curve c in the same figure this is an acceptable approximation if $v_p - v_{s,d}$ is not too large. The transfer characteristics in strong inversion remain close to a square law for a constant gate voltage v_g corresponding to a constant value of k_2. The gate-driven transfer characteristics depart more significantly from a square law since v_p and ψ_p increase linearly with v_g, thus modulating the value of k_1 and k_2 according to (8.9). This can be observed in the $\sqrt{i_f(v_p)}$ plot illustrated in Figure 8.3.

The source and drain transconductances at fixed voltages (depicted in Figure 5.1) are reduced proportionally to μ_z/μ_0. Hence from (8.8),

$$g_{ms,d} = \frac{\mu_z}{\mu_0}q_{s,d} = \frac{1}{k_1 + k_2/q_{s,d}}. \tag{8.13}$$

Their dependency on the forward and reverse currents can be obtained by first expressing $q_{s,d}(i_{f,r})$. Since expression (8.10) cannot be inverted, this can be done by inverting its second-order approximation (8.12). This yields

$$q_{s,d} = \frac{\sqrt{1 + 4k_2 i_{f,r}} - 1}{2}, \tag{8.14}$$

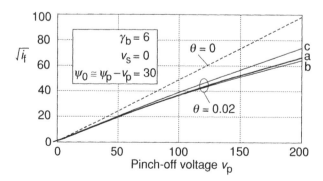

Figure 8.3   Transfer characteristics from the gate: (a) exact result (8.10); (b) third-order series expansion (8.11); (c) second-order series expansion (8.12)

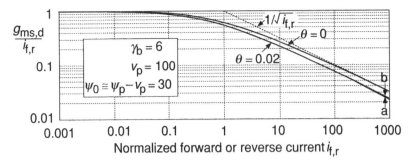

Figure 8.4 Effect of mobility reduction on the transconductance to current ratio: (a) using (8.15); (b) exact result

which is identical to (4.24) for $\theta = 0$ ($k_1 = 0$ and $k_2 = 1$). Introducing (8.14) in (8.13) results in

$$g_{\text{ms,d}} = \cfrac{1}{k_1 + \cfrac{2k_2}{\sqrt{1+4k_2 i_{\text{f,r}}}-1}}, \tag{8.15}$$

which is identical to (5.11) for $\theta = 0$. Using this result, the variation of $g_{\text{ms,d}}/i_{\text{f,r}}$ with $i_{\text{f,r}}$ is plotted as curve a in Figure 8.4 for a large value of θ. It departs only very slighly from the "exact" solution (curve b), obtained by using $q_{\text{s,d}}$ as a parameter in equation (8.13) of g_{ms} and in the full equation (8.10) of $i_{\text{f,r}}$. The gate transconductance g_{m} is still related to g_{ms} and g_{md} by (5.9) since this relationship was derived independent of the mobility. Therefore, all expressions of g_{m} are affected by mobility reduction.

It is worth noticing that the simple expression (5.17) of g_{m} in linear mode is *no longer valid*. The reason for it can be understood by examining Figure 8.5 that shows $\mu_z q_{\text{i}}/\mu_0(v)$ for various values of v_{p}, calculated from (8.8) and (3.48) (using q_{i} as a parameter). Indeed, changing v_{p} (or v_{g}) not only moves the curve vertically [as it did to $q_{\text{i}}(v)$ in Figure 5.2] but also modifies its slope. The figure also shows g_{ms}, g_{md}, and i_{d} for $v_{\text{p}} = 80$ and given values of v_{s} and v_{d}.

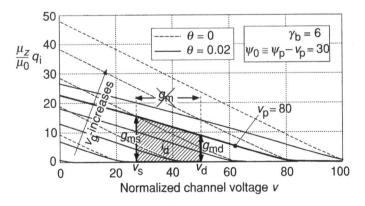

Figure 8.5 Effect of increasing v_{p}

8.3 NONUNIFORM VERTICAL DOPING

8.3.1 Introduction and General Case

In Part I, the doping concentration N_b of the local substrate (or bulk) was assumed to be uniform. As a consequence, the depletion charge Q_b, which is produced by repelling the holes before electrons can be attracted in the channel, was proportional to the square-root function (3.29) of the surface potential Ψ_s. The threshold function defined by (3.33) was therefore nonlinear, since it can be expressed as

$$V_{TB} = V_{FB} + \Psi_s + \frac{-Q_b(\Psi_s)}{C_{ox}}, \tag{8.16}$$

The slope n of this function was slightly decreasing with increasing Ψ_s, as shown by (3.34) and by Figure 3.7.

If the doping concentration is a function of depth z into the substrate, $Q_b(\Psi_s)$ is modified, and so are $V_{TB}(\Psi_s)$ and $n(\Psi_s)$. Let us consider the general case with the arbitrary profile $N_b(z)$ illustrated in Figure 8.6.

According to the classical depletion zone approximation, the density of holes is assumed to drop to a negligible value at depletion depth z_d, leaving a space charge $-qN_b$ per unit volume for $z < z_d$. Hence the total depletion charge density is given by

$$Q_b = -\epsilon_{si}E_z(z=0) = -q\int_0^{z_d} N_b\,dz, \tag{8.17}$$

which is a function of depletion depth z_d. Notice that the origin of the z-axis ($z = 0$) is positioned here just underneath the infinitely thin inverted charge sheet (see Section 3.4), so that E_z is not affected by Q_i.

The vertical field is zero for $z \geq z_d$. According to Poisson's equation, this field increases with the integral of the depleted charge for $z < z_d$:

$$E_z = \frac{q}{\epsilon_{si}}\int_z^{z_d} N_b\,dz. \tag{8.18}$$

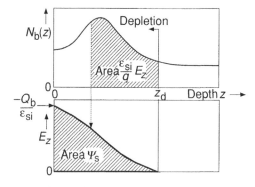

Figure 8.6   Calculation of $Q_b(\Psi_s)$ for an arbitrary doping profile $N_b(z)$

The surface potential is obtained by integrating the field (8.18) across the depletion region; hence,

$$\Psi_s = \frac{q}{\epsilon_{si}} \int_0^{z_d} \left(\int_z^{z_d} N_b \, dz \right) dz, \tag{8.19}$$

which is also a function of z_d.

Using z_d as the parameter, $Q_b(\Psi_s)$ can be calculated by means of parametric equations (8.17) and (8.19). Knowing $Q_b(\Psi_s)$, $V_{TB}(\Psi_s)$ and its derivative $n(\Psi_s)$ can then be obtained from (8.16).

This calculation will be carried out in the following subsections for two particular analytical profiles $N_b(z)$ that can be used as approximations of real profiles.

8.3.2 Constant Gradient Doping Profile

Consider the doping profile described by

$$N_b = N_{b0} \left(1 + S \frac{z}{z_c} \right) \tag{8.20}$$

and illustrated in Figure 8.7. For $S = 1$, the doping concentration increases linearly from its surface value N_{b0} and is doubled at characteristic depth z_c. For $S = -1$, it decreases to reach zero at z_c; the model is then valid only for a depletion depth $z_d < z_c$.

Introducing this profile into (8.17) and (8.19) yields the parametric equations of $\Psi_s(Q_b)$:

$$Q_b = -q N_{b0} \left(z_d + S \frac{z_d^2}{2z_c} \right), \tag{8.21}$$

$$\Psi_s = \frac{q N_{b0}}{\epsilon_{si}} \left(\frac{z_d^2}{2} + S \frac{z_d^3}{3z_c} \right), \tag{8.22}$$

These equations can be simplified by introducing a parameter

$$P \triangleq \frac{z_d}{z_c}, \tag{8.23}$$

and by defining a normalized characteristic depth

$$\zeta_c \triangleq \frac{z_c}{t_{ox}} \frac{\epsilon_{ox}}{\epsilon_{si}} = \frac{z_c C_{ox}}{\epsilon_{si}}, \tag{8.24}$$

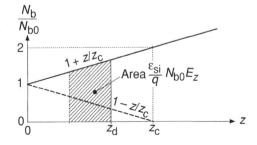

Figure 8.7   Constant gradient doping profile $N_b(z)$

and the substrate modulation factor at the surface

$$\Gamma_{b0} \overset{\Delta}{=} \frac{\sqrt{2qN_{b0}\epsilon_{si}}}{C_{ox}}. \tag{8.25}$$

They become

$$-\frac{Q_b}{C_{ox}} = \frac{\Gamma_{b0}^2\zeta_c}{2}\left(P + S\frac{P^2}{2}\right), \tag{8.26}$$

$$\Psi_s = \frac{\Gamma_{b0}^2\zeta_c^2}{2}\left(\frac{P^2}{2} + S\frac{P^3}{3}\right). \tag{8.27}$$

These two results can then be introduced in (8.16), giving

$$V_{TB} - V_{FB} = \frac{\Gamma_{b0}^2\zeta_c}{2}\left(P + \frac{\zeta_c + S}{2}P^2 + S\zeta_c\frac{P^3}{3}\right). \tag{8.28}$$

Now, (8.28) and (8.27) can be used as parametric equations of $V_{TB}(\Psi_s)$, with P as the independent parameter. This threshold function is plotted in Figure 8.8(a) for $\Gamma_{b0}^2 = 40U_T$ and for several values of the normalized characteristic depth ζ_c. Uniform doping corresponds to $\zeta_c = \infty$.

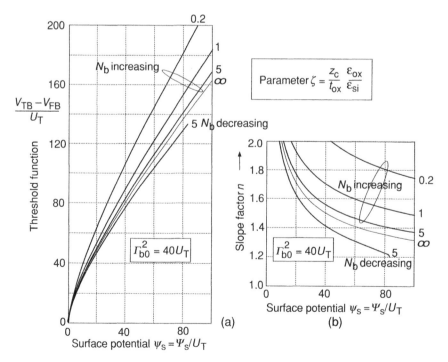

Figure 8.8   Effect of a constant gradient profile: (a) on threshold function $V_{TB}(\Psi_s)$; (b) on slope factor $n(\Psi_s)$

The slope factor can also be expressed as a function of parameter P by differentiating (8.28) and (8.27)

$$n = \frac{dV_{TB}}{d\Psi_s} = \frac{dV_{TB}}{dP} \Big/ \frac{d\Psi_s}{dP} = \frac{1 + P(\zeta_c + S) + \zeta_c SP^2}{\zeta_c P(1 + SP)}. \tag{8.29}$$

This equation can be associated with (8.27) to obtain $n(\Psi_s)$. This slope factor is plotted in Figure 8.8(b) with the same values of Γ_{b0} and ζ_c.

As could be expected, the slope factor for a given value of surface doping N_{b0} is increased for an increasing doping profile. However, this effect is significant only for $\zeta_c < 5$, corresponding to $z_c < 15t_{ox}$.

8.3.3 Step Profile

The step profile illustrated in Figure 8.9 is another approximation of real profiles that is tractable analytically. Introducing this profile into (8.17) and (8.19) yields, for $z_d \geq z_c$

$$Q_b = -q\,[N_{b0}z_c + N_{bc}(z_d - z_c)] \tag{8.30}$$

and

$$\Psi_s = \frac{q}{\epsilon_{si}} \left[N_{bc}\frac{(z_d - z_c)^2}{2} + N_{bc}z_c(z_d - z_c) + N_{b0}\frac{z_c^2}{2} \right]. \tag{8.31}$$

These equations can be simplified by introducing the parameter P defined by (8.23) and variables ζ_c and Γ_{b0} defined (8.24) and (8.25), and by defining the doping ratio

$$\nu \overset{\triangle}{=} \frac{N_{bc}}{N_{b0}}, \tag{8.32}$$

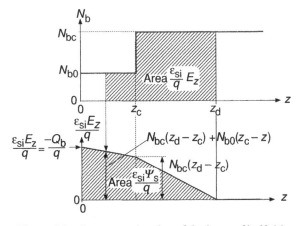

Figure 8.9   Step approximation of doping profile $N_b(z)$

resulting in

$$\frac{-Q_b}{C_{ox}} = \frac{\zeta_c \Gamma_{b0}^2}{2}[1 + \nu(P-1)] \tag{8.33}$$

and

$$\Psi_s = \frac{\zeta_c^2 \Gamma_{b0}^2}{2}\left[\frac{\nu}{2}(P-1)^2 + \nu(P-1) + \frac{1}{2}\right]. \tag{8.34}$$

These equations could be used as parametric equations of $\Psi_s(Q_b)$. Instead, parameter P can be calculated from its second-order equation (8.34):

$$P = \sqrt{1 - \frac{1}{\nu} + \frac{4\Psi_s}{\nu \zeta_c^2 \Gamma_{b0}^2}}, \tag{8.35}$$

and introduced in equation (8.33) of Q_b. According to (8.16), the threshold function then becomes

$$V_{TB} - V_{FB} = \Psi_s + \frac{\zeta_c \Gamma_{b0}^2}{2}\left[1 - \nu + \sqrt{\nu\left(\nu - 1 + \frac{4\Psi_s}{\zeta_c^2 \Gamma_{bo}^2}\right)}\right]. \tag{8.36}$$

This expression is plotted in Figure 8.10 for various values of normalized step depth ζ_c. Part (a) of the figure is for a 1 to 10 step up of doping concentration. The threshold is strongly increased if the step is shallow, since most of the depletion region extends within the region of higher doping concentration N_{bc}. This increase is attenuated when the step depth is increased.

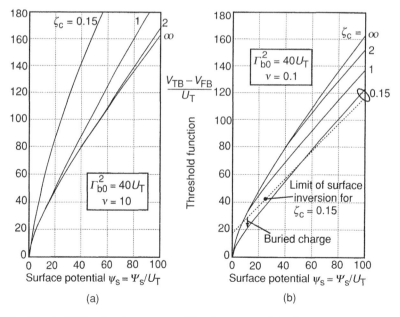

Figure 8.10   Threshold function for a step profile of normalized depth ζ_c: (a) for a step up 1 to 10; (b) for a step down 10 to 1

Expression (8.36) is valid only when the depletion depth extends beyond the step, corresponding to $P \geq 1$, and hence from (8.35) for

$$\Psi_s \geq \left(\frac{\zeta_c \Gamma_{b0}}{2} \right)^2 = \frac{q N_{b0} z_c^2}{2\epsilon_{si}}. \tag{8.37}$$

For smaller values of surface potential, the situation is reduced to that of uniform doping N_{b0} (corresponding to $\zeta_c = \infty$).

Figure 8.10(b) shows the case of a 10 to 1 step down of doping concentration for the same values of Γ_{b0} and ζ_c. The threshold function is reduced if the step is shallow, since the depletion region extends mostly in the region of lower doping.

Figure 8.11 is another plot of expression (8.36) of the threshold function, for a fixed value of step depth ($\zeta_c = 2$) and various values of the doping ratio ν. Below the limit given by (8.37), the depletion region does not reach the step depth z_c. The doping is therefore uniform with substrate modulation factor Γ_{b0}. Beyond this limit, the depletion region enters the region of different doping and the threshold function depends on the doping ratio ν.

When the gate voltage V_G starts exceeding the threshold function V_{TB} at some position along the channel, some inversion charge Q_i starts appearing. According to (3.5), the concentration of this inversion charge will dominate at the depth for which $\Psi - \Phi_F$ is maximum. If the doping concentration is uniform, then Fermi potential Φ_F is constant and the inversion charge appears at the surface ($z = 0$), since Ψ is always maximum at the surface. This is also true if the concentration increases with depth z, corresponding to an increase of Φ_F given by (3.8).

But if, on the contrary, the concentration decreases steeply with depth, the corresponding decrease of Φ_F can possibly overcome the decrease of Ψ. The inversion layer is then created below the surface, resulting in what is called a *buried channel*.

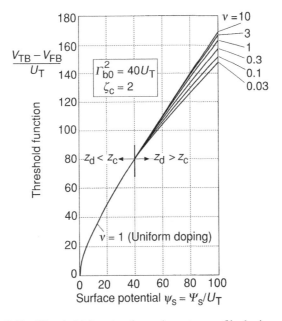

Figure 8.11   Threshold function for various step profile doping ratios ν

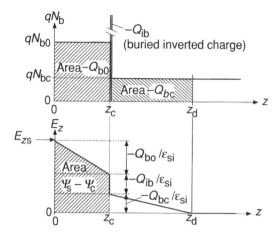

Figure 8.12   Charge densities and electrical field with a buried channel

For the step profile discussed here, the channel will be buried as long as the difference of Fermi potentials for the two doping concentrations exceeds the drop of potential across depth z_c:

$$\Phi_{F0} - \Phi_{Fc} > \Psi_s - \Psi_c, \tag{8.38}$$

or by introducing expression (3.8) of Φ_F

$$\Psi_s - \Psi_c < U_T \ln(1/\nu), \tag{8.39}$$

where $\Psi_c = \Psi(z = z_c)$.

The effect of this buried inverted charge Q_{ib} on the electric field is illustrated in Figure 8.12. The total charge Q_t underneath the surface can be split into three parts:

$$Q_t \overset{\triangle}{=} Q_b + Q_{ib} = Q_{b0} + Q_{bc} + Q_{ib}. \tag{8.40}$$

The buried inverted charge (supposed to be a very thin layer according to the charge sheet model) produces a step of field. The total voltage drop across z_c is easily obtained by inspection:

$$\Psi_s - \Psi_c = -\frac{z_c}{\varepsilon_{si}} \left(Q_{ib} + Q_{bc} + \frac{Q_{b0}}{2} \right). \tag{8.41}$$

Introducing (8.40) and the variables defined by (8.24) and (8.25) yields

$$\frac{-Q_t}{C_{ox}} = \frac{\zeta_c \Gamma_{b0}^2}{4} + \frac{\Psi_s - \Psi_c}{\zeta_c}. \tag{8.42}$$

This charge increases with the surface field. When the limit condition (8.39) for surface inversion is reached, it has a maximum value given by

$$\frac{-Q_{t\,max}}{C_{ox}} = \frac{\zeta_c \Gamma_{b0}^2}{4} + \frac{U_T}{\zeta_c} \ln(1/\nu) \tag{8.43}$$

that is independent of Ψ_s.

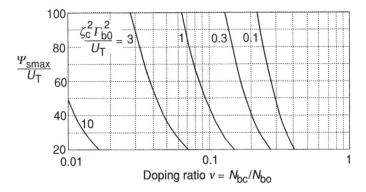

Figure 8.13 Maximum surface potential for which the channel remains buried

According to (8.16), the threshold for surface inversion can therefore be expressed as

$$V_{TBs} - V_{FB} = \Psi_s - \frac{Q_{t\,max}}{C_{ox}} = \Psi_s + \frac{\zeta_c \Gamma_{b0}^2}{4} + \frac{U_T}{\zeta_c} \ln(1/v).$$ (8.44)

This limit is plotted in dotted line in Figure 8.10(b) for $\zeta_c = 0.15$. Since the buried charge cannot be positive, this threshold is valid only when it is larger than the inversion threshold V_{TB}. This would not be the case for the deeper steps (larger ζ_c) shown in the same figure (except for very low values of Ψ_s).

To obtain a buried channel for a deeper step, the lower doping concentration should be further reduced, in order to increase the difference of Fermi potentials.

The maximum value of surface potential for which the channel remains buried can be obtained by introducing the value of $\Psi_s - \Psi_c$ into condition (8.39). It can easily be verified that this value is given by the last two terms in (8.34), since the first term correponds to Ψ_c. Hence, for this maximum value,

$$\frac{\zeta_c^2 \Gamma_{b0}^2}{2}\left[v(P-1) + \frac{1}{2}\right] = U_T \ln(1/\zeta_c).$$ (8.45)

Introducing expression (8.35) of parameter P and solving for Ψ_s yield

$$\Psi_{s\,max} = \frac{\zeta_c^2 \Gamma_{b0}^2}{4}\left[\frac{1}{v}\left(v - \frac{1}{2} - \frac{2}{\zeta_c^2 \Gamma_{b0}^2}\ln v\right)^2 + 1 - v\right].$$ (8.46)

This limit is plotted in Figure 8.13 as a function of the doping ratio v. It can be seen that, for a given value of parameter $\zeta_c^2 \Gamma_{b0}^2$ [that is proportional to $N_{b0} z_c^2$ according to (8.37)], the channel may remain buried up to large values of the surface potential if $v = N_{bc}/N_{b0}$ is sufficiently small. Such a situation is shown in Figure 8.14, with $\zeta_c^2 \Gamma_{b0}^2/U_T = 0.2$ and $v = 0.1$. As can be seen, $V_{TBs} > V_{TB}$ in the whole range represented in the diagram. If the surface potential is close to its pinch-off value Ψ_P, then all the inverted charge is buried. This will be true all along the channel if the transistor is in weak inversion (Q_i negligible). If the surface potential is lower than the value Ψ_{Ps} defined in the figure (surface pinch-off potential), then only a part Q_{ib} of the total inverted charge Q_i buried.

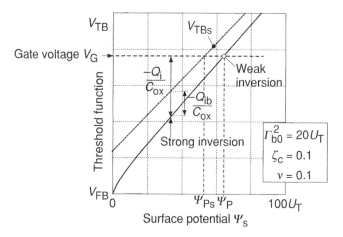

Figure 8.14 Inversion charge Q_i and its buried part Q_{ib}

The slope factor is easily calculated from (8.36) as

$$n = \frac{dV_{TB}}{d\Psi_s} = 1 + \sqrt{\frac{v}{\zeta_c^2(v-1) + 4\Psi_s/\Gamma_{b0}^2}}.$$ (8.47)

It is plotted in Figure 8.15 for various values of normalized step depth ζ_c, with the same parameters as in Figure 8.10.

Figure 8.15(a) shows the case of a 1 to 10 step up in doping concentration. Compared to uniform doping ($\zeta_c = \infty$), the slope is strongly increased if the step is very shallow, as can be expected from the fact that the depletion occurs mainly in the highly doped region. It remains higher when the depth is increased, but is more constant with the variation of the surface potential. Indeed, if the doping ratio v tends to infinity, (8.47) shows that the slope factor n remains constant at the value $1 + 1/\zeta_c$ as soon as the surface potential exceeds the limit given by (8.37) to have the depletion region reaching the step. For the example of the figure with $\zeta_c = 2$, this limit is reached for $\Psi_s = 40U_T$. Below this value, the slope factor is that of uniform doping ($\zeta_c = \infty$).

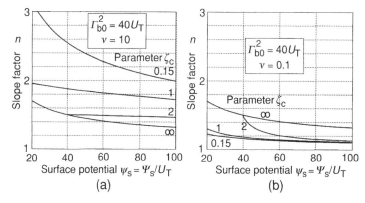

Figure 8.15 Slope factor for various values of normalized step depth ζ_c: (a) for a step up 1 to 10; (b) for a step down 10 to 1

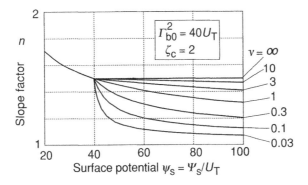

Figure 8.16   Slope factor for various step profile doping ratios v

Figure 8.15(b) shows the case of a 10 to 1 step down of doping concentration. The slope factor is lowered by a shallow step (most of the depletion occurring in the lightly doped region). However, it increases very steeply when the surface potential is reduced and approaches the limit given by (8.37).

In Figure 8.16, the slope factor is plotted for $\zeta_c = 2$ and various values of the doping ratio. It shows again that when the depletion region reaches the step, the way n varies with the surface potential strongly depends on the doping ratio.

8.3.4 Effect on the Basic Model

The nonuniformity of vertical doping affects only the shape of the threshold function, and hence the slope factor and the threshold voltage. Therefore, the whole model derived in Part I is still applicable with new values of these parameters. The threshold voltage at equilibrium V_{T0} was defined by (3.58) as the value of the threshold function V_{TB} for $V = 0$ (channel at equilibrium). Hence, from (3.56),

$$V_{T0} = V_{TB}(\Psi_s = \Psi_0),\tag{8.48}$$

where Ψ_0 is given by (3.66) with Φ_F calculated from (3.8) with $N_b = N_{b0}$.

The effect of the constant gradient profile as illustrated in Figure 8.8 can be approximated by fitting the substrate modulation factor Γ_b of a uniform profile. Both the threshold voltage and the slope factor are increased (decreased) if the doping is increasing (decreasing) with depth.

For the step profile approximation, the threshold voltage can be expressed explicitly from (8.36):

$$V_{T0} = V_{FB} + \Psi_0 + \frac{\zeta_c \Gamma_{b0}^2}{2}\left[1 - v + \sqrt{v\left(v - 1 + \frac{4\Psi_0}{\zeta_c^2 \Gamma_{b0}^2}\right)}\right].\tag{8.49}$$

The pinch-off potential Ψ_P was defined in Section 3.5 as the value of surface potential for which the inverted charge is zero; that is,

$$\Psi_P = \Psi_s(V_G = V_{TB}),\tag{8.50}$$

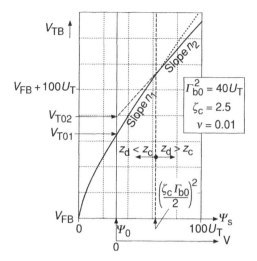

Figure 8.17 Dual slope threshold function

and is related to the pinch-off voltage by $V_P = \Psi_P - \Psi_0$ according to (3.66). But no simple relation exists here to replace expression (3.37) of Ψ_P. Instead, V_P can be approximated by (3.63).

For a given value of surface doping (and hence of Γ_{b0}), the effect of a step profile depends on the step depth z_c and the step doping ratio $\nu = N_{bc}/N_{b0}$. A step up ($\nu > 1$) can only increase V_{T0} and n, but it improves the linearity of threshold function V_{TB} and hence reduces the variation of n with the surface potential, as shown by Figure 8.11 and Figure 8.16.

A step down ($\nu < 1$) can only reduce V_{T0} and n, but it may result in an abrupt change of slope at the limit given by (8.37), as illustrated in Figure 8.16. Therefore, a dual slope threshold model might be needed as illustrated by the example of Figure 8.17.

Threshold V_{T01} and slope n_1 correspond to the region of uniform doping. They are thus given by (3.58) and (3.34) with $\Gamma_b = \Gamma_{b0}$ [or by (8.49) and (8.47) with $\nu = 1$]. According to Figure 8.16, the evaluation of n_2 should be made by (8.47) with a value of Ψ_s sufficiently larger than $\zeta_c \Gamma_{b0}^2/2$. Inspection of Figure 8.17 shows that the second threshold voltage can be calculated by

$$V_{T02} = V_{T01} + (n_1 - n_2) \left[(\zeta_c \Gamma_{b0}/2)^2 - \Psi_0 \right]. \tag{8.51}$$

8.4 POLYSILICON DEPLETION

8.4.1 Definition of the Effect

In the basic model discussed in Part I, we have assumed a constant potential V_G throughout the thickness of the gate electrode, which would always be true if the gate material was a metal. It is still true for a (poly)silicon gate, as long as the thickness of the layer of positive charge Q_G (that is concentrated at the lower face of the gate electrode) is so small that the voltage $\Delta\Psi_g$ across it is negligible. If the gate is N-type, we can assume that a positive gate charge Q_G is entirely produced by the depletion layer created at the lower face of the gate. If the gate is P-type, this depletion layer can possibly create a negative gate charge Q_G.

To calculate $\Delta\Psi_g$, we can further assume that the model of equations (3.29) and (3.30) giving

$$\Psi_s = \left(\frac{Q_b}{\Gamma_b C_{ox}}\right)^2 \quad \text{with} \quad \Gamma_b = \frac{\sqrt{2qN_b\epsilon_{si}}}{C_{ox}} \tag{8.52}$$

for the monocrystalline bulk with doping concentration N_b remains valid for the polysilicon gate with concentration N_g. Then by analogy [45]

$$\Delta\Psi_g = \pm\left(\frac{Q_g}{\Gamma_g C_{ox}}\right)^2 \quad \text{for} \quad \pm Q_g > 0, \tag{8.53}$$

where

$$\Gamma_g = \frac{\sqrt{2qN_g\epsilon_{si}}}{C_{ox}} \tag{8.54}$$

is the *gate modulation factor*. The *positive sign* in (8.53) corresponds to the positive voltage drop created through the depletion layer of an *N-type gate* (for $Q_g > 0$), whereas the *negative sign* corresponds to the negative voltage drop that might eventually be created in the depletion layer of a *P-type gate* (for $Q_g < 0$).

Comparing (8.52) with (8.53) and (8.54) shows that $\Delta\Psi_g$ remains negligible as long as $N_g/N_b \gg (Q_g/Q_b)^2$.

Now, while scaling-down process dimensions, N_b must be increased whereas N_g cannot be increased proportionally. Hence $\Delta\Psi_g$ may become nonnegligible, especially if Q_g is made much larger than $-Q_b$ by a large value of inverted charge $-Q_i$ (very strong inversion).

Although the original definition of the fixed interface charge Q_{fc} in Section 2.2 included the equivalent effect of the charge distributed throughout the oxide thickness, let us assume here that all this charge is physically located at the silicon-oxide interface. Hence, because of the overall charge neutrality

$$Q_g = -(Q_b + Q_i + Q_{fc}). \tag{8.55}$$

The depletion voltage at the gate then becomes

$$\Delta\Psi_g = \pm\left(\frac{Q_b + Q_i + Q_{fc}}{\Gamma_g C_{ox}}\right)^2. \tag{8.56}$$

8.4.2 Effect on the Mobile Inverted Charge

This voltage drop $\Delta\Psi_g$ given by (8.56) must be subtracted from V_G in the voltage to charge relation (3.19). This yields

$$\pm\frac{1}{\Gamma_g^2}\left(\frac{Q_i + Q_b + Q_{fc}}{C_{ox}}\right)^2 - \frac{Q_i + Q_b + Q_{fc}}{C_{ox}} - (V_G - \Phi_{ms} - \Psi_s) = 0. \tag{8.57}$$

The mobile charge density Q_i is then obtained by first solving this second-order equation in $(Q_i + Q_b + Q_{fc})/C_{ox}$, giving

$$Q_i + Q_b + Q_{fc} = C_{ox} \left[\pm \frac{\Gamma_g^2}{2} \left(1 - \sqrt{1 \pm \frac{4}{\Gamma_g^2}(V_G - \Psi_s - \Phi_{ms})} \right) \right]. \qquad (8.58)$$

Then, by introducing expression (3.29) of Q_b,

$$Q_i = C_{ox} \left[\Gamma_b \sqrt{\Psi_s} \pm \frac{\Gamma_g^2}{2} \left(1 - \sqrt{1 \pm \frac{4}{\Gamma_g^2}(V_G - \Psi_s - \Phi_{ms})} \right) \right] - Q_{fc}. \qquad (8.59)$$

8.4.3 Slope Factors and Pinch-Off Surface Potential

We can now compare (8.59) with (3.32), to which it reduces for Γ_g very large. As a first remark, Φ_{ms} and Q_{fc} cannot be lumped anymore into a flat-band voltage V_{FB}. But most important is the fact that Q_i is no longer proportional to the difference between the gate voltage V_G and a threshold function $V_{TB}(\Psi_s)$. The diagram of Figure 3.7 is therefore no longer applicable. As a consequence, the slope of $\frac{Q_i}{C_{ox}}(\Psi_s)$ is no longer identical to that of $V_G(\Psi_P)$. Therefore, the single slope factor n introduced in Part I must be replaced by two distinct slope factors:

$$n_v = \frac{dV_G}{d\Psi_P} = \frac{dV_G}{dV_P} \quad \text{and} \quad n_q = \frac{dQ_i/C_{ox}}{d\Psi_s}. \qquad (8.60)$$

In order to obtain an expression for n_v, we have to first calculate $V_G(\Psi_P)$ that replaces (3.33) plotted in Figure 3.7. Introducing $Q_i = 0$ and $\Psi_s = \Psi_P$ in (8.59) results in

$$V_G = \Phi_{ms} - \frac{Q_{fc}}{C_{ox}} + \Psi_P + \Gamma_b\sqrt{\Psi_P} \pm \frac{1}{\Gamma_g^2} \left(\frac{Q_{fc}}{C_{ox}} - \Gamma_b\sqrt{\Psi_P} \right)^2, \qquad (8.61)$$

which reduces to expression (3.33) of $V_{TB}(\Psi_s)$ for Γ_g very large.

It must be reminded that, according to (3.66), Ψ_P is related to the pinch-off voltage V_P by

$$\Psi_P = V_P + \Psi_0 = V_P + 2\Phi_F + V_{sh}. \qquad (8.62)$$

Equation (8.61) is plotted in Figure 8.18 for particular values of $\gamma_b = \Gamma_b/U_T$ and $\gamma_g = \Gamma_g/U_T$, and for five different values of fixed charge Q_{fc}/C_{ox} ranging from -40 to 40.

Now, polydepletion is possible only if the charge Q_G on the gate is positive for an N-type gate or negative for a P-type gate. At pinch-off $Q_i = 0$, hence from (8.55) polydepletion at pinch-off occurs only for

$$\pm(Q_b + Q_{fc}) < 0. \qquad (8.63)$$

By introducing expression (3.29) of Q_b with $\Psi_s = \Psi_P$, this condition becomes

$$\pm \left(\Gamma_b\sqrt{\Psi_P} - \frac{Q_{fc}}{C_{ox}} \right) > 0. \qquad (8.64)$$

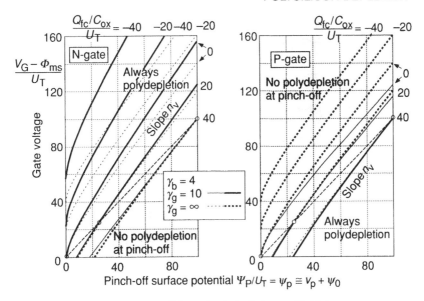

Figure 8.18   Effect of polydepletion on $V_G(\Psi_p)$

Introduced in (8.61), condition (8.64) becomes

$$\pm(V_G - \Phi_{ms} - \Psi_P) > 0. \tag{8.65}$$

This limit is also represented in Figure 8.18. No polydepletion occurs at pinch-off if condition (8.65) is not fulfilled; the $V_G(\Psi_P)$ function is then that for γ_g very large. The valid curve for a given value of Q_{fc} is shown in thick line. Its slope n_v is always *increased* by polydepletion. Notice that polydepletion can occur below the limit given by (8.65) if $\Psi_s < \Psi_P$ (Q_i no longer negligible in (8.63)).

We see that, for the N-channel transistor considered here, a large positive value of fixed charge Q_{fc} may prevent polydepletion for an N-type gate, and make it possible for an P-type gate.

The flat-band voltage V_{FB} was defined as the value of gate voltage V_G for which $\Psi_s = 0$; hence, $Q_{si} = Q_b + Q_i = 0$. Using (8.57), we obtain

$$V_{FB} = \Phi_{ms} - \frac{Q_{fc}}{C_{ox}} \pm \left(\frac{Q_{fc}}{\Gamma_g C_{ox}}\right)^2 \quad \text{for} \quad \pm Q_g = \mp Q_{fc} > 0. \tag{8.66}$$

The flat-band voltage is increased by polydepletion in an N-type gate for a positive fixed charge; it is decreased by a negative fixed charge for a P-type gate. However, we have seen before in (8.59) that the fixed charge does not simply contribute to a shift V_{FB} of gate voltage, as is the case without polydepletion.

The pinch-off surface potential can be calculated from the gate voltage by inverting (8.61), with the result of a very complicated expression for the general case. This expression is simplified if $Q_{fc} = 0$ hence $V_{FB} = \Phi_{ms}$. As can be seen in Figure 8.18, there is then no polydepletion for a P-type gate; $\Psi_P(V_G)$ is therefore given by (3.37). For a *N-type gate*, inverting (8.61) yields

$$\Psi_P = \left\{ \frac{\Gamma_g}{2(\Gamma_b^2 + \Gamma_g^2)} \left[\sqrt{4(\Gamma_b^2 + \Gamma_g^2)(V_G - V_{FB}) + \Gamma_b^2\Gamma_g^2} - \Gamma_b\Gamma_g \right] \right\}^2 \quad \text{(for } Q_{fc} = 0\text{)},$$

(8.67)

which also provides $V_P(V_G)$ by introducing (8.62). It can be verified that this expression is also reduced to (3.37) for $\Gamma_g \gg \Gamma_b$.

8.4.4 Voltage Slope Factor n_v

According to definition (8.60), n_v is the slope of $V_G(\Psi_P)$ plotted in Figure 8.18. Differentiation of (8.61) gives

$$n_v = \frac{dV_G}{d\Psi_P} = 1 + \frac{\Gamma_b}{2\sqrt{\Psi_P}} \pm \left(\frac{\Gamma_b}{\Gamma_g}\right)^2 \left(1 - \frac{Q_{fc}/C_{ox}}{\Gamma_b\sqrt{\Psi_P}}\right),$$

(8.68)

which reduces to n given by (3.68) for $\Gamma_b \ll \Gamma_g$. This slope factor is plotted in Figure 8.19 for numerical values identical to those of Figure 8.18.

Condition (8.64) is used to identify the valid part of the curve (shown in thick line) for each value of Q_{fc}. Notice that n_v cannot be lower than its value for very large γ_g/γ_b, that is independent of Q_{fc} as shown by (8.68). This confirms that the voltage slope factor n_v is always *increased by polydepletion*.

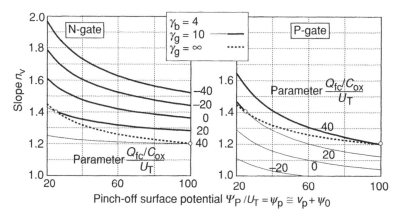

Figure 8.19   Variation of slope factor n_v with pinch-off surface potential Ψ_P

8.4.5 Charge Slope Factor n_q

Slope factor n_q is obtained by differentiation of (8.59):

$$n_q = \frac{dQ_i/C_{ox}}{d\Psi_s} = \frac{\Gamma_b}{2\sqrt{\Psi_s}} + \frac{1}{\sqrt{1 \pm \frac{4}{\Gamma_g^2}(V_G - \Psi_s - \Phi_{ms})}}. \tag{8.69}$$

Unlike n (to which it reduces for Γ_g very large) this slope is not only dependent on surface potential Ψ_s but also dependent on gate voltage V_G. This dependency on V_G may be replaced by that on Ψ_P by introducing (8.61), giving

$$n_q = \frac{\Gamma_b}{2\sqrt{\Psi_s}} + \frac{1}{\sqrt{\left[1 \pm \frac{2}{\Gamma_g^2}(\Gamma_b\sqrt{\Psi_P} - \frac{Q_{fc}}{C_{ox}})\right]^2 \pm \frac{4}{\Gamma_g^2}(\Psi_P - \Psi_s)}}. \tag{8.70}$$

Here again, Ψ_P may be replaced by $\Psi_0 + V_P$.

As done without polydepletion, this slope factor may be evaluated at the value of surface potential Ψ_s that is best adapted to the mode of operation.

In weak inversion, or close to it, it may be evaluated at $\Psi_s = \Psi_P$; (8.70) then becomes

$$n_q = n_{qw} = \frac{\Gamma_b}{2\sqrt{\Psi_P}} + \frac{1}{1 \pm \frac{2}{\Gamma_g^2}\left(\Gamma_b\sqrt{\Psi_P} - Q_{fc}C_{ox}\right)}. \tag{8.71}$$

This variation of n_{qw} with Ψ_P is plotted in Figure 8.20 with the same numerical values as those of Figures 8.18 and 8.19.

Condition (8.64) is used to identify the valid part of the curve (shown in thick line) for each value of Q_{fc}. Notice that n_q cannot be higher than its value for very large γ_g/γ_b, that is independent of Q_{fc} as shown by (8.70). This shows that the charge slope factor n_q is always *decreased by polydepletion*.

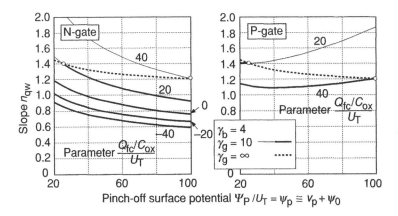

Figure 8.20   Variation of slope factor n_{qw} with pinch-off surface potential Ψ_P

For strong inversion, a better evaluation of n_q given by (8.70) may be at $\Psi_s = \Psi_0 + V_P/2$, resulting in

$$n_q = n_{qs} = \frac{\Gamma_b}{2\sqrt{\Psi_0 + V_P/2}} + \frac{1}{\sqrt{\left[1 \pm \frac{2}{\Gamma_g^2}(\Gamma_b\sqrt{\Psi_0 + V_P} - \frac{Q_{fc}}{C_{ox}})\right]^2 \pm \frac{2}{\Gamma_g^2}V_P}}. \tag{8.72}$$

8.4.6 Effect on $Q_i(V)$, Currents, and Transconductances

The continuous expression of $V(Q_i)$ (3.48) and its approximations (3.49) and (3.50) in weak and strong inversion remain valid if n is replaced by n_q given by (8.70) in the normalization of Q_i. Hence the specific charge (3.42) used for this normalization must be replaced by

$$Q_{spec} = -2n_q U_T C_{ox}, \tag{8.73}$$

The same remark applies to the normalized drain current and its forward and reverse components derived in Sections 4.4.1 and 4.4.6 provided that the specific current is redefined as

$$I_{spec} = 2n_q \beta U_T^2. \tag{8.74}$$

In particular, results (4.25) and (4.39) and the corresponding characteristics in Figure 4.5 remain unchanged.

As explained in Chapter 5, the source transconductance G_{ms} is proportional to the density Q_{iS} of mobile charge at the source (in normalized form, $g_{ms} = q_s$). Symmetrically, the drain transconductance G_{md} is proportional to the density Q_{iD} of mobile charge at the drain (in normalized form, $g_{md} = q_d$). Therefore, all expressions of $G_{ms,d}$ as functions of $I_{F,R}$ or $V_P - V_{S,D}$ remain valid, provided n is replaced by n_q. The specific conductance used for normalization becomes

$$G_{spec} = 2n_q \beta U_T. \tag{8.75}$$

Now, since by definition (8.60) the slope of $V_G(V_P)$ is n_v, the gate transconductance is given by

$$G_m = \frac{G_{ms} - G_{md}}{n_v}, \tag{8.76}$$

instead of (5.9).

It can be pointed out that, since polydepletion reduces n_q, it reduces $G_{ms,d}$ (except in weak inversion at a fixed current). Since it increases n_v, G_m is further reduced proportionally.

The effect of polydepletion can be combined with that of mobility reduction by simply applying relations (8.73) to (8.76) in the analysis of Section 8.2.

8.4.7 Strong Inversion Approximation

According to (3.50) and (8.73), the mobile charge in strong inversion can be approximated by

$$-\frac{Q_i}{C_{ox}} = n_q(V_P - V).$$
(8.77)

This approximation is illustrated in Figure 8.21. Assuming constant mobility (no mobility reduction), the forward and reverse components of drain current as well as the various transconductances are also shown on the curve.

As discussed in Section 3.6.3, the surface potential can be assumed to be independent of the gate voltage and equal to $\Psi_0 + V$. The (equilibrium) threshold voltage V_{T0} was defined as the value of the gate voltage for which the mobile charge density is zero at equilibrium ($V = 0$). Hence, according to (8.77),

$$V_{T0} = V_G(V_P = V = 0).$$
(8.78)

Now, since $V_P = \Psi_P - \Psi_0$ (3.66), then

$$V_{T0} = V_G(\Psi_P = \Psi_0)$$
(8.79)

or, from (8.61),

$$V_{T0} = \Phi_{ms} - \frac{Q_{fc}}{C_{ox}} + \Psi_0 + \Gamma_b\sqrt{\Psi_0} \pm \frac{1}{\Gamma_g^2}\left(\frac{Q_{fc}}{C_{ox}} - \Gamma_b\sqrt{\Psi_0}\right)^2,$$
(8.80)

with the condition that the term in parentheses must be positive for an N-type gate and negative for a P-type gate. Otherwise no polydepletion occurs, corresponding to an infinite value of Γ_g.

The threshold V_{T0} can also be found in Figure 8.18 as the value of V_G for $\Psi_P = \Psi_0$. As we can see, V_{T0} is always increased by polydepletion in an N-type gate. For a P-type gate, it might be decreased, but only with very large positive values of fixed charge Q_{fc}.

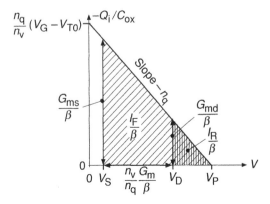

Figure 8.21   Effect of polydepletion on strong inversion approximation

According to (8.60), n_v is the slope of $V_G(V_P)$. Furthermore, $V_{T0} = V_G(V_P = 0)$ as expressed by (8.78). Therefore, the pinch-off voltage can be approximated by

$$V_P = \frac{V_G - V_{T0}}{n_v}. \tag{8.81}$$

This approximation can be used to obtain $V_P = \Psi_P - \Psi_0$, instead of (8.67), which was anyhow valid only for $Q_{fc} = 0$.

Introducing (8.81) into (8.77) provides a simple expression of $Q_i(V_G, V)$:

$$-\frac{Q_i}{C_{ox}} = n_q \left(\frac{V_G - V_{T0}}{n_v} - V \right). \tag{8.82}$$

Hence, a variation ΔV_G of gate voltage results in a vertical shift $\frac{n_q}{n_v} \Delta V_G < \Delta V_G$ of $Q_i/C_{ox}(V)$ as depicted in Figure 8.21. This explains the value of gate transconductance in linear mode

$$G_m = \frac{n_q}{n_v} \beta (V_D - V_S). \tag{8.83}$$

8.5 BAND GAP WIDENING

8.5.1 Introduction

In modern deep submicron processes, the oxide thickness is reduced to very small values, whereas the gate voltage is not reduced proportionally. The electric field at the silicon surface is therefore increased, and confined states originating from a quantum treatment can no longer be ignored. As a consequence, the highest allowed energy level for holes is slightly below the top of the valence band by an amount $q\Delta\Psi_v$, and the lowest allowed energy level for electrons is slightly above the bottom of the conduction band by an amount $q\Delta\Psi_c$. With this *band gap widening* effect, expression (3.5) of the concentration of electrons must be modified to [47,108]

$$n_p = n_i \exp \frac{\Psi - \Phi_F - V - \Delta\Psi_c}{U_T}. \tag{8.84}$$

In inversion, it can be assumed that $\Delta\Psi_c$ is primarily a function of the effective vertical field at the silicon-oxide interface. Since this field is proportional to the charge in the silicon, $\Delta\Psi_c$ can be expressed as [47]

$$\Delta\Psi_c = A_{qm}(-Q_b - \eta Q_i)^{2/3}. \tag{8.85}$$

The physical constant A_{qm} is given by

$$A_{qm} = \left(\frac{1}{2m^*q} \right)^{1/3} \left(\frac{9}{16} \frac{h}{\epsilon_{si}} \right)^{2/3} = 3.53 \frac{Vm^{4/3}}{A^{2/3}s^{2/3}}, \tag{8.86}$$

where h is the Planck's constant and m^* is the effective mass of the electron (equal to 98% of its free mass for <100> substrate orientation).

The constant η accounts for the effective value of the surface field. Its value is typically 3/4, but it can be used as a fitting parameter.

For a uniformly doped substrate, the depletion charge linearized around its value at $\Psi_s = \Psi_P$ is given by (3.54) and is not affected by possible polydepletion. It can further be related to the inverted charge by means of (3.39) (with $n = n_q$ to account for polydepletion), resulting in

$$Q_b = -\Gamma_b C_{ox} \sqrt{\Psi_P} - \frac{n-1}{n_q} Q_i, \tag{8.87}$$

where n is given by (3.34) and n_q is given by (8.70). Introducing this expression of $Q_b(Q_i)$ into (8.85) gives

$$\Delta\Psi_c = A_{qm} \left[\Gamma_b C_{ox} \sqrt{\Psi_P} + \left(\frac{1-n}{n_q} + \eta \right) (-Q_i) \right]^{2/3}, \tag{8.88}$$

or, with the normalized variables defined by (3.41), (3.43), and (8.73),

$$\delta\psi_c = \frac{A_{qm} C_{ox}^{2/3}}{U_T^{1/3}} \left[\gamma_b \sqrt{\psi_P} + (1 - n + \eta n_q) 2q_i \right]^{2/3}, \tag{8.89}$$

which is plotted in Figure 8.22 for several values of the pinch-off potential, $\eta = 3/4$, and $n_q = n = n_w$ given by (3.68).

As given by (8.88), $\Delta\Psi_c$ is a nonlinear function of both Ψ_P and Q_i. It can be linearized with respect to Q_i around $Q_i = 0$, giving

$$\Delta\Psi_c = \underbrace{A_{qm} \left(\Gamma_b C_{ox} \sqrt{\Psi_P} \right)^{2/3}}_{\Delta\Psi_{cP}}$$

$$+ \underbrace{\frac{2}{3} A_{qm} C_{ox} (\eta n_q + 1 - n) \left(\Gamma_b C_{ox} \sqrt{\Psi_P} \right)^{-1/3} \frac{-Q_i}{n_q C_{ox}}}_{\delta_{qm}}, \tag{8.90}$$

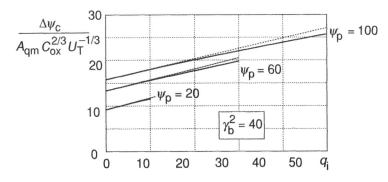

Figure 8.22   Dependency of band gap widening $\Delta\psi_c$ on inverted charge q_i; linear approximation in dotted line

or simply

$$\Delta \Psi_c = \Delta \Psi_{cP} + \delta_{qm} \frac{-Q_i}{n_q C_{ox}}. \tag{8.91}$$

The comparison with (8.89) in Figure 8.22 shows that this linearization is an acceptable approximation.

8.5.2 Extension of the General Charge–Voltage Expression

According to (8.84), the effect of band gap widening on the electron concentration is corrected for by adding $\delta \Psi_c$ to the channel voltage V. With the charge sheet approximation, the same correction can be introduced in expression (3.38) of $Q_i(\Psi_s)$, where the contribution of holes is neglected, and in equation (3.40) resulting from the linearization around $\Psi_s = \Psi_P$. Replacing n by n_q to also account for polydepletion, the latter can hence be rewritten as

$$\frac{\Psi_P - 2\Phi_F - V}{U_T} = 2\frac{-Q_i(1 + \delta_{qm})}{2n_q U_T C_{ox}} + \ln \frac{-Q_i(1 + \delta_{qm})}{2n_q U_T C_{ox}}$$

$$+ \frac{\Delta \Psi_{cP}}{U_T} + \ln \left[\frac{2n_q \sqrt{U_T}}{\Gamma_b(1 + \delta_{qm})} \left(\frac{-Q_i}{C_{ox}\Gamma_b \sqrt{U_T}} + 2\sqrt{\frac{\Psi_P}{U_T} + \frac{Q_i}{n_q C_{ox} U_T}} \right) \right]. \tag{8.92}$$

To obtain this form, $(1 + \delta_{qm})$ has been introduced in the numerator of the first logarithmic term and in the denominator of the second logarithmic term for compensation.

The normalized charge q_i defined by (3.41) can then be introduced, with an extended definition of the specific charge

$$Q_{spec} = -2n_q U_T C_{ox}/(1 + \delta_{qm}), \tag{8.93}$$

which reduces to (8.73) if band gap widening is negligible [and to (3.42) if polydepletion is also negligible].

Equation (8.92) then becomes

$$\psi_p - 2\phi_f - v = 2q_i + \ln q_i + \ln \left[\frac{2n_q}{\gamma_b(1 + \delta_{qm})} \left(\frac{2n_q q_i}{\gamma_b(1 + \delta_{qm})} \right. \right.$$

$$\left. \left. + 2\sqrt{\psi_p - \frac{2q_i}{(1 + \delta_{qm})}} \right) \right] + \delta \psi_{cp}. \tag{8.94}$$

It has been shown in Figure 3.10 that the second logarithmic term is practically independent of q_i. The terms in q_i can therefore be neglected, providing the simplified result

$$2q_i + \ln q_i + \underbrace{\ln \left(\frac{4n_q}{\gamma_b} \sqrt{\psi_p} \right) + \delta \psi_{cp} - \ln(1 + \delta_{qm})}_{v_{sh}} = \psi_p - 2\phi_f - v, \tag{8.95}$$

which is identical to (3.45) with an expression of voltage shift v_{sh} extended to include the effect of band gap broadening.

The general charge–voltage expression (3.48) *does therefore include quantum effects*, provided specific charge Q_{spec} (that is used to normalize mobile inverted charge Q_i) is extended to (8.93), and v_{sh} [that relates pinch-off voltage v_p to pinch-off potential ψ_p according to (3.47)] is extended to

$$v_{sh} = \ln\left(\frac{4n_q}{\gamma_b}\sqrt{\psi_p}\right) + \delta\psi_{cp} - \ln(1 + \delta_{qm}). \tag{8.96}$$

The correction term δ_{qm} is given by (8.90), and can be expressed by introducing definition (3.30) of the substrate modulation factor Γ_b:

$$\delta_{qm} = \frac{2}{3}A_{qm}C_{ox}(\eta n_q + 1 - n)(2q\epsilon_{si}N_b\Psi_P)^{-1/6}. \tag{8.97}$$

For a fixed doping concentration N_b, δ_{qm} is approximately proportional to C_{ox}. Even though N_b is also increased while scaling-down the process features, the dependency on C_{ox} dominates. Expression (8.97) is plotted in Figure 8.23 as a function of Ψ_P for $n_q = n = n_w$ (no polydepletion, evaluation of n at pinch-off) and for several combinations of values of N_b and C_{ox}.

As can be seen, δ_{qm} is a very weak function of ψ_p. It can therefore be considered constant and evaluated at a particular value of Ψ_P; for example, $\Psi_P = 2\Phi_F$. It is also a very weak function of N_b. Indeed, for the whole set of values considered in Figure 8.23 and for $\Psi_P > 0.7$ V,

$$\frac{\delta_{qm}}{C_{ox}} = 10 \text{ to } 20\,\text{m}^2/\text{F} = 10^{-2} \text{ to } 2 \times 10^{-2}\,\mu\text{m}^2/\text{fF}. \tag{8.98}$$

For a given value of $V_P - V$, the inverted charge Q_i is *reduced* since Q_{spec} is reduced according to (8.93).

Even for a very large value of C_{ox}, δ_{qm} remains smaller than unity and can therefore be neglected in (8.96). Hence, the increase ΔV_{sh} of V_{sh} due to band gap widening is reduced to

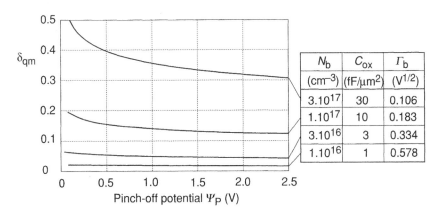

N_b	C_{ox}	Γ_b
(cm^{-3})	(fF/μm^2)	(V$^{1/2}$)
3.10^{17}	30	0.106
1.10^{17}	10	0.183
3.10^{16}	3	0.334
1.10^{16}	1	0.578

Figure 8.23   Correction term δ_{qm} according to (8.97)

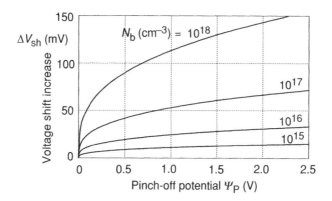

Figure 8.24   Contribution of band gap widening to voltage shift V_{sh}

$\Delta\Psi_{cP}$ defined in (8.90). Introducing definition (3.30) of Γ_b, it can be expressed as

$$\Delta V_{sh} \cong \Delta\Psi_{cP} = A_{qm}\,(2q\epsilon_{si}N_b\Psi_P)^{1/3}, \tag{8.99}$$

which is *independent of* C_{ox} and depends only on the doping concentration N_b and the pinch-off potential Ψ_P. It is represented in Figure 8.24 for various values of N_b. It can be noticed that it still has a nonnegligible value even for the very low doping concentration $N_b = 10^{15}\,\mathrm{cm}^{-3}$.

Using nonnormalized variables, (8.96) then becomes

$$V_{sh} = U_T \ln\left(\frac{4n_q}{\Gamma_b}\sqrt{\Psi_P}\right) + \Delta V_{sh}. \tag{8.100}$$

Although this increase of V_{sh} is proportional to $(\Psi_P)^{1/3}$, it may again be considered as a constant evaluated at some value of Ψ_P, for example $2\Phi_F$.

The effect of band gap widening is to increase the voltage shift V_{sh}, thereby increasing $\Psi_P - V_P$ and Ψ_0 according to (3.66). Hence the threshold V_{T0} defined by (3.58) is *increased*, and the pinch-off voltage V_P is *decreased* at constant Ψ_P (or constant V_G).

8.5.3 Extension of the General Current–Voltage Expression

Due to quantum effects, the electric field experienced by the mobile carriers in the channel is no longer $-\mathrm{d}\Psi_s/\mathrm{d}x$ but

$$E_x = -\frac{\mathrm{d}(\Psi_s - \Delta\Psi_c)}{\mathrm{d}x}. \tag{8.101}$$

Expression (4.2) of the drain current must therefore be modified to

$$I_D = \mu W\left(\underbrace{-Q_i\frac{\mathrm{d}(\Psi_s - \Delta\Psi_c)}{\mathrm{d}x}}_{\text{drift}} + \underbrace{U_T\frac{\mathrm{d}Q_i}{\mathrm{d}x}}_{\text{diffusion}}\right). \tag{8.102}$$

However, according to (8.84), the same correction must be included in the density of inverted charge, so that (4.4) becomes

$$\frac{dQ_i}{dx} = \frac{Q_i}{U_T}\left(\frac{d(\Psi_s - \Delta\Psi_c)}{dx} - \frac{dV}{dx}\right), \tag{8.103}$$

which when introduced in (8.102) *gives exactly* (4.5).

Hence, all the results derived in Chapter 4 *remain valid*, provided the value of V_{sh} used in expression (3.66) of Ψ_0 is increased by ΔV_{sh} given by (8.99) [increasing thereby the value of V_{T0} according to (3.58)], and the expression of the specific current defined by (4.14) is extended with that of the specific charge (8.93), giving

$$I_{spec} = \mu U_T \frac{W}{L}(-Q_{spec}) = \frac{2n_q\mu C_{ox}}{1 + \delta_{qm}}\frac{W}{L}U_T^2, \tag{8.104}$$

which reduces to (8.74) if band gap widening is negligible [and to (4.14) if polydepletion is also negligible].

The same is true for all other results derived so far.

8.6 GATE LEAKAGE CURRENT

Silicon dioxide used to isolate the gate electrode from the channel is an excellent dielectric material. Due to its large band gap, its intrinsic leakage current is negligible as long as its thickness t_{ox} is larger than 3 nm and the voltage V_{ox} across it does not exceed a few volts.

Increasing V_{ox} may result in field-induced (Fowler–Nordheim) tunneling of carriers [109], even with larger values of t_{ox}. This "high-voltage" leakage current is exploited to charge or discharge an isolated gate in EPROM [110] and E^{2}PROM [111] nonvolatile memories.

Now, when t_{ox} is reduced below 3 nm in aggressively scaled-down processes, a gate leakage current starts to appear even at low V_{ox}, as the result of direct tunneling of carriers through the oxide.

For an N-channel transistor operated in inversion ($\Psi_s > 0$), this intrinsic gate current consists of electrons tunneling from the inversion layer to the gate (holes for a P-channel transistor). The resulting current density can be expressed as [112, 113]

$$J_G = \frac{K_G}{\epsilon_{ox}}\frac{V_{ox}}{t_{ox}}(-Q_i)P_{tun}. \tag{8.105}$$

Practical values for constant K_G are [113] 3×10^{-5} A/V^2 for electrons and 4×10^{-5} A/V^2 for holes, and P_{tun} is the tunneling probability.

A suitable formulation of this probability, which covers both direct and Fowler–Nordheim tunneling, is given by [113]

$$P_{tun} = \begin{cases} \exp\left(-\dfrac{E_B t_{ox}}{V_{ox}}\left[1 - \left(1 - \dfrac{V_{ox}}{X_B}\right)^{3/2}\right]\right) & \text{for } V_{ox} \leq X_B \text{ (direct)} \\[4mm] \exp\left(-\dfrac{E_B t_{ox}}{V_{ox}}\right) & \text{for } V_{ox} \geq X_B \text{ (Fowler–Nordheim).} \end{cases} \tag{8.106}$$

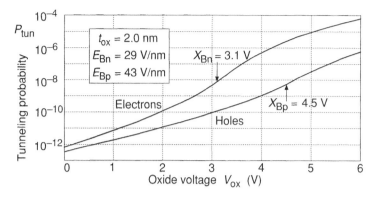

Figure 8.25   Tunneling probability for $t_{ox} = 2$ nm

where X_B is the oxide-channel voltage barrier and E_B is a characteristic electric field. Their values for electrons and holes are given in Figure 8.25, which shows the corresponding plots of (8.106) for $t_{ox} = 2$ nm. As can be seen, the probability of direct tunneling is a strong function of V_{ox}. Its dependency on t_{ox} is even stronger, as illustrated in Figure 8.26 for two values of V_{ox}. Indeed, for small values of V_{ox}, P_{tun} increases by about 12 orders of magnitude when t_{ox} is reduced from 3 nm to 1 nm.

Using normalized variables $q_i = Q_i/Q_{spec}$, $v_{ox} = V_{ox}/U_T$, and $\xi = x/L$, the gate current is obtained by integrating (8.105) along the channel:

$$I_G = I_{G0} \int_0^1 q_i v_{ox} P_{tun}(v_{ox})\, d\xi, \tag{8.107}$$

with

$$I_{G0} = \frac{2 n_q K_G U_T^2 W L}{t_{ox}^2}. \tag{8.108}$$

Since the very thin oxide is always associated with large substrate concentration, polysilicon depletion must be taken into account and n_q is the charge slope factor given by (8.69) or (8.70). The voltage across the oxide can be obtained directly from (3.19) with $Q_{si} = Q_b + Q_i$:

$$V_{ox} = E_{ox} t_{ox} = -\frac{Q_b + Q_i + Q_{fc}}{C_{ox}}. \tag{8.109}$$

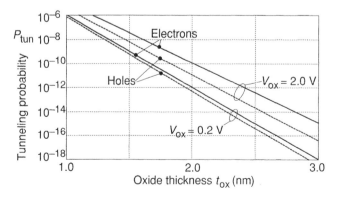

Figure 8.26   Probability of direct tunneling as a function of t_{ox}

After normalization of Q_i and Q_{fc} to Q_{spec} given by (8.73) and introduction of expression (3.29) of Q_b, we obtain

$$v_{ox} = 2n_q(q_i + q_{fc}) + \gamma_b \sqrt{\psi_s},$$ (8.110)

where voltages are normalized according to (3.43).

Now, with the charge–potential linearization introduced in Section 3.6, and using (3.66) and (3.48),

$$\psi_s = \psi_p - 2q_i = \psi_0 + v_p - 2q_i = \psi_0 + v + \ln q_i,$$ (8.111)

which when introduced in (8.110) finally yields

$$v_{ox} = 2n_q(q_i + q_{fc}) + \gamma_b \sqrt{\psi_0 + v + \ln q_i}.$$ (8.112)

The argument of integral (8.107) is a complicated function of q_i, which is itself a function of ξ in the general case, and no analytic solution can be found. For the particular case of $v = v_s = v_d$ (equipotential channel), the situation is much simpler since all three terms inside the integral are constant and the integral is just their product:

$$\frac{I_G}{I_{G0}} = q_i v_{ox} P_{tun}(v_{ox}) \quad \text{(for equipotential channel)}.$$ (8.113)

We shall assume that the gate current remains sufficiently small to have no effect on the potential.

Using q_i as a parameter, this product can be calculated to obtain $I_G(q_i)$, whereas $(v_p - v)(q_i)$ is given by (3.48). An example of the resulting plot of $I_G(v_p - v)$ is represented in Figure 8.27

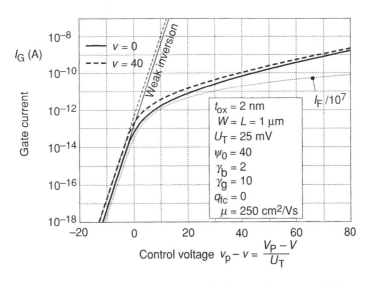

Figure 8.27   Gate leakage current for an equipotential channel ($v = v_s = v_d$); the corresponding variation of forward current I_F is also shown for comparison

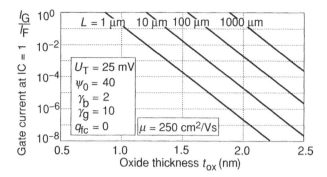

Figure 8.28 Variation with oxide thickness of the relative gate current at $IC = 1$ (equipotential channel)

for two values of (equipotential) channel voltage v. As can be seen, $I_G(v_p - v)$ is slightly dependent of the value of v, due to the presence of v in (8.112).

For comparison, the forward current, $I_F(v_p - v)$, has been calculated from parameter q_i, using (4.19) and (8.74). It is also represented in Figure 8.27 after division by 10^7 to fit in the same scale. As can be seen, the gate current is approximately proportional to I_F in weak inversion, but it increases faster in strong inversion.

Due to the very steep function $P_{tun}(t_{ox})$, I_G/I_F is very strongly dependent on t_{ox}. In Figure 8.28 the variation of this ratio is represented for $I_F = I_{spec}$ (inversion coefficient $IC = 1$), for which $q_i = (\sqrt{5} - 1)/2$ according to (4.24). The ratio I_G/I_F for a given value of inversion coefficient does not depend on the channel width, but it increases with the square of the channel length (since $I_G \propto L$ and $I_F \propto 1/L$).

In weak inversion, $q_i \ll 1$ and $\psi_s = \psi_p = \psi_0 + v_p$. Equation (8.110) then becomes

$$v_{ox} = 2n_q q_{fc} + \gamma_b \sqrt{\psi_0 + v_p}. \tag{8.114}$$

Hence v_{ox} and P_{tun} are constant along the channel *even if v is not constant* ($v_d \neq v_s$). In this case, the gate current is obtained by replacing q_i in (8.113) by its average value $\overline{q_i}$. Since in weak inversion $i_d = -dq_i/d\xi$ according to (4.21), q_i changes with a constant slope between its values q_s at the source and q_d at the drain (as long as the gate current remains a small perturbation, $I_G \ll I_F$), as illustrated in Figure 8.29. Thus, using (3.49), the average charge density can be expressed as

$$\overline{q_i} = \frac{e^{v_p}}{2}(e^{-v_s} + e^{-v_d}). \tag{8.115}$$

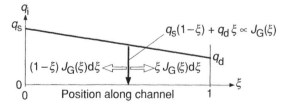

Figure 8.29 Weak inversion: profile of mobile charge and source–drain repartition of local gate current

The gate current in weak inversion is obtained by introducing (8.114), (8.115), and (8.106) into (8.113). The result for $v_d = v_s = v$ is also plotted in Figure 8.27. In saturation ($v_d \gg v_s$), the gate current would be divided by 2.

Since in weak inversion V_{ox} is constant, the local density of gate current $J_G(\xi)$ injected in the channel is simply proportional to q_i. Now, the two parts of the channel may be considered as separate transistors with normalized channel lengths ξ and $1 - \xi$. According to the concept of pseudo-resistor introduced in Section 4.5.4, the elementary local contribution $J_G(\xi)\,d\xi$ of gate current splits proportionally to the inverse of the channel lengths (assuming constant width), as indicated in Figure 8.29. Hence the fraction of gate current flowing to the source is given by

$$\frac{I_{GS}}{I_G} = \frac{\int_0^1 [q_s(1 - \xi) + q_d\xi](1 - \xi)\,d\xi}{(q_s + q_d)/2} = \frac{2q_s + q_d}{3(q_s + q_d)}, \tag{8.116}$$

or, by introducing (3.49) to express the charge densities from the corresponding voltages:

$$\frac{I_{GS}}{I_G} = 1 - \frac{I_{GD}}{I_G} = \frac{1 + 2\exp(v_d - v_s)}{3[1 + \exp(v_d - v_s)]}. \tag{8.117}$$

This result is plotted in Figure 8.30. The gate current splits evenly to source and drain for $v_d = v_s$ to reach ratios 2/3 and 1/3 in saturation.

As was pointed out before, the calculation of the gate current in moderate and strong inversion is much more complicated for $v_d \neq v_s$ because v_{ox} and P_{tun} are then variable along the channel. No exact analytical solution can be found, but approximations show that the overall gate current I_G does not change much with v_d [113]. Its source–drain repartition saturates for $v_d > v_p$ to a value of I_{GS}/I_G that increases with v_p.

Since the gate current depends on the pinch-off voltage, it creates a dc gate conductance

$$G_g = \frac{1}{n_v}\frac{dI_G}{dV_P}, \tag{8.118}$$

which is split into a gate-to-source conductance G_{gs} and a gate-to-drain conductance G_{gd} proportionally to the splitting of current. We have seen that, in weak inversion, the gate current is approximately proportional to the drain current; hence, $G_g \cong I_G/(nU_T)$. In strong inversion G_g/I_G decreases, but not as fast as the gate transconductance.

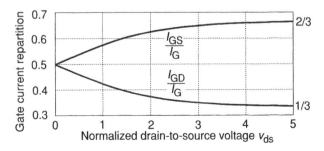

Figure 8.30   Source–drain repartition of gate current in weak inversion

The gate current also exhibits shot noise of spectral density $2q I_G$ [114]. Its contribution may dominate at low frequencies, when the gate noise induced from the channel becomes negligible.

In addition to the gate to channel current discussed above, tunneling also produces a current in the small areas where the gate overlaps the source and drain diffusions. This component of gate leakage may dominate for short channel lengths or for large drain voltages [112, 113].

9 Short-Channel Effects

In the previous chapters, the channel was assumed to be long, allowing for a one-dimensional analysis using the gradual channel approximation. This chapter is devoted to the effects appearing when reducing the length of the transistor to dimensions that get close to the depletion width. In such a situation, the one-dimensional approach is no longer valid and a two-dimensional analysis is required. Furthermore, since the terminal voltages are not scaled-down proportionally to the device length reduction, the longitudinal electric field increases beyond a certain critical field above which the carrier velocity starts to saturate. This velocity saturation (VS) effect is presented in Section 9.1 using different velocity-field relations. Another short-channel effect that strongly limits the performance (particularly the voltage gain) of analog circuits is the channel length modulation (CLM) described in Section 9.2. When the device length gets small the surface potential in the channel region is no longer defined uniquely by the vertical field, but becomes influenced by the drain (or source) voltage. This effect is called the drain-induced barrier lowering (DIBL) and is presented in Section 9.3. A pseudo two-dimensional analysis is used for CLM and DIBL to derive analytical expressions for the channel length reduction and the surface potential. Short-channel effects such as VS and CLM not only impact the current, but also impact the thermal noise. A short-channel thermal noise model including the effects of VS, CLM, but also carrier heating and mobility reduction due to the vertical field is presented in Section 9.4.

9.1 VELOCITY SATURATION

An important effect that appears when the longitudinal electric field E_x within the device starts to become large is the saturation of the drift velocity. The drift velocity v_{drift} of electrons and holes in bulk silicon is plotted versus E_x in Figure 9.1. At low longitudinal electric field, the velocity is proportional to the electric field with a proportionality factor equal to the mobility μ_z at low longitudinal field. When E_x approaches the *critical field* E_c, the velocity starts to saturate towards a maximum value v_{sat}. The shape of the velocity-field relation in silicon is slightly different for electrons and holes. Typical values for E_c and v_{sat} at room temperature for electrons and holes are given in Table 9.1. The critical field E_c is related to the saturated

Charge-Based MOS Transistor Modeling: The EKV Model for Low-Power and RF IC Design   C. Enz and E. Vittoz
© 2006 John Wiley & Sons, Ltd.

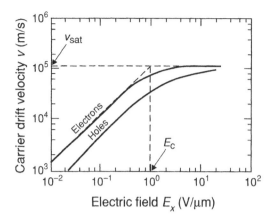

Figure 9.1 Drift velocity in silicon for electrons and holes versus electric field

drift velocity and the mobility at low longitudinal field by

$$E_c \triangleq \frac{v_{sat}}{\mu_z}.$$
(9.1)

As shown in Figure 9.2(a) and discussed in Section 8.2, the mobility at low longitudinal field μ_z actually depends on the vertical field E_z. The larger the vertical field, the smaller μ_z and the larger the critical field, since v_{sat} is constant for a given type of silicon. More generally, any factor reducing the low longitudinal field mobility μ_z, pushes the limit of velocity saturation to higher values of the longitudinal field.

Two other mobilities can be defined as illustrated in Figure 9.2(b). The *effective mobility* μ_{eff} combining the effects of reduction due to the vertical field and VS and defined as

$$\mu_{eff} \triangleq \frac{v_{drift}}{|E_x|}.$$
(9.2)

μ_{eff} is also called the *cord mobility*, since it actually corresponds to the secant between the origin and the operating point $v(E_x)$, as shown in Figure 9.2(b). Another mobility that will be used in Section 9.4 is the *differential mobility* defined as

$$\mu_{diff} \triangleq \frac{dv_{drift}}{dE_x}.$$
(9.3)

Different velocity-field models will be considered below in order to analyze the effect of velocity saturation on the drain current and on the transconductances.

Table 9.1 Typical values of the saturated drift velocity and the critical field for bulk silicon at room temperature [67]

	v_{sat}	E_c
Electrons	10^5 m/s	1 V/μm
Holes	8×10^4 m/s	3 V/μm

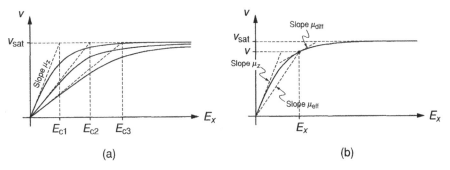

Figure 9.2 (a) Dependence of the mobility μ_z at low longitudinal field with the vertical field E_z. (b) Illustration of the different mobilities

9.1.1 Velocity-Field Models

9.1.1.1 Model 1

The simplest model to describe the velocity-field dependence is to consider the velocity proportional to the longitudinal field with a constant slope up to the critical field, above which it stays constant and equal to the saturated velocity. This model is described by the piecewise linear model defined by

$$v_{\text{drift}}(E_x) = \begin{cases} \mu_z|E_x| & \text{for } |E_x| < E_c \\ v_{\text{sat}} & \text{for } |E_x| \geq E_c, \end{cases} \tag{9.4}$$

which can be normalized to the saturation value v_{sat} according to

$$v(e) \triangleq \frac{v_{\text{drift}}}{v_{\text{sat}}} = \begin{cases} e & \text{for } e < 1 \\ 1 & \text{for } e \geq 1, \end{cases} \tag{9.5}$$

where e is the longitudinal electric field E_x normalized to the critical field E_c:

$$e \triangleq \frac{|E_x|}{E_c}. \tag{9.6}$$

The corresponding effective mobility μ_{eff} is then simply given by

$$\mu_{\text{eff}}(E_x) \triangleq \frac{v_{\text{drift}}}{|E_x|} = \begin{cases} \mu_z & \text{for } E_x < E_c \\ v_{\text{sat}}/|E_x| & \text{for } E_x \geq E_c, \end{cases} \tag{9.7}$$

or in a normalized form

$$u(e) \triangleq \frac{\mu_{\text{eff}}}{\mu_z} = \begin{cases} 1 & \text{for } e < 1 \\ 1/e & \text{for } e \geq 1. \end{cases} \tag{9.8}$$

The main advantage of this model is obviously its simplicity, allowing to get a first understanding of the physical phenomenon. But on the other hand, it has discontinuous derivatives at $e = 1$ that may induce a bad behavior of the model.

9.1.1.2 Model 2

A continuous model also accounting for the difference between the velocity-field characteristics of electrons and holes as illustrated in Figure 9.1 is defined by

$$v_{\text{drift}}(E_x) = v_{\text{sat}} \frac{|E_x|/E_c}{\left[1 + \left(\frac{|E_x|}{E_c}\right)^\alpha\right]^{\frac{1}{\alpha}}} = \mu_z \frac{|E_x|}{\left[1 + \left(\frac{|E_x|}{E_c}\right)^\alpha\right]^{\frac{1}{\alpha}}}, \tag{9.9}$$

where $\alpha = 2$ for electrons and $\alpha = 1$ for holes. Equation (9.9) can be written in a normalized form as

$$v(e) \triangleq \frac{v_{\text{drift}}}{v_{\text{sat}}} = \frac{e}{(1 + e^\alpha)^{\frac{1}{\alpha}}}. \tag{9.10}$$

Note that this continuous model has been used in many compact models with the approximation that α is equal to unity for both holes and electrons. For the sake of simplicity, we will also make the same assumption in the following development.

The effective mobility is then given by

$$\mu_{\text{eff}}(E_x) \triangleq \frac{v_{\text{drift}}}{|E_x|} = \frac{\mu_z}{1 + |E_x|/E_c}, \tag{9.11}$$

which can also be written in normalized form as

$$u(e) \triangleq \frac{\mu_{\text{eff}}}{\mu_z} = \frac{1}{1 + e}. \tag{9.12}$$

9.1.1.3 Model 3

Although the continuous velocity-field model given by (9.9) and (9.10) insures the continuity of the current and the output conductance versus the drain voltage, it requires the electric field to become infinity at the drain for the velocity and hence the current to saturate, which is not physical. Another velocity-field model that will also be used subsequently in Section 9.2 for the derivation of the CLM model is given by

$$v_{\text{drift}}(E_x) = \begin{cases} v_{\text{sat}} \dfrac{|E_x|/E_c}{1 + |E_x|/(2E_c)} & \text{for } |E_x| < 2E_c \\ v_{\text{sat}} & \text{for } |E_x| \geq 2E_c, \end{cases} \tag{9.13}$$

or in normalized form

$$v(e) \triangleq \frac{v_{\text{drift}}}{v_{\text{sat}}} = \begin{cases} \dfrac{e}{1 + e/2} & \text{for } e < 2 \\ 1 & \text{for } e \geq 2. \end{cases} \tag{9.14}$$

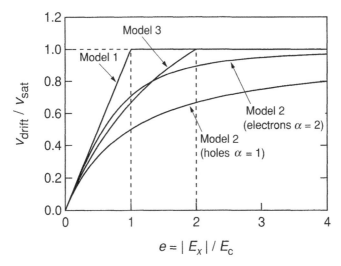

Figure 9.3   The different velocity-field models

The corresponding effective mobility-field model is then given by

$$\mu_{\text{eff}}(E_x) \triangleq \frac{v_{\text{drift}}}{|E_x|} = \begin{cases} \dfrac{\mu_z}{1 + |E_x|/(2E_c)} & \text{for } |E_x| < 2E_c \\ \dfrac{v_{\text{sat}}}{|E_x|} & \text{for } |E_x| \geq 2E_c, \end{cases} \tag{9.15}$$

or in normalized form

$$u(e) \triangleq \frac{\mu_{\text{eff}}}{\mu_z} = \begin{cases} \dfrac{1}{1 + e/2} & \text{for } e < 2 \\ \dfrac{1}{e} & \text{for } e \geq 2. \end{cases} \tag{9.16}$$

This last model does not require the field to become infinity for the velocity to saturate. Actually, the velocity saturates at $E_{\text{sat}} \triangleq 2E_c$. This allows to define a point in the channel where the field becomes equal to twice the critical field E_c and where the carrier velocity saturates. The channel can then be split into a nonvelocity saturation region on the direction from that point toward the source and a VS region on the direction from that point toward the drain where the velocity is equal to v_{sat}. The length of this VS region will then be used in the CLM model.

The three velocity-field models presented above are plotted in Figure 9.3. Note that the approximation defined in (9.13) (corresponding to Model 3) is closer to the velocity-field model of electrons given by (9.9) with $\alpha = 2$. Also note that the shape of the velocity-field curve will strongly affect the current and output conductance versus drain voltage. All three velocity-field models will be used hereafter to evaluate the effect of velocity saturation on the profile of the inversion charge, the drain current, and the transconductances.

9.1.2 Effect of VS on the Drain Current

In weak inversion the current is carried by diffusion only and the surface potential gradient along the channel is zero and therefore the longitudinal electric field is null. Velocity saturation

Table 9.2   Typical values of the VS parameter λ_c

L	1 μm	0.5 μm	0.17 μm	0.1 μm
λ_c	0.05	0.1	0.3	0.5

can therefore be neglected in weak inversion. Although a complete model including both the diffusion component and drift component including the effect of velocity saturation can be derived, in the following derivation, the diffusion current is neglected for the sake of simplicity. Note that this does not introduce a significant error in strong inversion.

The drift component of the drain current is proportional to the velocity and can be written as

$$I_D = W(-Q_i)v_{\text{drift}} = W(-Q_i)\mu_{\text{eff}}|E_x| = W(-Q_i)\mu_{\text{eff}}\frac{d\Psi_s}{dx}, \tag{9.17}$$

or using the normalized variables defined above

$$i_d = 2q_i\frac{v}{\lambda_c} = 2q_i\frac{ue}{\lambda_c} = uq_i\frac{d\psi_s}{d\xi}, \tag{9.18}$$

where the normalized variables i_d, q_i, ψ_s, and ξ have their usual meaning. The parameter λ_c accounts for the VS effect and depends on the transistor length L according to

$$\lambda_c \triangleq \frac{2\mu_z U_T}{v_{\text{sat}}L} = \frac{2U_T}{E_c L}. \tag{9.19}$$

Note that λ_c tends to zero for very long-channel devices. Setting it to zero corresponds to ignore the effect of VS. Typical values of λ_c for different channel lengths are given in Table 9.2.

As discussed in Section 3.6.1 and according to (3.39), the inversion charge can be linearized with respect to Ψ_s giving a relation between the surface potential gradient and the inversion charge gradient

$$\frac{d\Psi_s}{dx} = \frac{1}{nC_{\text{ox}}}\frac{dQ_i}{dx}, \tag{9.20}$$

or in normalized form

$$\frac{d\psi_s}{d\xi} = -2\frac{dq_i}{d\xi}. \tag{9.21}$$

Replacing (9.21) into (9.18) results in

$$i_d = -2uq_i\frac{dq_i}{d\xi}. \tag{9.22}$$

The main difference between (9.22) and the long-channel model (4.21) (in strong inversion) discussed in Section 4.4.1 is the field-dependent mobility term u which depends on the chosen velocity-field model.

The inversion charge density at the drain $-Q_{\mathrm{iD}}$ (or its normalized form q_{d}) plays a particular role in the presence of VS. When the drift velocity saturates right at the drain, $v = 1$ at $\xi = 1$. The drain current cannot increase anymore and is then limited to the saturation value $i_{\mathrm{d\,sat}}$ given by (9.18) evaluated at $\xi = 1$ with $v = 1$ and $q_{\mathrm{i}} = q_{\mathrm{d\,sat}}$:

$$i_{\mathrm{d\,sat}} \triangleq \frac{2}{\lambda_{\mathrm{c}}} q_{\mathrm{d\,sat}}, \tag{9.23}$$

where $q_{\mathrm{d\,sat}}$ is the value of the inversion charge density at the drain which is required to sustain the drain current when carrier velocity is saturated at the drain. The main difference compared to the long-channel situation is that the charge density at the drain does not vanish to zero as it does at the onset of saturation for long-channel devices when $v_{\mathrm{d}} = v_{\mathrm{p}}$, but it has to be finite in order for the current to flow despite the saturated velocity. To account for this saturation of the drain charge, the normalized drain charge in strong inversion given by (3.50) has to be modified according to

$$q_{\mathrm{d}} = \begin{cases} \dfrac{v_{\mathrm{p}} - v_{\mathrm{d}}}{2} & \text{for } v_{\mathrm{d}} < v_{\mathrm{d\,sat}} \\[2mm] q_{\mathrm{d\,sat}} & \text{for } v_{\mathrm{d}} \geq v_{\mathrm{d\,sat}}, \end{cases} \tag{9.24}$$

where the *drain saturation voltage* $v_{\mathrm{d\,sat}}$ is the value of the drain voltage at which the current and the drain charge saturate. It is obtained by replacing q_{d} and v_{d} in $q_{\mathrm{d}} = (v_{\mathrm{p}} - v_{\mathrm{d}})/2$ by $q_{\mathrm{d\,sat}}$ and $v_{\mathrm{d\,sat}}$ respectively, resulting in

$$v_{\mathrm{d\,sat}} = v_{\mathrm{p}} - 2q_{\mathrm{d\,sat}}. \tag{9.25}$$

Note that, due to VS, $v_{\mathrm{d\,sat}}$ is always smaller than the pinch-off voltage v_{p} corresponding to the saturation voltage when velocity saturation is not present.

9.1.2.1 Model 1

In a first step, the simple piecewise linear velocity-field model given by (9.5) will be used to derive the current. The latter is obtained by integrating (9.18) or (9.22) from source to drain, assuming that the drain voltage is sufficiently low for the longitudinal electric field to remain smaller than the critical field at any point along the channel. The carrier velocity is then not saturated along the channel and hence $v = e$ and $u = 1$, resulting in

$$i_{\mathrm{d}} = q_{\mathrm{s}}^2 - q_{\mathrm{d}}^2, \tag{9.26}$$

where q_{s} is the inversion charge density taken at the source, which in strong inversion and assuming there is no velocity saturation at the source is related to the source voltage according to

$$q_{\mathrm{s}} \triangleq q_{\mathrm{i}}(\xi = 0) = \frac{v_{\mathrm{p}} - v_{\mathrm{s}}}{2}, \tag{9.27}$$

and q_{d} is the inversion charge density taken at the drain and given by (9.24).

To ensure current continuity, from (9.26), the drain current in saturation can also be written as

$$i_{d\,sat} = q_s^2 - q_{d\,sat}^2. \tag{9.28}$$

If there was no VS, like in the long-channel case, the drain current would saturate as soon as $q_d = 0$ which occurs when $v_d = v_p$. When VS is present, the drain current saturates as soon as q_d reaches $q_{d\,sat}$, which occurs when v_d becomes equal to $v_{d\,sat}$.

The saturation current $i_{d\,sat}$, saturation drain charge density $q_{d\,sat}$, and saturation voltage $v_{d\,sat}$ are depending on the source charge density q_s. They can be expressed in terms of q_s by solving equations (9.23), (9.28), and (9.25) for $q_{d\,sat}$, $i_{d\,sat}$, and $v_{d\,sat}$, resulting in

$$q_{d\,sat} = \frac{1}{\lambda_c}\left[\sqrt{1 + (\lambda_c q_s)^2} - 1\right], \tag{9.29a}$$

$$i_{d\,sat} = \frac{2}{\lambda_c}q_{d\,sat} = \frac{2}{\lambda_c^2}\left[\sqrt{1 + (\lambda_c q_s)^2} - 1\right], \tag{9.29b}$$

$$v_{d\,sat} = v_p - 2q_{d\,sat} = v_p - \frac{2}{\lambda_c}\left[\sqrt{1 + (\lambda_c q_s)^2} - 1\right]. \tag{9.29c}$$

The drain saturation voltage is plotted versus the pinch-off voltage in Figure 9.4(a), which shows that $v_{d\,sat}$ can be substantially smaller than v_p.

The drain current including VS is plotted versus v_d in Figure 9.4(b) together with the current which does not include the effect of VS. For $v_d < v_{d\,sat}$, the drain current follows the current that does not include VS up to $v_{d\,sat}$, above which it saturates to the value $i_{d\,sat}$ given by (9.29b). The output conductance obtained with this model is clearly discontinuous due to the discontinuity of the derivative of the velocity-field relation given by (9.5). A continuous model will be derived below (Model 2).

The profile of the inversion charge density can be obtained by integrating (9.18) or (9.22) with $u = 1$ from the source to a point x along the channel. This leads to

$$i_d\xi = q_s^2 - q_i^2, \tag{9.30}$$

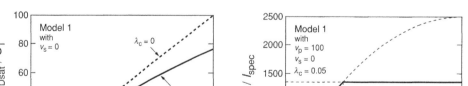
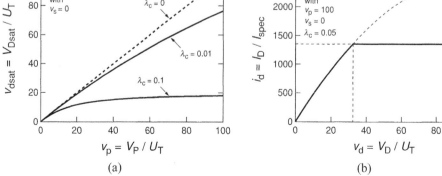

Figure 9.4   (a) Drain saturation voltage versus pinch-off voltage for Model 1 with $\lambda_c = 0$, 0.01, 0.1. (b) Drain current versus drain voltage for Model 1 with $\lambda_c = 0.05$

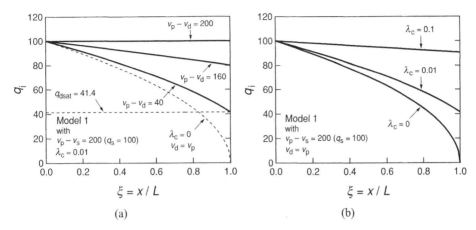

Figure 9.5 Inversion charge density versus position along the channel for Model 1; (a) for $v_p - v_s = 200$ ($q_s = 100$), $v_p - v_d = 200$, 160, 40, and $\lambda_c = 0.01$; (b) for $v_p - v_s = 200$, $v_d = v_p$ and $\lambda_c = 0$, 0.01, 0.1

which can be solved for q_i, resulting in

$$q_i(\xi) = \sqrt{q_s^2 - i_d \xi},$$
(9.31)

where i_d accounts for VS with q_d given by (9.24).

As Figure 9.5 shows, the profile of the inversion charge along the channel is affected by the VS occurring right at the drain. Indeed, the inversion charge at the drain cannot decrease down to zero, but remains clamped at $q_{d\,sat}$ when v_d gets larger than $v_{d\,sat}$. The higher the product $\lambda_c q_s$, the stronger the effect. Figure 9.5(a) shows the inversion charge profile for different values of $v_p - v_d$ and for $\lambda_c = 0.01$. When $v_p - v_d = v_p - v_s = 200$ (or $v_d = v_s$), the lateral field is zero and the profile is not affected by VS and hence remains uniform from source to drain. When v_d increases but remains smaller than $v_{d\,sat}$ (for example, the curve labeled $v_p - v_d = 160$ in Figure 9.5(a)), the profile decreases at the drain like in the long-channel case. As soon as v_d becomes larger then $v_{d\,sat}$, the charge at the drain cannot decrease anymore and remains clamped to $q_{d\,sat}$. In saturation, i.e., for $v_d > v_{d\,sat}$ the actual charge at the drain is no longer set by $v_p - v_d$, but by the current $i_{d\,sat}$ which has to flow even though the carrier velocity is saturated close to the drain. Figure 9.5(b) shows the inversion charge profile in saturation (obtained by setting $v_d = v_p > v_{d\,sat}$) for different values of λ_c and for a constant value of q_s. Increasing λ_c increases $q_{d\,sat}$ which tends to q_s making the profile become almost uniform from source to drain, even though the transistor is in saturation. This situation is somehow similar to what happens when $v_d \cong v_s$, but with the difference that the current is equal to the saturation current $i_{d\,sat}$, which depends only on q_s as stated by (9.29b).

Knowing the inversion charge profile allows to also get the longitudinal electric field. Indeed, setting $u = 1$ in (9.18) and solving for e, we get

$$e(\xi) = \frac{\lambda_c i_d}{2 q_i} = \frac{\lambda_c i_d}{2\sqrt{q_s^2 - i_d \xi}}.$$
(9.32)

In strong VS condition, i.e., for $\lambda_c q_s \gg 1$, the saturation current given by (9.29b) reduces to

$$i_{d\,sat} \cong \frac{2q_s}{\lambda_c} \cong \frac{v_p - v_s}{\lambda_c},$$ (9.33)

which after denormalization simplifies to

$$I_{D\,sat} \cong W v_{sat}(-Q_{iS}) \cong nWC_{ox}v_{sat}(V_P - V_S)$$
$$\cong WC_{ox}v_{sat}(V_G - V_{T0} - nV_S).$$ (9.34)

Equation (9.34) shows that when the channel is under strong VS conditions, the drain current in saturation does not depend on the channel length anymore and varies linearly instead of quadratically with respect to the overdrive voltage. This can be explained by the fact that in such condition on one hand the inversion charge becomes almost uniform along the channel from source to drain and equal to the value taken at the source and on the other hand that the carriers are moving at their maximum velocity which is constant along the channel. The charge moving from source to drain per unit time corresponding to the drain current is therefore constant and independent of the channel length L for a given W and $V_P - V_S$.

9.1.2.2 Model 2

The discontinuity problem inherent to the simple piecewise linear velocity-field model of (9.4) or (9.5) can be avoided by using the continuous velocity-field model given by (9.9) or (9.10) with $\alpha = 1$. The longitudinal field E_x can be expressed in terms of the inversion charge density gradient by using (9.20)

$$|E_x| = \frac{d\Psi_s}{dx} = \frac{1}{nC_{ox}} \frac{dQ_i}{dx},$$ (9.35)

or in a normalized form

$$e \triangleq \frac{|E_x|}{E_c} = \frac{U_T}{E_c L} \frac{d\psi_s}{d\xi} = -\frac{2U_T}{E_c L} \frac{dq_i}{d\xi} = -\lambda_c \frac{dq_i}{d\xi}.$$ (9.36)

The normalized velocity and mobility can then be written as

$$v \triangleq \frac{v_{drift}}{v_{sat}} = \frac{-\lambda_c \frac{dq_i}{d\xi}}{1 - \lambda_c \frac{dq_i}{d\xi}},$$ (9.37a)

$$u \triangleq \frac{\mu_{eff}}{\mu_z} = \frac{1}{1 - \lambda_c \frac{dq_i}{d\xi}}.$$ (9.37b)

Replacing v in (9.18) by (9.37a) or equivalently u in (9.22) by (9.37b) results in

$$i_d = \frac{-2q_i}{1 - \lambda_c \frac{dq_i}{d\xi}} \frac{dq_i}{d\xi}.$$ (9.38)

Rearranging (9.38) leads to

$$i_d = -(2q_i - \lambda_c i_d)\frac{dq_i}{d\xi}. \tag{9.39}$$

Integrating (9.39) from source (where $\xi = 0$ and $q_i = q_s$) to drain (where $\xi = 1$ and $q_i = q_d$) leads to

$$
\begin{aligned}
i_d &= -\int_{q_s}^{q_d}(2q_i - \lambda_c i_d)\,dq_i \\
&= \int_{q_d}^{q_s}2q_i\,dq_i - \lambda_c i_d\int_{q_d}^{q_s}dq_i \\
&= q_s^2 - q_d^2 - \lambda_c i_d(q_s - q_d).
\end{aligned}
\tag{9.40}
$$

Solving (9.40) for i_d results in the current expression accounting for VS using the continuous velocity-field model (Model 2):

$$i_d = \frac{q_s^2 - q_d^2}{1 + \lambda_c(q_s - q_d)}. \tag{9.41}$$

The drain current given by (9.41) is plotted in Figure 9.6(b) and compared to the current without any VS effect. Note that the drain charge density q_d in (9.41) should be taken equal to (9.24) to account for the saturation of q_d to $q_{d\,sat}$ when $v_d \geq v_{d\,sat}$. If q_d is taken equal to $(v_p - v_d)/2$ instead (without accounting for the saturation), the current reaches a maximum at $v_d = v_{d\,sat}$ and then decreases as shown by the dashed line in Figure 9.6(b). Now, the current cannot actually decrease, but must saturate to $i_{d\,sat}$ for $v_d \geq v_{d\,sat}$. The charge at the drain must also saturate to $q_d = q_{d\,sat}$ for $v_d \geq v_{d\,sat}$. Also note that at the onset of saturation, the electric field right at the drain has to tend to infinity in order for the velocity to tend to the saturation velocity v_{sat}.

The saturation current is obtained by replacing i_d and q_d in (9.41) by $i_{d\,sat}$ and $q_{d\,sat}$, respectively, resulting in

$$i_{d\,sat} = \frac{q_s^2 - q_{d\,sat}^2}{1 + \lambda_c(q_s - q_{d\,sat})}. \tag{9.42}$$

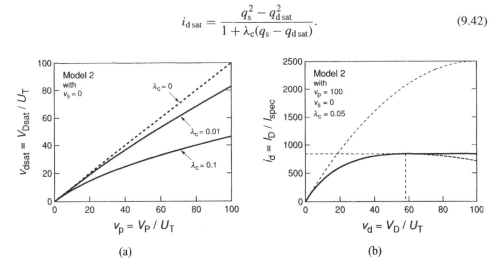

Figure 9.6 (a) Drain saturation voltage versus pinch-off voltage for Model 2 with $\lambda_c = 0,\ 0.01,\ 0.1$. (b) Drain current versus drain voltage for Model 2 with $\lambda_c = 0.05$

Equations (9.23), (9.42), and (9.25) can then be solved for $q_{d\,sat}$, $i_{d\,sat}$, and $v_{d\,sat}$, resulting in

$$q_{d\,sat} = \frac{1}{\lambda_c}\left(1 + \lambda_c q_s - \sqrt{1 + 2\lambda_c q_s}\right), \tag{9.43a}$$

$$i_{d\,sat} = \frac{2}{\lambda_c}q_{d\,sat} = \frac{2}{\lambda_c^2}\left(1 + \lambda_c q_s - \sqrt{1 + 2\lambda_c q_s}\right), \tag{9.43b}$$

$$v_{d\,sat} = v_p - 2q_{d\,sat} = v_p - \frac{2}{\lambda_c}\left(1 + \lambda_c q_s - \sqrt{1 + 2\lambda_c q_s}\right). \tag{9.43c}$$

The drain saturation voltage given by (9.43c) is plotted versus the pinch-off voltage in Figure 9.6(a) for two different values of λ_c. It clearly shows the reduction of the saturation voltage with respect to the pinch-off voltage, due to VS.

The profile of the inversion charge along the channel is obtained by integrating (9.39) from the source, where $q_i = q_s$, to a point in the channel, resulting in

$$i_d\xi = \frac{q_s^2 - q_i^2}{1 + \lambda_c(q_s - q_i)}. \tag{9.44}$$

Solving (9.44) for q_i results in

$$q_i(\xi) = \frac{\lambda_c}{2}\xi i_d + \sqrt{\left(q_s - \frac{\lambda_c}{2}\xi i_d\right)^2 - \xi i_d}, \tag{9.45}$$

where the current i_d is given by (9.41). Equation (9.45) is plotted in Figure 9.7(a) for different values of $v_p - v_d$ and in saturation (i.e., for $v_d > v_{d\,sat}$) for different values of λ_c in Figure 9.7(b).

The longitudinal electrical field is obtained by solving (9.18) and (9.12), resulting in

$$e(\xi) = \frac{\lambda_c i_d}{2q_i(\xi) - \lambda_c i_d}. \tag{9.46}$$

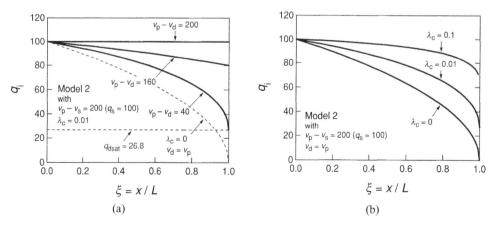

Figure 9.7 Inversion charge density versus position along the channel for Model 2: (a) for $v_p - v_s = 200$ ($q_s = 100$), $v_p - v_d = 200$, 160, 40, and $\lambda_c = 0.01$; (b) for $v_p - v_s = 200$, $v_d = v_p$ and $\lambda_c = 0$, 0.01, 0.1

9.1.2.3 Model 3

Finally, the effect on the drain current can also be evaluated for the velocity-field model given by (9.13) or its normalized form (9.14). The drain current can be derived in a similar way than for the Model 2 described above, resulting in

$$i_d = \frac{q_s^2 - q_d^2}{1 + \frac{\lambda_c}{2}(q_s - q_d)},$$ (9.47)

for $v_d < v_{d\,sat}$, whereas

$$i_d = i_{d\,sat} = \frac{q_s^2 - q_{d\,sat}^2}{1 + \frac{\lambda_c}{2}(q_s - q_{d\,sat})},$$ (9.48)

for $v_d \geq v_{d\,sat}$.

Equations (9.23), (9.48), and (9.25) can then be solved for $q_{d\,sat}$, $i_{d\,sat}$ and $v_{d\,sat}$, resulting in

$$q_{d\,sat} = \frac{\frac{\lambda_c}{2}q_s^2}{1 + \frac{\lambda_c}{2}q_s},$$ (9.49a)

$$i_{d\,sat} = \frac{2}{\lambda_c}q_{d\,sat} = \frac{q_s^2}{1 + \frac{\lambda_c}{2}q_s},$$ (9.49b)

$$v_{d\,sat} = v_p - 2q_{d\,sat} = v_p - \frac{\lambda_c q_s^2}{1 + \frac{\lambda_c}{2}q_s}.$$ (9.49c)

The drain saturation voltage $v_{d\,sat}$ for Model 3 given by (9.49c) and the drain saturation current for Model 3 given by (9.49b) are plotted in Figures 9.8(a) and 9.8(b), respectively.

The profile of the inversion charge for Model 3 is obtained in a similar way than for Model 2, resulting in

$$q_i(\xi) = \frac{\lambda_c}{4}\xi i_d + \sqrt{\left(q_s - \frac{\lambda_c}{4}\xi i_d\right)^2 - \xi i_d},$$ (9.50)

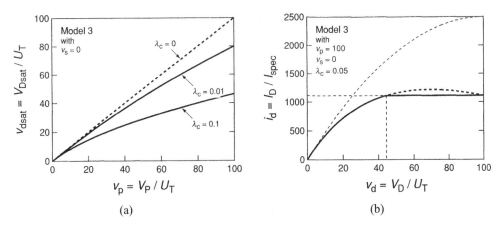

Figure 9.8 (a) Drain saturation voltage versus pinch-off voltage for Model 3 with $\lambda_c = 0$, 0.01, 0.1. (b) Drain current versus drain voltage for Model 3 with $\lambda_c = 0.05$

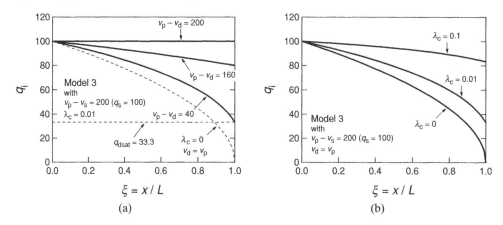

Figure 9.9   Inversion charge density versus position along the channel for Model 3: (a) for $v_p - v_s = 200$ $(q_s = 100)$, $v_p - v_d = 200,\ 160,\ 40$, and $\lambda_c = 0.01$; (b) for $v_p - v_s = 200$, $v_d = v_p$ and $\lambda_c = 0$, $0.01,\ 0.1$.

where i_d is given by (9.47). The inversion charge profile for Model 3 given by (9.50) is plotted in Figure 9.9.

The longitudinal electrical field is obtained by solving (9.18) and (9.16), resulting in

$$e(\xi) = \frac{2\lambda_c i_d}{4q_i(\xi) - \lambda_c i_d}. \tag{9.51}$$

9.1.2.4 Model comparison

The drain saturation voltages for the three different velocity-field models are plotted versus the pinch-off voltage in Figure 9.10(a). The piecewise linear model (Model 1) predicts the smallest saturation voltage, the continuous model (Model 2) the highest, and the third model is in between.

The drain currents using the three different velocity-field models are plotted in Figure 9.10(b). The piecewise linear model (Model 1) gives the highest saturation current, the continuous model (Model 2) gives the smallest, whereas the third model again lies in between. Again note that the CLM effect has not been accounted for and hence the currents remain constant in saturation. The CLM effect will be analyzed in Section 9.2.

The inversion charge profiles in saturation for the three different models are plotted in Figure 9.11(a). The three models look very similar, starting at $q_s = 100$ on the source and decreasing to the value of $q_{d\,sat}$ at the drain.

Finally, the longitudinal normalized electric field profiles in saturation are plotted for the three velocity-field models in Figure 9.11(b). For Model 1, the field reaches the critical field right at the drain. For Model 2, the field becomes infinity at the drain in order for the velocity to saturate. Finally, for Model 3, the field reaches twice the critical field at the drain, which for this model is the value at which the velocity saturates.

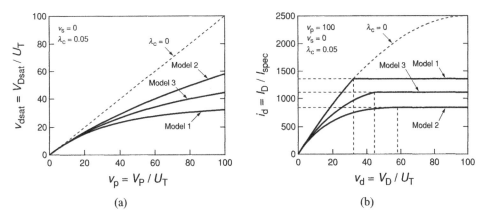

Figure 9.10 Comparison of the drain saturation voltages and of the drain saturation currents for the three velocity-field models for $\lambda_c = 0.05$

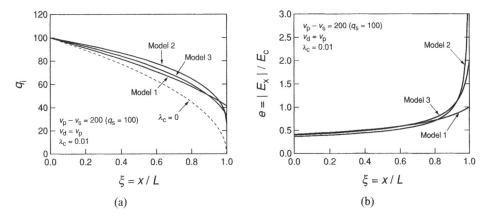

Figure 9.11 Comparison of the inversion charge and longitudinal field profile in saturation for the three velocity-field models

9.1.3 Effect of VS on the Transconductances

The transconductances are defined by (5.2). The normalized transconductances can be written as

$$g_{ms} \triangleq \frac{G_{ms}}{G_{spec}} = -\frac{\partial i_d}{\partial v_s} = -\frac{\partial i_d}{\partial q_s} \frac{\partial q_s}{\partial v_s}, \tag{9.52a}$$

$$g_{md} \triangleq \frac{G_{md}}{G_{spec}} = \frac{\partial i_d}{\partial v_d} = -\frac{\partial i_d}{\partial q_d} \frac{\partial q_d}{\partial v_d}, \tag{9.52b}$$

$$g_m \triangleq \frac{G_m}{G_{spec}} = \frac{\partial i_d}{\partial v_g} = \frac{\partial i_d}{\partial v_p} \frac{\partial v_p}{\partial v_g} = \frac{1}{n} \frac{\partial i_d}{\partial v_p}$$

$$= \frac{1}{n} \left(\frac{\partial i_d}{\partial q_s} \frac{\partial q_s}{\partial v_p} + \frac{\partial i_d}{\partial q_d} \frac{\partial q_d}{\partial v_p} \right), \tag{9.52c}$$

where G_{spec} is defined as

$$G_{\text{spec}} \triangleq \frac{I_{\text{spec}}}{U_T}. \tag{9.53}$$

From (9.27) and (9.24), below saturation the partial derivatives of q_s and q_d with respect to v_s and v_d respectively are given by

$$\frac{\partial q_s}{\partial v_s} = -\frac{1}{2}, \tag{9.54a}$$

$$\frac{\partial q_d}{\partial v_d} = -\frac{1}{2}, \tag{9.54b}$$

and therefore the normalized transconductances in strong inversion are simply given by

$$g_{\text{ms}} = \frac{1}{2} \frac{\partial i_d}{\partial q_s}, \tag{9.55a}$$

$$g_{\text{md}} = -\frac{1}{2} \frac{\partial i_d}{\partial q_d}. \tag{9.55b}$$

Note that the general relation (5.9) between G_m, G_{ms}, and G_{md} still holds, even though due to VS. the drain current cannot be split into a forward and reverse component that depend only on $v_p - v_s$ and $v_p - v_d$, respectively. But as long as the source and drain charges q_s and q_d depend only on the differences $v_p - v_s$ and $v_p - v_d$ respectively, the derivatives of the charges with respect to the pinch-off voltage are then given by

$$\frac{\partial q_s}{\partial v_p} = -\frac{\partial q_s}{\partial v_s} = \frac{1}{2} \tag{9.56a}$$

$$\frac{\partial q_d}{\partial v_p} = -\frac{\partial q_d}{\partial v_d} = \frac{1}{2}, \tag{9.56b}$$

where (9.54) has been used. It can then be shown from (9.52c), (9.55), and (9.56) that the gate transconductance is given by

$$g_m = \frac{g_{\text{ms}} - g_{\text{md}}}{n}, \tag{9.57}$$

even when VS occurs. In saturation, $g_{\text{md}} = 0$ and hence (9.57) reduces to

$$g_{\text{m sat}} = \frac{g_{\text{ms sat}}}{n}. \tag{9.58}$$

9.1.3.1 Model 1

With the piecewise linear velocity-field model, (9.55a) and (9.55b), for $v_d < v_{\text{d sat}}$, are then simply given by the long-channel values

$$g_{\text{ms}} = q_s, \tag{9.59a}$$

$$g_{\text{md}} = q_d. \tag{9.59b}$$

In saturation, i.e., for $v_d \geq v_{d\,sat}$, neglecting the effect of CLM, the drain current is clamped and held constant with respect to v_d to the value $i_{d\,sat}$. The drain transconductance G_{md} is therefore zero in saturation. The normalized source transconductance in saturation $g_{ms\,sat}$ is obtained by differentiation of (9.29b), resulting in

$$g_{ms\,sat} \triangleq \frac{\partial i_{d\,sat}}{\partial v_s} = \frac{1}{2}\frac{\partial i_{d\,sat}}{\partial q_s} = \frac{q_s}{\sqrt{1+(\lambda_c q_s)^2}}. \tag{9.60}$$

Due to VS, the simple piecewise model shows that the source transconductance is lowered by $\sqrt{1+(\lambda_c q_s)^2}$ compared to the value without VS. As mentioned above in the case of strong VS conditions, the saturation drain current becomes linear with q_s and $V_P - V_S$, resulting in a constant value of the saturation transconductance $g_{ms\,sat}$ inversely proportional to λ_c

$$g_{ms\,sat} \cong \frac{1}{\lambda_c} = \frac{v_{sat}L}{2\mu_z U_T} \qquad \text{for } \lambda_c q_s \gg 1, \tag{9.61}$$

or in denormalized form

$$G_{ms\,sat} = G_{spec} \cdot g_{ms\,sat} \cong \frac{G_{spec}}{\lambda_c} = n \cdot C_{ox} \cdot W \cdot v_{sat}. \tag{9.62}$$

The transconductance-to-current ratio in saturation can be derived from (9.60) and (9.29b) as

$$\frac{g_{ms\,sat}}{i_{d\,sat}} = \frac{\lambda_c^2 q_s}{2\left[1+(\lambda_c q_s)^2 - \sqrt{1+(\lambda_c q_s)^2}\right]}, \tag{9.63}$$

For $\lambda_c q_s \gg 1$, (9.63) reduces to

$$\frac{g_{ms\,sat}}{i_{d\,sat}} \cong \frac{1}{2q_s} = \frac{1}{\lambda_c i_{d\,sat}}, \tag{9.64}$$

which decreases inversely proportional to $i_{d\,sat}$ instead of $\sqrt{i_{d\,sat}}$ as it would when VS is not present.

9.1.3.2 Model 2

The normalized source, drain, and gate transconductances for $v_d < v_{d\,sat}$ in the case of the continuous velocity-field model (9.9) or (9.10) are given by

$$g_{ms} = \frac{q_s + \frac{\lambda_c}{2}(q_s - q_d)^2}{[1+\lambda_c(q_s - q_d)]^2} = \frac{\frac{v_p - v_s}{2} + \frac{\lambda_c}{8}(v_d - v_s)^2}{\left[1+\frac{\lambda_c}{2}(v_d - v_s)\right]^2}, \tag{9.65a}$$

$$g_{md} = \frac{q_d - \frac{\lambda_c}{2}(q_s - q_d)^2}{[1+\lambda_c(q_s - q_d)]^2} = \frac{\frac{v_p - v_d}{2} - \frac{\lambda_c}{8}(v_d - v_s)^2}{\left[1+\frac{\lambda_c}{2}(v_d - v_s)\right]^2}, \tag{9.65b}$$

$$g_m = \frac{q_s - q_d}{n}\frac{1}{1+\lambda_c(q_s - q_d)} = \frac{v_d - v_s}{2n}\frac{1}{1+\frac{\lambda_c}{2}(v_d - v_s)}. \tag{9.65c}$$

Similar to Model 1, in saturation, the drain current $i_{d\,sat}$ is clamped to a constant value with respect to v_d corresponding to its maximum value taken at $v_d = v_{d\,sat}$ and remains constant for $v_d \geq v_{d\,sat}$. The drain transconductance G_{md} is therefore zero and the normalized source transconductance in saturation $g_{ms\,sat}$ becomes

$$
g_{ms\,sat} \triangleq \frac{\partial i_{d\,sat}}{\partial v_s} = \frac{1}{2}\frac{\partial i_{d\,sat}}{\partial q_s} = \frac{1}{\lambda_c}\left[1 - \frac{1}{\sqrt{1 + 2\lambda_c q_s}}\right]
$$

$$
= \frac{2q_s}{1 + 2\lambda_c q_s + \sqrt{1 + 2\lambda_c q_s}} \cong \frac{1}{\lambda_c} \quad \text{for } q_s \gg 1,
$$

(9.66)

which can be approximated by

$$
g_{ms\,sat} \cong \frac{q_s}{1 + \frac{3}{2}\lambda_c q_s}.
$$

(9.67)

As mentioned above, for $\lambda_c q_s \gg 1$, $g_{ms\,sat}$ saturates to a constant value $1/\lambda_c$. The normalized gate transconductance is then simply given by (9.58).

The transconductance-over-current ratio in saturation is then given from (9.43b) and (9.66) as

$$
\frac{g_{ms\,sat}}{i_{d\,sat}} = \frac{\lambda_c}{1 + 2\lambda_c q_s - \sqrt{1 + 2\lambda_c q_s}} \cong \frac{1}{2q_s} = \frac{1}{\lambda_c i_{d\,sat}} \quad \text{for } q_s \gg 1,
$$

(9.68)

which has the same asymptote than for Model 1.

9.1.3.3 Model 3

The normalized source and drain transconductances for the third velocity-field model (9.13) or (9.14) and for $v_d < v_{d\,sat}$ are given by

$$
g_{ms} = \frac{q_s + \frac{\lambda_c}{4}(q_s - q_d)^2}{\left[1 + \frac{\lambda_c}{2}(q_s - q_d)\right]^2} = \frac{\frac{v_p - v_s}{2} + \frac{\lambda_c}{16}(v_d - v_s)^2}{\left[1 + \frac{\lambda_c}{4}(v_d - v_s)\right]^2},
$$

(9.69a)

$$
g_{md} = \frac{q_d - \frac{\lambda_c}{4}(q_s - q_d)^2}{\left[1 + \frac{\lambda_c}{2}(q_s - q_d)\right]^2} = \frac{\frac{v_p - v_d}{2} - \frac{\lambda_c}{16}(v_d - v_s)^2}{\left[1 + \frac{\lambda_c}{4}(v_d - v_s)\right]^2},
$$

(9.69b)

$$
g_m = \frac{q_s - q_d}{n}\frac{1}{1 + \frac{\lambda_c}{2}(q_s - q_d)} = \frac{v_d - v_s}{2n}\frac{1}{1 + \frac{\lambda_c}{4}(v_d - v_s)}.
$$

(9.69c)

As for Models 1 and 2, in saturation, the drain current is constant with respect to v_d and hence the drain transconductance is zero. The source transconductance is then obtained by differentiating (9.49b), resulting in

$$
g_{ms\,sat} \triangleq \frac{\partial i_{d\,sat}}{\partial v_s} = \frac{1}{2}\frac{\partial i_{d\,sat}}{\partial q_s} = \frac{q_s\left(1 + \frac{\lambda_c}{4}q_s\right)}{\left(1 + \frac{\lambda_c}{2}q_s\right)^2} \cong \frac{1}{\lambda_c} \quad \text{for } q_s \gg 1.
$$

(9.70)

The transconductance-over-current ratio in saturation is then given from (9.49b) and (9.70) as

$$\frac{g_{\text{ms sat}}}{i_{d\,\text{sat}}} = \frac{1 + \frac{\lambda_c}{4}q_s}{q_s \left(1 + \frac{\lambda_c}{2}q_s\right)} \simeq \frac{1}{2q_s} = \frac{1}{\lambda_c i_{d\,\text{sat}}} \quad \text{for } q_s \gg 1. \tag{9.71}$$

9.1.3.4 Model comparison

The normalized source transconductances in saturation for the three models are plotted versus the normalized pinch-off voltage in Figure 9.12(a) for $\lambda_c = 0.1$. As mentioned at the end of Section 9.1.2.1, the drain current in saturation becomes linear instead of quadratic with respect to the overdrive voltage $V_P - V_S$ (c.f. equation (9.34)) and hence the source transconductance in saturation saturates to a constant value equal to $1/\lambda_c$ as soon as velocity saturates.

The degradation of the source transconductance due to VS can be evaluated by defining the ratio of the source transconductance in saturation to the source transconductance in saturation without the effect of VS (which actually is simply equal to q_s)

$$\chi_{\text{ms sat}} \triangleq \frac{g_{\text{ms sat}}}{g_{\text{ms sat}}|_{\lambda_c=0}} = \frac{g_{\text{ms sat}}}{q_s}. \tag{9.72}$$

For the first piecewise linear model (Model 1), the degradation is given by

$$\chi_{\text{ms sat}} = \frac{1}{\sqrt{1 + (\lambda_c q_s)^2}}, \tag{9.73}$$

whereas for the continuous model (Model 2), it is given by

$$\chi_{\text{ms sat}} = \frac{2}{1 + 2\lambda_c q_s + \sqrt{1 + 2\lambda_c q_s}}$$
$$\simeq \frac{1}{1 + \frac{3}{2}\lambda_c q_s} \simeq \frac{1}{1 + \frac{3}{4}\lambda_c(v_p - v_s)}. \tag{9.74}$$

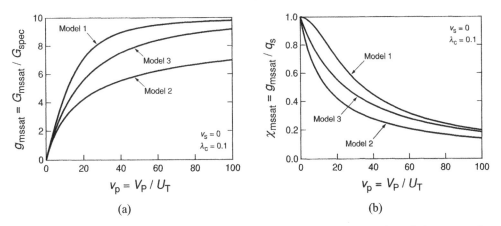

Figure 9.12   (a) Source transconductance and (b) source transconductance degradation versus the pinch-off voltage for the three velocity-field models

The transconductance degradation function for the third velocity-field model is given by

$$\chi_{\text{ms sat}} = \frac{1 + \frac{\lambda_c}{4} q_s}{\left(1 + \frac{\lambda_c}{2} q_s\right)^2}. \tag{9.75}$$

Figure 9.12(b) illustrates the degradation of the source transconductance in saturation due to VS with respect to the pinch-off voltage for the three velocity-field models. They all show quite a dramatic impact of VS on the transconductance. As an example, for a channel length $L = 0.1$ μm, corresponding to $\lambda_c \cong 0.5$, the reduction can be larger than a factor 10 for a pinch-off voltage of about 1 V (corresponding to $v_p \cong 40$ and $q_s \cong 20$). This obviously has a significant impact on power consumption for a given cutoff frequency. As explained in Section 9.4, it also has an important impact on the thermal noise.

9.2 CHANNEL LENGTH MODULATION

The CLM effect was already introduced in Section 4.6 using a very simple model that ignored the effect of VS. But obviously CLM is tightly linked to the effect of VS since carriers enter into velocity in the high longitudinal field region close to the drain. As shown in Figure 9.13(a), this effect can be explained simply by splitting the source to drain region into a nonsaturated region (the *channel region*) and a velocity saturation region (VSR) close to the drain [115]. The length of the saturation region ΔL where the carrier travels at saturated velocity depends on the longitudinal field and hence on the drain and gate bias voltages. The effective length where the carrier velocity is not saturated is therefore smaller than the length L between the source and drain junctions.

The first-order model introduced in Section 4.6 ignored VS and estimated the channel length reduction based on the length of the abrupt depletion regions on the source and drain sides of the channel. In this section we will derive a more accurate model that also accounts for the VS effect described in the previous section. To do this, we will use the third velocity-field model since the continuous model requires the field to become infinity for the carrier velocity to reach saturation, whereas the piecewise linear model is not accurate enough.

A schematic diagram of the VS region is shown in Figure 9.13(b). The VS region length can be derived by applying the Gauss' law to the ABCD box. Following the derivation made

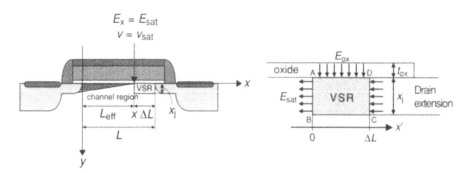

Figure 9.13   (a) VS region and channel region definition. (b) Schematic diagram of the VS region [115]

in [115], we will assume that (a) the carriers in the VS region are traveling with a saturated velocity; (b) the junction on the drain extension region is abrupt and the drain extension is heavily doped and hence is perfectly conducting; and (c) the current flows no deeper than the junction depth and is confined in the box. To simplify the derivation, we also change the coordinate system as indicated in Figure 9.13(b). To apply Gauss' law, we will further assume that the field is independent of y and that the field lines crossing the BC boundary contribute very little. This results in

$$-\epsilon_{si}x_j W E_{sat} + \epsilon_{si}x_j W E_x(x') + \epsilon_{ox}W \int_0^{x'} E_{ox}(x') \, dx' = Q_{box}Wx', \qquad (9.76)$$

where $E_{ox}(x')$ is simply given by

$$E_{ox}(x') = \frac{V_G - V_{FB} - \Psi_0 - V(x')}{t_{ox}}, \qquad (9.77)$$

and $Q_{box} \triangleq qn_px_j + qN_bx_j$ corresponds to the charge per unit area in the VSR with n_p being the electron concentration and N_b the fixed charge concentration.

Since the current and velocity are constant in the VSR, n_p is also constant in the VSR and hence differentiating (9.76) with respect to x' results in

$$\epsilon_{si}x_j \frac{dE_x(x')}{dx'} + \epsilon_{ox}E_{ox}(x') = Q_{box}. \qquad (9.78)$$

Replacing E_{ox} in (9.78) by (9.77) results in

$$\epsilon_{si}x_j \frac{dE_x(x')}{dx'} + C_{ox}[V_G - V_{FB} - \Psi_0 - V(x')] = Q_{box}. \qquad (9.79)$$

On the left side of the VS region, the gradual channel approximation applies and the gradient of the longitudinal field can be ignored compared to the gradient of the vertical field. That is, in this region, all the silicon charges are controlled by the vertical field only. This approximation also applies to the boundary at $x' = 0$, where the channel voltage $V(x' = 0)$ is equal to the saturation voltage $V_{D\,sat}$. The gradient of the longitudinal field $dE_x(x')/dx'$ in (9.79) can hence be neglected and (9.79) then simplifies to

$$\epsilon_{ox}E_{ox}(x' = 0) = C_{ox}(V_G - V_{FB} - \Psi_0 - V_{D\,sat}) = Q_{box}. \qquad (9.80)$$

The saturation voltage $V_{D\,sat}$ normalized to U_T is given by (9.49c). Replacing Q_{box} in (9.79) by (9.80) results in

$$\frac{dE_x(x')}{dx'} = \frac{V(x') - V_{D\,sat}}{\ell^2}, \qquad (9.81)$$

where

$$\ell \triangleq \sqrt{\frac{\epsilon_{si}}{\epsilon_{ox}}t_{ox}x_j} = \sqrt{\frac{\epsilon_{si}x_j}{C_{ox}}}. \qquad (9.82)$$

Differential equation (9.81) can be solved by applying the boundary conditions $E(x' = 0) = E_{sat} = 2E_c$ and $V(x' = 0) = V_{D\,sat}$, resulting in

$$E_x(x') = E_{sat} \cosh\left(\frac{x'}{\ell}\right) \tag{9.83}$$

and

$$V(x') = V_{D\,sat} + \ell E_{sat} \sinh\left(\frac{x'}{\ell}\right). \tag{9.84}$$

Equation (9.83) indicates that the field increases almost exponentially close to the drain where it becomes maximum at the end of the VS region and is given by

$$E_x(x' = \Delta L) = E_{max} = E_{sat} \cosh\left(\frac{\Delta L}{\ell}\right), \tag{9.85}$$

whereas

$$V(x' = \Delta L) = V_D = V_{D\,sat} + \ell E_{sat} \sinh\left(\frac{\Delta L}{\ell}\right). \tag{9.86}$$

Equations (9.85) and (9.86) can then be solved for ΔL and E_{max}

$$\Delta L = \ell \operatorname{asinh}(u) = \ell \ln\left(u + \sqrt{u^2 + 1}\right), \tag{9.87a}$$

$$E_{max} = E_{sat}\sqrt{u^2 + 1}, \tag{9.87b}$$

where

$$u \triangleq \frac{V_D - V_{D\,sat}}{\ell E_{sat}}. \tag{9.88}$$

The channel length reduction normalized to ℓ is plotted versus u in Figure 9.14.

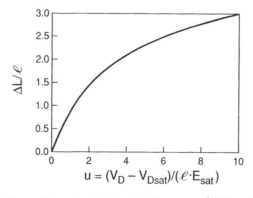

Figure 9.14   Channel length reduction $\Delta L/\ell$ versus $u \triangleq (V_D - V_{D\,sat})/(\ell E_{sat})$

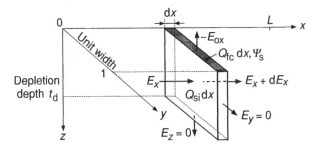

Figure 9.15 Application of Gauss' law to calculate Q_{si}

9.3 DRAIN INDUCED BARRIER LOWERING

9.3.1 Introduction

With the gradual channel approximation used to develop the long-channel model in Part I, the second derivative of the channel potential (first derivative of the electric field) was assumed to be negligible along the channel. As a consequence, the charge density in silicon Q_{si} depended only on the vertical surface field as illustrated in Figure 3.2. This approximation is no longer acceptable for short-channel transistors, and the variation of the horizontal field E_x must be included in the calculation of Q_{si}, as illustrated in Figure 9.15. This figure depicts an elementary volume of unit width and of length dx. Its depth is t_d, the depletion depth (or depletion thickness), beyond which the potential is assumed to be constant ($\Psi = 0$).

The negative elementary charge $Q_{si}\,dx$ enclosed in this volume is increased in absolute value by the difference $dE_x < 0$ of the horizontal field created by the drain voltage. Alternately, the value of E_{ox} needed to obtain a given charge density is reduced, corresponding to an increase of surface potential. Hence the name *drain induced barrier lowering* (DIBL) given to this effect.

9.3.2 Evaluation of the Surface Potential

Following the analysis carried out in [116], the application of Gauss' law to the elementary volume depicted in Figure 9.15 results in

$$(Q_{si} + Q_{fc})\,dx = -\epsilon_{ox} E_{ox}\,dx + \epsilon_{si} t_d\,dE_x, \tag{9.89}$$

where dE_x is assumed to be constant across the depletion depth t_d, with

$$\frac{dE_x}{dx} = -\frac{d^2\Psi_s}{dx^2} < 0. \tag{9.90}$$

By introducing expression (2.3) of E_{ox} and definition (3.22) of V_{FB}, we obtain a second-order differential equation of Ψ_s:

$$-L_c^2 \frac{d^2\Psi_s}{dx^2} + \Psi_s = V_G - V_{FB} + \frac{Q_{si}}{C_{ox}}, \tag{9.91}$$

where L_c is a characteristic length defined by

$$L_c \triangleq \sqrt{\frac{\epsilon_{si}t_d}{C_{ox}}}. \tag{9.92}$$

Now, t_d is itself slightly dependent on the surface potential Ψ_s; hence, equation (9.91) is nonlinear. This dependency is weak; therefore, we shall assume L_c constant in order to integrate the equation.

When the surface potential is constant along the channel, this equation has the particular solution,

$$\Psi_s = \Psi_{sl} \triangleq V_G - V_{FB} + \frac{Q_{si}(\Psi_{sl})}{C_{ox}}, \tag{9.93}$$

which is also the solution for a long channel corresponding to (3.19). Constant Ψ_s is possible only in weak inversion, or in strong inversion with $V_D = V_S$.

The general solution of (9.91) is then

$$\Psi_s(x) = \Psi_{sl} + [\Psi_s(0) - \Psi_{sl}] \frac{\sinh \frac{L-x}{L_c}}{\sinh \frac{L}{L_c}} + [\Psi_s(L) - \Psi_{sl}] \frac{\sinh \frac{x}{L_c}}{\sinh \frac{L}{L_c}}, \tag{9.94}$$

where $\Psi_s(0)$ and $\Psi_s(L)$ are the source and drain potentials. According to Figures 4.12 and 4.13, these can be expressed as

$$\Psi_s(0) = \Phi_B + V_S, \qquad \Psi_s(L) = \Phi_B + V_D, \tag{9.95}$$

where Φ_B is the junction potential barrier given by (4.55).

By introducing the normalized variables

$$\xi = x/L \quad \text{and} \quad \lambda = L/L_c, \tag{9.96}$$

equation (9.94) becomes

$$\Psi_s(\xi) = \Psi_{sl} + (\Phi_B + V_S - \Psi_{sl}) \frac{\sinh[\lambda(1-\xi)]}{\sinh \lambda} + (\Phi_B + V_D - \Psi_{sl}) \frac{\sinh(\lambda\xi)}{\sinh \lambda}. \tag{9.97}$$

It can be pointed out that the particular solution (9.93) used to integrate (9.91) corresponds to an equipotential channel ($V_S = V_D = V$) with $\Psi_s = \Psi_{sl} = \Phi_B + V$.

Now, for λ larger than a few units, this result can be approximated by

$$\Psi_s(\xi) \cong \Psi_{sl} + \underbrace{(\Phi_B + V_S - \Psi_{sl})e^{-\lambda\xi} + (\Phi_B + V_D - \Psi_{sl})e^{-\lambda(1-\xi)}}_{\triangleq \Delta\Psi_s(\xi)}, \tag{9.98}$$

where $\Delta\Psi_s(\xi)$ is the increase of surface potential with respect to the long-channel approximation.

Although L_c given by (9.92) has been assumed to be constant while integrating (9.91), it is slightly dependent on Ψ_s through t_d. By introducing (9.92) in (9.96) with t_d given by (3.26), we obtain

$$\lambda = \frac{L}{L_c} = L\sqrt[4]{\frac{q N_b C_{ox}^2}{2\Psi_s \epsilon_{si}^3}}, \qquad (9.99)$$

showing that λ is only a weak function of Ψ_s.

At the onset of inversion for $V_S = V_D = 0, \Psi_s = \Psi_P = \Psi_0$, which corresponds to a particular value of λ:

$$\lambda_0 \triangleq \frac{L}{L_{c0}} = L\sqrt[4]{\frac{q N_b C_{ox}^2}{2\Psi_0 \epsilon_{si}^3}} = L\sqrt{\frac{q N_b}{\epsilon_{si} \Gamma_b}} \Psi_0^{-1/4}. \qquad (9.100)$$

For the general case, we shall assume that the depletion depth is essentially controlled by $\Psi_s = \Psi_P$ and use the following approximation:

$$\lambda = \frac{L}{L_c} \cong \lambda_0 \left(\frac{\Psi_0}{\Psi_P}\right)^{1/4}. \qquad (9.101)$$

In reality, even in weak inversion, the surface potential with DIBL is not constant all along the channel. This could be accounted for by a fitting parameter in (9.101) [116] which we shall not introduce here.

In weak inversion, the surface potential for a long channel is constant: $\Psi_{sl} = \Psi_P$. The variation of surface potential due to DIBL given by (9.97) is plotted in Figure 9.16 for two values of pinch-off voltage and two values of drain voltage.

At both ends of the channel, the surface potential drops from $\Phi_B + V_{S,D}$ to $\Psi_P = \Psi_0 + V_P$ within a distance characterized by λ. For a long channel ($L \gg L_c$), this distance is negligible, and the surface potential is equal to Ψ_P over most of the channel length. On the contrary, for a short channel, the surface potential never reaches Ψ_P; the barrier for the current carriers is lowered and the current is increased. Furthermore, the amount of lowering depends on the drain voltage, as we can see in the figure. Therefore, even if the device is saturated (as is the case in the figure since $V_D - V_S \gg U_T$), the drain current keeps increasing with the drain voltage. It should be noticed that, for the values of N_b and C_{ox} used in the example of Figure 9.16, a channel length $L = 0.5\,\mu m$ is already too short to avoid DIBL.

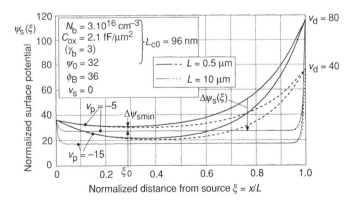

Figure 9.16 Surface potential in weak inversion: short channel ($L = 0.5\,\mu m$) compared with long channel ($L = 10\,\mu m$). All voltages are normalized to U_T

The increase of surface potential, $\Delta\Psi_s$, has a minimum at a position ξ_0 that can be calculated by differentiating (9.98). Neglecting again the variation of λ, we obtain

$$\xi_0 = \frac{1}{2}\left(1 - \frac{1}{\lambda}\ln\frac{\Phi_B + V_D - \Psi_{sl}}{\Phi_B + V_S - \Psi_{sl}}\right), \tag{9.102}$$

which introduced in (9.98) gives

$$\Delta\Psi_{s\,min} = 2e^{-\lambda/2}\sqrt{(\Phi_B + V_S - \Psi_{sl})(\Phi_B + V_D - \Psi_{sl})}. \tag{9.103}$$

In a long-channel transistor, when the pinch-off voltage $V_P = \Psi_P - \Psi_0$ is increased and approaches V_S, the inverted charge at the source end of the channel is no longer negligible and the device enters moderate inversion. The local surface potential stops the following Ψ_P to finally saturate at $\Psi_0 + V_S$. As V_P keeps increasing, this saturation (to the local value of channel voltage V) extends progressively to the whole channel. Hence, an increasing part of the channel has a variable surface potential (except for $V_D = V_S$), and the validity of the previous analysis is progressively lost.

Since (9.93) remains a solution of (9.91) as long as its second derivative is negligible, we shall extend (9.103) to moderate inversion by limiting Ψ_{sl} to its value at the source

$$\Psi_{sl} = \Psi_P + \frac{Q_{iS}}{nC_{ox}}, \tag{9.104}$$

or, with normalized variables

$$\psi_{sl} = \psi_p - 2q_s, \tag{9.105}$$

according to (3.39). The dependency of q_s on v_s is given by (3.48), the general relation between voltages and charge. Unfortunately, this relation cannot be inverted to provide the charge from the voltages, but we can use the following approximation:

$$q_i \cong \ln\left(1 + \exp\frac{v_p - v - 1}{2}\right), \tag{9.106}$$

which is good in strong and moderate inversion, when the charge cannot be neglected, as shown in Figure 9.17.

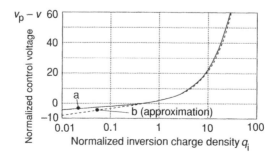

Figure 9.17 Normalized charge–voltage characteristics; (curve a) original relation (3.48); (curve b) approximation (9.106)

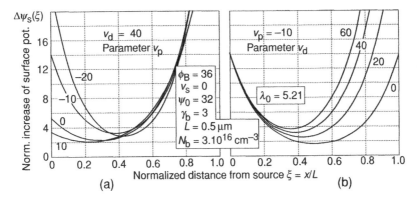

Figure 9.18 Increase of surface potential due to DIBL

Introducing this approximation in (9.105) with $\psi_p = \psi_0 + v_p$ gives

$$\psi_{sl} = \psi_0 + v_p - 2\ln\left(1 + \exp\frac{v_p - v_s - 1}{2}\right). \tag{9.107}$$

As required, this expression tends to ψ_p in weak inversion, and to $\psi_0 + v_s$ in strong inversion.

The increase of surface potential $\Delta\psi_s$ can now be calculated by introducing (9.107) in the normalized form of (9.98). It is plotted in Figure 9.18(a) with the same parameters as in Figure 9.16, and for a fixed value of v_d and several values of v_p ranging from weak inversion to moderate inversion.

In Figure 9.18(b), v_p is fixed in weak inversion and $\Delta\psi_s$ is plotted for various values of v_d.

The increase of surface potential goes through a minimum given by (9.103). This minimum is plotted in Figure 9.19 for the same numerical values. As can be expected, this minimum increases with the drain voltage. As the pinch-off voltage increases from very negative values (corresponding to very small currents), $\Delta\psi_{s\,min}$ first increases until it reaches a maximum. It then decreases as moderate inversion is approached.

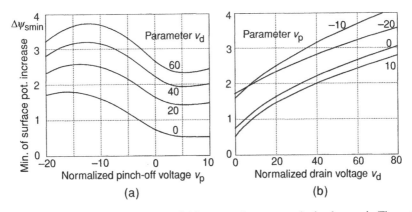

Figure 9.19 Minimum of surface potential increase. Same numerical values as in Figure 9.18

9.3.3 Effect on the Drain Current

In weak inversion, the surface potential for a long channel is constant and equal to Ψ_P. Before diffusing to the drain, the current carriers must pass a potential barrier of height $\Phi_B + V_S - \Psi_P$. As shown in Figure 9.16, when the channel is shortened and L approaches the characteristic length L_{c0} defined by (9.100), the surface potential cannot reach Ψ_P and its minimum is $\Delta\Psi_{s\,min}$ higher. This corresponds to a lowering of the barrier by $\Delta\Psi_{s\,min}$, which (in weak inversion) has an exponential effect on Q_i, the density of mobile charge. The position of the barrier also moves away from the source, thereby reducing the effective channel length, but this linear effect can be neglected in a first approximation.

Hence, the effect of DIBL on weak inversion current can be accounted for by adding $\Delta\Psi_{s\,min}$ given by (9.103) to Ψ_P or V_P in the equation of the current for a long channel.

In strong inversion, the surface field increases with the inversion coefficient; hence, the relative importance of the lateral electric flux in Figure 9.15 is progressively reduced. Hence, although the previous analysis loses its validity, we shall also apply it to strong inversion for the calculation of current, as a very first approximation. The general current–voltages relation (4.25) and the alternative continuous voltages–current approximation (4.39) then become respectively:

$$v_p - v_{s,d} + \Delta\psi_{s\,min} = \sqrt{1 + 4i_{f,r}} + \ln\left(\sqrt{1 + 4i_{f,r}} - 1\right) - (1 + \ln 2), \tag{9.108}$$

$$i_{f,r} = \ln^2\left(1 + \exp\frac{v_p - v_{s,d} + \Delta\psi_{s\,min}}{2}\right). \tag{9.109}$$

Using this last equation, the gate-to-drain characteristics in saturation ($i_d = i_f$ since $v_p < v_d$) are plotted in Figure 9.20(a) for several values of λ_0. The relative increase of current is plotted in Figure 9.20(b). It is large in weak inversion and decreases at the approach of moderate inversion. As mentioned before, the results lose their validity for $V_P > 0$, although the qualitative tendency is certainly correct. Furthermore, other short-channel effects (for example, VS) are then combined with DIBL.

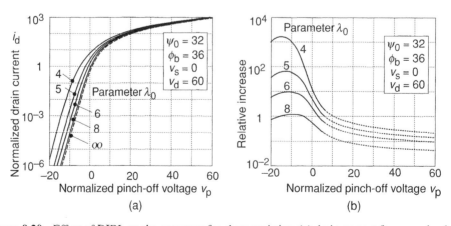

Figure 9.20 Effect of DIBL on the gate transfer characteristics: (a) drain current for several values of λ_0; (b) relative increase of current

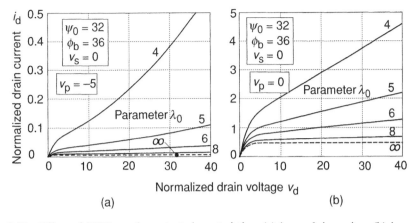

Figure 9.21 Effect of DIBL on the output characteristics: (a) in weak inversion; (b) in moderate inversion

As can be seen, the current increases very abruptly with the reduction of length: 1 to 1000 for λ_0 reduced from 8 to 4. It would be almost negligible (less than 10%) for $\lambda_0 > 12$.

It must be pointed out that, since $\Delta\Psi_{s\,min}$ is a function of all the bias voltages (V_S, V_D, and V_G), I_F becomes dependent on V_D, and I_R becomes dependent on V_S. As predicted in Section 4.5.3, the fundamental property is degraded. As a result, the output conductance in saturation is drastically increased, as we can see in the output characteristics plotted in Figures 9.21(a) and 9.21(b) for two different values of v_p. Again, this increase is maximum in weak inversion. As a matter of fact, since $\Delta\Psi_{s\,min}$ increases approximately linearly with the drain voltage (see Figure 9.19(b)), the drain current in weak inversion increases exponentially with the drain voltage, as was already pointed out in 1973 [10].

The characteristics of Figure 9.21 have been obtained by $i_d = i_f - i_r$, although the superposition property is no longer valid here. However, it affects only the low-voltage part of the characteristics ($v_d - v_s < 5$) for which i_r is not negligible.

Figure 9.22 shows the source-to-drain transfer characteristics also plotted from (9.109) (the curves correspond to saturation: $i_d = i_f$ except for $v_s > v_d - 5$). As for the gate-to-drain

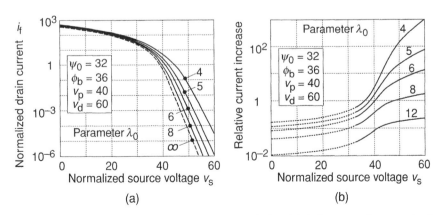

Figure 9.22 Effect of DIBL on source transfer characteristics: (a) drain current for several values of λ_0; (b) relative increase of current

characteristics of Figure 9.20, the current is significantly increased in weak inversion for $\lambda_0 < 8$.

9.3.4 Effect on Small-Signal Parameters in Weak Inversion

As already pointed out, the previous analysis of DIBL is valid only in weak inversion, where its effect is maximum. Therefore, the discussion of the related small-signal parameters will be limited to weak inversion. In addition, we shall consider only the saturated transistor with $i_d = i_f$. This forward component of current given by (9.109) can be approximated by

$$i_f = \exp{(v_p - v_s + \Delta\psi_{s\,min})}, \tag{9.110}$$

where $\Delta\psi_{s\,min}$ is given by (9.103) with $\psi_{sl} = \psi_p = \psi_0 + v_p$.

The differentiation of (9.110) with respect to v_s provides the source transconductance in weak inversion (normalized to G_{spec} given by (5.6)):

$$g_{ms} = i_f\left(1 - \frac{d\Delta\psi_{s\,min}}{dv_s}\right) = i_f\left(1 - e^{-\lambda/2}\sqrt{\frac{\phi_b - \psi_0 - v_p + v_d}{\phi_b - \psi_0 - v_p + v_s}}\right). \tag{9.111}$$

Due to DIBL, I_F also depends on the drain voltage. The drain current therefore keeps increasing in saturation, as shown in Figure 9.21. The corresponding output conductance (which is actually a component of the drain transconductance that is proportional to I_F) is obtained by differentiating (9.110) with respect to v_d:

$$g_{md\,sat} = i_f\frac{d\Delta\psi_{s\,min}}{dv_d} = i_f\,e^{-\lambda/2}\sqrt{\frac{\phi_b - \psi_0 - v_p + v_s}{\phi_b - \psi_0 - v_p + v_d}}. \tag{9.112}$$

Finally, the gate transconductance in weak inversion is obtained by differentiating (9.110) with respect to v_p. Knowing that $dv_p/dv_g = n$ and that λ also depends on v_p according to (9.101), we obtain

$$
\begin{aligned}
ng_m &= i_f\left(1 + \frac{d\Delta\psi_{s\,min}}{dv_p}\right) \\
&\quad i_f\left[1 + e^{-\lambda/2}\left(\frac{\lambda R(v_p, v_s, v_d)}{4(\psi_0 + v_p)} - \frac{2(\phi_b - \psi_0 - v_p) + v_s + v_d}{R(v_p, v_s, v_d)}\right)\right],
\end{aligned} \tag{9.113}
$$

where

$$R(v_p, v_s, v_d) = \sqrt{(\phi_b - \psi_0 - v_p + v_s)(\phi_b - \psi_0 - v_p + v_d)}. \tag{9.114}$$

The three transconductance to current ratios given by (9.111), (9.112), and (9.113) are plotted in Figure 9.23 as functions of the drain voltage for two values of λ_0. Their variation with the source voltage (at constant $v_p - v_s$) is shown in Figure 9.24.

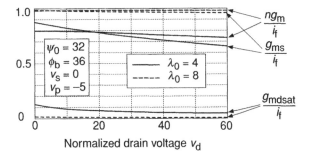

Figure 9.23   Effect of DIBL on transconductance in weak inversion: dependency on drain voltage

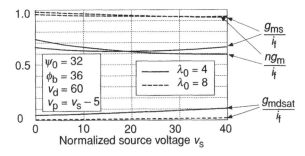

Figure 9.24   Effect of DIBL on transconductance in weak inversion: dependency on source voltage (with constant $v_p - v_s$)

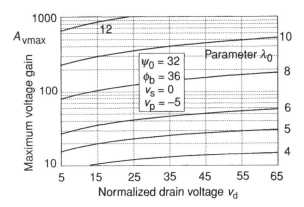

Figure 9.25   Effect of DIBL on maximum voltage gain

Although DIBL may increase the current in weak inversion by several orders of magnitude, g_m/i_f and g_{ms}/i_f are reduced only by less than 40%, even with λ_0 as low as 4. The reduction is negligible for $\lambda_0 > 8$. Hence, if the device is biased at a constant current (as should always be the case in weak inversion), the effect of DIBL on source and gate transconductance is not very significant.

As could be expected from the output characteristics of Figure 9.21, the main problem lies in $g_{md\,sat}$, the residual conductance in saturation. For a long channel, the conductance g_{ds} due to CLM is much smaller than g_{ms}; therefore, the small-signal voltage gain given by (5.24) can be very large. With the DIBL occurring in a short channel, the voltage gain is limited to

$$A_{v\,max} = \frac{G_{ms}}{G_{md\,sat}} = \frac{g_{ms}}{g_{md\,sat}}. \tag{9.115}$$

As shown by Figure 9.25, this maximum gain becomes very small for $\lambda_0 = 4$. It increases rapidly for larger values of λ_0 to be limited by CLM for λ_0 larger than 10–12.

9.4 SHORT-CHANNEL THERMAL NOISE MODEL

The long-channel thermal noise model derived in Section 6.2 assumed that the mobility was constant along the channel and that the channel length was sufficiently long so that VS and CLM could be neglected. These assumptions are obviously not valid anymore for short-channel devices where VS and CLM effects have to be accounted for. Mobility reduction due to the vertical field also greatly influences the thermal noise and has also to be included. This will be done in the next sections.

9.4.1 Thermal Noise Drain Conductance

In order to account for the CLM, the region between the source and the drain is split into the channel region of length L_{eff} on the source side, where the carrier velocity is not saturated, and the VS region on the drain side, where the carrier velocity saturates to v_{sat} (Figure 9.26). Since in the VS region the carrier travel at their maximum saturated velocity, they will not respond to the local change of the electric field caused by the noise voltage fluctuations. Therefore, the noise fluctuations generated in the VS region do not propagate to the drain since the conductance on the drain side is zero [117]. Therefore, the dominant contribution to the drain noise mainly comes from the channel region between $x = 0$ and $x = L_{eff}$.

Since the effects of VS, mobility reduction due to the vertical field, and CLM are predominant in strong inversion, we will derive the PSD (power spectral density) of the drain

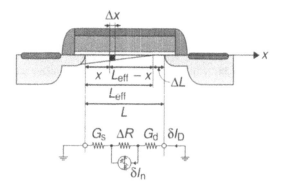

Figure 9.26 Cross section of the MOS channel with a thermal noise source at position x and with the VS region to account for the effect of CLM

current fluctuations assuming that the transistor is biased in strong inversion. The model can be extended to cover all regions of operation as described in [55].

We will reuse the general approach presented in Section 6.1. The PSD of the drain current fluctuations due to the single elementary noise source δI_n is given by (6.3), which is repeated here for convenience:

$$S_{\delta I_D^2} = G_{ch}^2 \Delta R^2 S_{\delta I_n^2},$$

where ΔR is the resistance across the channel slice and G_{ch} is the channel conductance at point x along the channel and given by (6.2)

$$\frac{1}{G_{ch}} \triangleq \frac{1}{G_s} + \frac{1}{G_d}. \tag{9.116}$$

Conductances G_s and G_d are the channel conductances seen by the local thermal noise current source δI_n on the source and drain sides respectively. They can be derived by splitting the total transistor into transistor M_1 and M_2 on the source and drain side, respectively, as shown in Figures 6.2 and 6.3. Conductance G_s actually corresponds to the drain transconductance of transistor M_1 after having isolated it from transistor M_2:

$$G_s = G_{md1} \triangleq \frac{dI_D}{dV}, \tag{9.117}$$

where current I_D is redefined as the current entering the drain of transistor M_1:

$$I_D = \frac{W}{x} \int_{V_S}^{V} (-Q_i) \mu_{eff} \, dV'. \tag{9.118}$$

Notice that for clarity, V' is used as dummy variable for integration in (9.118) to distinguish it from variable V. In order to differentiate (9.118) for calculating conductance G_s, we must remember that μ_{eff} is a function of the electric field E_x which depends on the position x along the channel or equivalently on the channel voltage V at that position. In the most general case, we do not know μ_{eff} as a function of the channel voltage. Even though we would know it, it is not sure we could integrate (9.118). A work around is to notice that the current I_D given by

$$I_D = -W[-Q_i(V)]\mu_{eff}(E_x)E_x \tag{9.119}$$

is constant along the channel and (9.119) can be solved for E_x, which now becomes a function of V and I_D. Replacing $E_x = E_x(V, I_D)$ in the expression of μ_{eff} makes μ_{eff} become a function of V and I_D and (9.119) becomes

$$I_D = -W[-Q_i(V)]\mu_{eff}(V, I_D)E_x$$
$$= W[-Q_i(V)]\mu_{eff}(V, I_D)\frac{dV}{dx}. \tag{9.120}$$

The current I_D is then given by a function $\mathcal{F}(V, I_D)$ of variables V and I_D defined by

$$I_D = \mathcal{F}(V, I_D) \triangleq \frac{W}{x} \int_{V_S}^{V} [-Q_i(V')]\mu_{eff}(V', I_D) \, dV'. \tag{9.121}$$

The total differential of current I_D is then given by

$$dI_D = d\mathcal{F} = \frac{\partial \mathcal{F}}{\partial V} dV + \frac{\partial \mathcal{F}}{\partial I_D} dI_D, \tag{9.122}$$

and the total derivative corresponding to G_s writes

$$G_s = \frac{dI_D}{dV} = \frac{\partial \mathcal{F}}{\partial V} + \frac{\partial \mathcal{F}}{\partial I_D} \frac{dI_D}{dV}, \tag{9.123}$$

which can be solved for $G_s = dI_D/dV$ as

$$G_s = \frac{dI_D}{dV} = \frac{\frac{\partial \mathcal{F}}{\partial V}}{1 - \frac{\partial \mathcal{F}}{\partial I_D}}. \tag{9.124}$$

The partial derivatives of $\mathcal{F}$ can be evaluated from (9.121) as

$$\frac{\partial \mathcal{F}}{\partial V} = \frac{W}{x}(-Q_i)\mu_{\text{eff}}, \tag{9.125a}$$

$$\frac{\partial \mathcal{F}}{\partial I_D} = \frac{W}{x} \int_{V_S}^{V} (-Q_i) \frac{\partial \mu_{\text{eff}}}{\partial I_D} dV'. \tag{9.125b}$$

Replacing (9.125a) and (9.125b) in (9.124) results in [54]

$$G_s = \frac{W(-Q_i)\mu_{\text{eff}}}{x - W \int_{V_S}^{V}(-Q_i)\frac{\partial \mu_{\text{eff}}}{\partial I_D} dV'}. \tag{9.126}$$

In order to evaluate $\partial\mu_{\text{eff}}/\partial I_D$, we notice that

$$\frac{\partial \mu_{\text{eff}}}{\partial I_D} = \frac{\partial \mu_{\text{eff}}}{\partial E_x} \frac{\partial E_x}{\partial I_D} = \mu'_{\text{eff}} \frac{\partial E_x}{\partial I_D}, \tag{9.127}$$

where

$$\mu'_{\text{eff}} \triangleq \frac{\partial \mu_{\text{eff}}}{\partial E_x}. \tag{9.128}$$

From (9.120), we have

$$\frac{\partial I_D}{\partial E_x} = -W(-Q_i)\left(\mu_{\text{eff}} + \mu'_{\text{eff}}E_x\right) = -W(-Q_i)\mu_{\text{diff}}, \tag{9.129}$$

where μ_{diff} is the differential mobility defined by (9.3) and which can be written as

$$\mu_{\text{diff}} = \frac{dv_{\text{drift}}}{dE_x} = \frac{d(\mu_{\text{eff}}E_x)}{dE_x} = \mu_{\text{eff}} + \mu'_{\text{eff}}E_x. \tag{9.130}$$

Hence

$$\frac{\partial \mu_{\text{eff}}}{\partial I_D} = \frac{\mu'_{\text{eff}}}{-W(-Q_i)\mu_{\text{diff}}}. \tag{9.131}$$

Replacing (9.131) in (9.126) finally results in [54]

$$G_s = \frac{W(-Q_i)\mu_{\text{eff}}}{x + \int_{V_S}^{V} \frac{\mu'_{\text{eff}}}{\mu_{\text{diff}}} \, dV'}. \tag{9.132}$$

Conductance G_d corresponds to the source transconductance of transistor M_2 and is defined as

$$G_d = G_{\text{ms2}} \triangleq -\frac{dI_D}{dV}. \tag{9.133}$$

with the current I_D defined as

$$I_D = \frac{W}{L_{\text{eff}} - x} \int_{V}^{V_{\text{Deff}}} (-Q_i)\mu_{\text{eff}} \, dV', \tag{9.134}$$

where saturation at the drain and CLM were accounted for by defining L_{eff} and V_{Deff} as

$$L_{\text{eff}} = \begin{cases} L & \text{for } V_D < V_{D\,\text{sat}} \\ L - \Delta L & \text{for } V_D \geq V_{D\,\text{sat}}, \end{cases} \tag{9.135}$$

where ΔL is the channel reduction due CLM, and

$$V_{\text{Deff}} = \begin{cases} V_D & \text{for } V_D < V_{D\,\text{sat}} \\ V_{D\,\text{sat}} & \text{for } V_D \geq V_{D\,\text{sat}}. \end{cases} \tag{9.136}$$

Conductance G_d is then obtained in a similar way than G_s leading to

$$G_d = \frac{W(-Q_i)\mu_{\text{eff}}}{(L_{\text{eff}} - x) + \int_{V}^{V_{\text{Deff}}} \frac{\mu'_{\text{eff}}}{\mu_{\text{diff}}} \, dV'}. \tag{9.137}$$

The channel conductance G_{ch} at a position x is then easily obtained from (9.116), (9.132), and (9.137) as [54]

$$G_{\text{ch}} = \frac{W(-Q_i)\mu_{\text{eff}}}{L_{\text{eff}} + \int_{V_S}^{V_{\text{Deff}}} \frac{\mu'_{\text{eff}}}{\mu_{\text{diff}}} \, dV'}. \tag{9.138}$$

The channel slice resistance ΔR can be calculated in a similar way than G_d, but integrating from x to $x + \Delta x$ instead of L (or L_{eff} in saturation). After noticing that the voltage drop between x and $x + \Delta x$ is simply equal to $\Delta V = -E_x \Delta x$, we get [54]

$$\frac{1}{\Delta R} = \Delta G = \frac{W(-Q_i)\mu_{\text{eff}}}{\Delta x - \frac{\mu'_{\text{eff}}}{\mu_{\text{diff}}} E_x \Delta x} = \frac{W(-Q_i)\mu_{\text{diff}}}{\Delta x}. \tag{9.139}$$

Here, we would like to point out that in case of a long-channel MOS transistor, μ_{eff} is independent of the electric field E_x and hence $\mu'_{\text{eff}} = 0$. Conductance G_{ch} and resistance ΔR then reduce to

$$G_{\text{ch}} = \mu_{\text{eff}}(-Q_i)\frac{W}{L_{\text{eff}}} = \frac{W(-Q_i)\mu_{\text{eff}}}{L_{\text{eff}}}, \tag{9.140a}$$

$$\Delta R = \frac{\Delta x}{W\mu_{\text{eff}}(-Q_i)}, \tag{9.140b}$$

which correspond to the expressions (6.9) and (6.10), respectively, which were derived in Section 6.1 for the thermal noise of a long-channel device.

Calculation of the current noise source between x and $x + \Delta x$ is a difficult task. This is because, in presence of an electric field, the segment is no longer in equilibrium and in nonequilibrium, the Einstein relation is no longer valid. The method to model device noise in nonequilibrium is to assume that even in nonequilibrium an Einstein-like relation holds between the mobility and the diffusivity D_n (of electrons). One needs to be careful with definitions when making this transition because it often acts as a source of error. The procedure is to define [118–121]

$$D_n = \frac{kT_n\mu_{\text{diff}}}{q}, \tag{9.141}$$

where T_n is defined as the *noise temperature*. Since D_n is unknown in nonequilibrium, this relationship in general provides nothing new but a definition of T_n. Note that in most of the cases the noise temperature T_n is different from the carrier temperature T_C [118].

It is shown in [118] that T_n becomes equal to T_C when the velocity distribution is heated Maxwellian. It is to be noted that we are considering the inversion layer of the MOS transistor where the carrier density is in the order of 10^{18} cm^{-3} and that kind of carrier concentration thermalize the distribution function and enforces it to be heated Maxwellian (we are assuming the channel to be nondegenerate) [122–124]. This observation provides a great simplification by expressing diffusivity D_n in terms of known quantities. It allows to write the expression of $S_{\delta I_n^2}$ as [54, 119]

$$S_{\delta I_n^2} = 4q\,W(-Q_i)\frac{D_n}{\Delta x} = 4kT_C\frac{W(-Q_i)\mu_{\text{diff}}}{\Delta x} = 4kT_C\Delta G, \tag{9.142}$$

where the expression of ΔG given by (9.139) has been used.

It is important to point out that many publications about numerical noise simulation [125–127] as well as [128] use the cord mobility μ_{eff} instead of the differential mobility μ_{diff} when replacing the diffusivity. Ref. [128] has also made an additional assumption that the decrease in cord mobility is exactly balanced by the increase in temperature leaving the diffusivity about constant. Device simulations made for a typical 0.18 μm N-channel MOS transistor show that the product of the cord mobility times the carrier temperature is actually not constant but increases monotonically from source to drain [54]. This indicates that the increase in temperature is not fully compensated by the decrease in cord mobility.

In order to define a noise conductance, (9.142) can be written as

$$S_{\delta I_n^2} = 4k T_L \frac{T_C}{T_L} \Delta G, \tag{9.143}$$

where T_L is the lattice temperature. The PSD of the drain current fluctuations due to δI_n is then given by

$$S_{\delta I_D^2} = 4k T_L \frac{T_C}{T_L} G_{ch}^2 \Delta R. \tag{9.144}$$

Replacing G_{ch} by (9.138) and ΔR by the inverse of (9.139) results in [54]

$$S_{\delta I_D^2} = 4k T_L M \frac{W}{L_{eff}^2} \frac{T_C}{T_L} \frac{\mu_{eff}^2}{\mu_{diff}} (-Q_i) \Delta x, \tag{9.145}$$

where [54]

$$M \triangleq \frac{1}{\left(1 + \frac{1}{L_{eff}} \int_{V_s}^{V_{Deff}} \frac{\mu'_{eff}}{\mu_{diff}} dV \right)^2}. \tag{9.146}$$

The PSD of the total noise current fluctuation at the drain $S_{\Delta I_D^2}$ can be derived by integrating the PSD due to an elementary contribution $S_{\delta I_{D_i}^2}$ at position x over the channel assuming that the contribution of each slice at different positions along the channel remains uncorrelated. This leads to [54]

$$S_{\Delta I_D^2} = 4k T_L M \frac{W}{L_{eff}^2} \int_0^{L_{eff}} \frac{T_C}{T_L} \frac{\mu_{eff}^2}{\mu_{diff}} (-Q_i) \, dx, \tag{9.147}$$

which can be written as

$$S_{\Delta I_D^2} \triangleq 4k T_L G_{nD}, \tag{9.148}$$

and where G_{nD} is the thermal noise conductance at the drain given by

$$G_{nD} = M \frac{W}{L_{eff}^2} \int_0^{L_{eff}} \frac{T_C}{T_L} \frac{\mu_{eff}^2}{\mu_{diff}} (-Q_i) \, dx. \tag{9.149}$$

Equation (9.149) is very general and does not depend on a particular velocity-field (or mobility-field) model. Nevertheless, the relation between the carrier temperature T_C and the lattice temperature T_L has to be consistent with the field-dependent mobility model. It can be shown that the third velocity-field model (Model 3), which is also used for deriving the channel length reduction ΔL in Section 9.2, actually arises as an approximation of [122, 123, 129]

$$\mu_{eff} = \mu_z \sqrt{\frac{T_L}{T_C}}, \tag{9.150}$$

which gives the relation between carrier temperature and mobility as

$$\frac{T_C}{T_L} = \left(\frac{\mu_z}{\mu_{eff}}\right)^2 = \begin{cases} \left(1 + \frac{|E_x|}{2E_c}\right)^2 & \text{for } |E_x| < 2E_c \\ \left(\frac{|E_x|}{E_c}\right)^2 & \text{for } |E_x| \geq 2E_c. \end{cases} \tag{9.151}$$

Using that same mobility-field relation (9.15), we can deduce the following relations:

$$\frac{\mu'_{eff}}{\mu_{diff}} = \frac{\mu'_{eff}}{\mu_{eff} + \mu'_{eff}E_x} = -\frac{1}{2E_c} \tag{9.152}$$

and

$$\frac{\mu^2_{eff}}{\mu_{diff}} = \frac{\mu^2_{eff}}{\mu_{eff} + \mu'_{eff}E_x} = \mu_z. \tag{9.153}$$

Note that relations (9.152) and (9.153) are valid only for $|E_x| < 2E_c$ and can be applied only in the nonsaturation region comprised between x and L_{eff} where $|E_x|$ reaches $2E_c$ and v_{drift} becomes equal to v_{sat}. Anyway, as was pointed out earlier, the contribution of the VS region to the total drain noise is null since in the third velocity-field model, the differential mobility is null and therefore, according to (9.139), no noise is produced.

Introducing (9.152) into (9.146), the M factor then simply reduces to

$$M = \frac{1}{\left(1 - \frac{V_{Deff} - V_S}{2L_{eff} \cdot E_c}\right)^2}, \tag{9.154}$$

whereas using (9.153) in (9.149), G_{nD} becomes

$$G_{nD} = M\frac{W}{L^2_{eff}} \int_0^{L_{eff}} \mu_z \frac{T_C}{T_L}(-Q_i(x))dx \tag{9.155}$$

again valid for $|E_x| < 2E_c$, i.e. in the non-saturation region. The effect of mobility reduction due to the vertical field has been discussed in Section 8.2 and is modelled by (8.5). Although the mobility given by (8.5) is a local mobility and includes q_i which depends on the position x along the channel, an effective mobility can be approximated by replacing q_i in (8.5) by an average charge $(q_s + q_d)/2$. The mobility μ_z can then be taken out of the integral in (9.155), leading to

$$\begin{aligned} G_{nD} &= M\mu_z\frac{W}{L^2_{eff}} \int_0^{L_{eff}} \frac{T_C}{T_L}(-Q_i(x))dx \\ &= M\mu_z\frac{W}{L^2_{eff}} \int_0^{L_{eff}} \left(\frac{\mu_z}{\mu_{eff}}\right)^2(-Q_i(x))dx \\ &= M\mu_z\frac{W}{L^2_{eff}} \int_0^{L_{eff}} \left(1 + \frac{|E_x|}{2E_c}\right)^2(-Q_i(x))dx, \end{aligned} \tag{9.156}$$

where (9.151) has been used.

Notice that for a long-channel MOS transistor, μ_{eff} is independent of the electric field E_x and equal to the mobility at low longitudinal field $\mu_{\text{eff}} = \mu_z$ and hence $\mu'_{\text{eff}} = 0$ and the differential mobility is equal to the cord mobility $\mu_{\text{diff}} = \mu_{\text{eff}} = \mu_z$. The carrier temperature T_C is then equal to the lattice temperature T_L and the effective length L_{eff} is approximately equal to the source-to-drain length L. Factor M is then simply equal to unity and (9.149) and (9.155) then simplify to

$$G_{\text{nD}} = \mu_z \frac{W}{L^2} \int_0^L [-Q_{\text{i}}(x)] \, dx, \qquad (9.157)$$

which corresponds to the long-channel expression (6.15) obtained earlier in Section 6.2.

As shown in (9.150), the effect of VS cannot be considered without the effect of carrier heating. On the other hand, the effect of mobility reduction due to the vertical field and CLM can be considered separately. This will be done in the next sections.

9.4.2 Effect of VS and Carrier Heating on Thermal Noise

In this section we will assume that the effects of mobility reduction due to the vertical field and CLM can be neglected and we will concentrate on the combined effect of VS and carrier heating. This means that $\mu_z = \mu_0$ and $L_{\text{eff}} = L$. To this purpose we will again use the third velocity-field model (Model 3) defined in (9.15) or its normalized form (9.16).

The drain thermal noise conductance accounting for VS and carrier heating only is given by (9.155) with $L_{\text{eff}} = L$ and $\mu_z = \mu_0$[1]

$$G_{\text{nD(VS+CH)}} = M\mu_0 \frac{W}{L^2} \int_0^L \left(1 + \frac{|E_x|}{2E_c}\right)^2 [-Q_{\text{i}}(x)] \, dx, \qquad (9.158)$$

or in normalized form

$$g_{\text{nD(VS+CH)}} \triangleq \frac{G_{\text{nD(VS+CH)}}}{G_{\text{spec}}} = M \int_0^1 \left[1 + \frac{e(\xi)}{2}\right]^2 q_{\text{i}}(\xi) \, d\xi, \qquad (9.159)$$

where e is the normalized longitudinal field defined by (9.6) and

$$M = \frac{1}{\left(1 - \frac{V_{\text{Deff}} - V_{\text{S}}}{2LE_c}\right)^2}. \qquad (9.160)$$

$g_{\text{nD(VS+CH)}}$ can be split into two components $g_{\text{nD(VS)}}$ and $g_{\text{nD(CH)}}$:

$$g_{\text{nD(VS+CH)}} = g_{\text{nD(VS)}} + g_{\text{nD(CH)}}, \qquad (9.161)$$

where $g_{\text{nD(VS)}}$ accounts for the effects of VS only and is given by

$$g_{\text{nD(VS)}} = M \int_0^1 q_{\text{i}} d\xi = M q_{\text{I}}, \qquad (9.162)$$

[1] The fact that only the VS and carrier heating effects are accounted for is indicated by the sign (VS + CH) at the end of the subscript.

whereas $g_{nD(CH)}$ accounts for the additional effect of carrier heating and is given by

$$g_{nD(CH)} = M \int_0^1 \left[e(\xi) + \left(\frac{e(\xi)}{2} \right)^2 \right] q_i(\xi) \, d\xi, \tag{9.163}$$

Note that even though the integral in (9.162) is the same expression as obtained for the long-channel case, the inversion charge distribution along the channel when VS occurs can be quite different from that without VS as in the long-channel case. Therefore $g_{nD(VS)}$ is not simply equal to M times the long-channel value of g_{nD} as (9.161) might wrongly suggest.

For $v_d \leq v_{d\,sat}$, the normalized noise conductances $g_{nD(VS)}$ and $g_{nD(CH)}$ can be found by operating a change of variable from ξ to q_i using (9.22) and (9.36) with the definitions of the normalized velocity v and mobility u given by (9.14) and (9.16) respectively. This results in expressions for $dq_i/d\xi$ and e given by

$$\frac{dq_i}{d\xi} = \frac{-i_d}{2q_i - \frac{\lambda_c}{2} i_d}, \tag{9.164a}$$

$$e = \frac{\lambda_c i_d}{2q_i - \frac{\lambda_c}{2} i_d}, \tag{9.164b}$$

where i_d is given by (9.47). $g_{nD(VS)}$ then becomes

$$\begin{aligned}
g_{nD(VS)} &= \frac{M}{i_d} \int_{q_d}^{q_s} q_i \left(2q_i + \frac{\lambda_c}{2} i_d \right) dq_i \\
&= \frac{M}{i_d} \left[\frac{2}{3} (q_s^3 - q_d^3) - \frac{\lambda_c i_d}{4} (q_s^2 - q_d^2) \right]. \tag{9.165}
\end{aligned}$$

Similarly, $g_{nD(CH)}$ is obtained by replacing in (9.163) e by (9.164b) and $d\xi$ by dq_i given by (9.164a), resulting in

$$\begin{aligned}
g_{nD(CH)} = M \frac{\lambda_c}{2} \Bigg[q_s^2 - q_d^2 + \frac{\lambda_c i_d}{4} (q_s - q_d) \\
+ \left(\frac{\lambda_c i_d}{4} \right)^2 \ln \left(\frac{q_s - \lambda_c/4i_d}{q_d - \lambda_c/4i_d} \right) \Bigg]. \tag{9.166}
\end{aligned}$$

For $v_d \geq v_{d\,sat}$, $g_{nD(VS)} = g_{nD\,sat(VS)}$, and $g_{nD(CH)} = g_{nD\,sat(CH)}$ which can be obtained by replacing i_d and q_d in (9.165) and (9.166) by (9.49b) and (9.49a) respectively.

The corresponding thermal noise parameter can also be split into two components according to

$$\delta_{nD(VS+CH)} = \delta_{nD(VS)} + \delta_{nD(CH)}, \tag{9.167}$$

where $\delta_{nD(VS)} \triangleq g_{nD(VS)}/q_s$ and $\delta_{nD(CH)} \triangleq g_{nD(CH)}/q_s$.

Similarly, the thermal noise excess factor including both VS and carrier heating in saturation is given by

$$\gamma_{nD\,sat(VS+CH)} = \gamma_{nD\,sat(VS)} + \gamma_{nD\,sat(CH)},\qquad(9.168)$$

where $\gamma_{nD\,sat(VS)} \triangleq g_{nD\,sat(VS)}/g_{msat}$ and $\gamma_{nD\,sat(CH)} \triangleq g_{nD\,sat(CH)}/g_{m\,sat}$ with $g_{m\,sat}$ given by (9.70).

The thermal noise parameter $\delta_{nD(VS)}$ is plotted versus the normalized drain voltage $v_d \triangleq V_D/U_T$ and for $v_s = 0$ in Figure 9.27(a). It clearly shows that for $\lambda_c > 0$, the thermal noise conductance and therefore the thermal noise parameter is lower than the long-channel

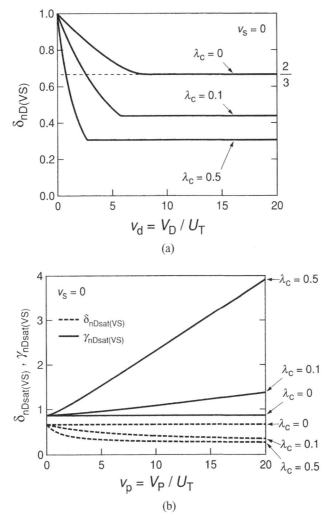

Figure 9.27   (a) Thermal noise parameter $\delta_{nD(VS)}$ accounting for VS only versus normalized drain voltage v_d. (b) Thermal noise parameter $\delta_{nD\,sat(VS)}$ and noise excess factor $\gamma_{nD\,sat(VS)}$ in saturation versus normalized pinch-off voltage v_p and accounting for VS only

value. At first glance, this might be surprising, since there are more inversion charges in the channel due to VS at the drain. It can be explained by the fact that the noise at the drain is transferred from the local noise sources in the channel to the drain through the square of the magnitude of the (trans)conductance, which is proportional to the mobility. Therefore even though there are more charges in the channel, they produce less noise at the drain compared to the situation where the transistor is biased at $V_{DS} = 0$, which is taken as the reference for the definition of the thermal noise parameter δ_{nD}.

Because the transconductance degrades faster than the noise conductance, the situation is different for the noise excess factor accounting for VS $\gamma_{nD\,sat(VS)}$. As opposed to $\delta_{nD\,sat(VS)}$, $\gamma_{nD\,sat(VS)}$ deteriorates as the product $\lambda_c q_s$ increases when VS is present. This is illustrated in Figure 9.27(b), where $\delta_{nD\,sat(VS)}$ and $\gamma_{nD\,sat(VS)}$ are plotted versus v_p for $v_s = 0$ and for different λ_c. For $v_p = 20$ ($V_G - V_{T0} \cong 400\,\text{mV}$) and $\lambda_c = 0.50$ ($L \cong 0.1\,\mu\text{m}$), $\gamma_{nD\,sat(VS)}$ reaches about 4.

$\delta_{nD\,sat(VS)}$ and $\delta_{nD\,sat(CH)}$ are plotted in Figure 9.28(a) versus v_p and for $v_s = 0$. As already mentioned above, the $\delta_{nD\,sat(VS)}$ noise parameter in saturation is smaller than the long-channel value $2/3$ obtained when VS is not present. On the other hand, the term due to carrier heating $\delta_{nD\,sat(CH)}$ is increasing from zero, compensating the reduction of $\delta_{nD\,sat(VS)}$ so that the sum $\delta_{nD\,sat(VS+CH)}$ finally remains slightly above the long-channel value $2/3$.

$\gamma_{nD\,sat(VS)}$, $\gamma_{nD\,sat(CH)}$, and $\gamma_{nD\,sat(VS+CH)}$ are plotted versus v_p and for $v_s = 0$ in Figure 9.28(b). On the contrary to $\delta_{nD\,sat(VS+CH)}$, the effect of carrier heating does not compensate the effect of VS, but it deteriorates $\gamma_{nD\,sat(VS)}$ even further by increasing it significantly from the value without carrier heating.

9.4.3 Effects of Vertical Field Mobility Reduction and Channel Length Modulation

In addition to VS and carrier heating, the reduction of mobility due to the vertical field and the effect of CLM have also to be accounted for. Mobility reduction due to the vertical field will affect both γ_{nD} and δ_{nD} only through parameter λ_c because they are expressed as the ratio of conductances.

The CLM effect is discussed in Section 9.2. The effective channel length used in (9.149) can be approximated by

$$L_{\text{eff}} = L - \Delta L, \tag{9.169}$$

where ΔL is given by (9.87a). CLM will affect γ_{nD} and δ_{nD} in different ways. It will affect γ_{nD} only through parameter λ_c, but for δ_{nD}, in addition to effecting through λ_c, it will increase it by a factor of $\frac{1}{1-\Delta L/L}$.

Some results from the above model are presented in Figures 9.29 and 9.30 for a typical 0.18 μm MOSFET (with $E_c = 2$ V/μm, $\theta = 0.3$, and $\ell = 30$ nm) with different levels of approximation. The noise parameter δ_{nD} is plotted versus v_d in Figure 9.29(a). When CLM is absent, δ_{nD} gets saturated for $v_d \geq v_{d\,sat}$. CLM causes the noise conductance to increase with respect to the drain voltage because the channel length decreases with the drain voltage. Since G_{ds0} is defined at $V_{DS} = 0$, it is not affected by CLM. As a result δ_{nD} increases with the drain voltage. Note that this increase is much less when the effect of mobility reduction due to the

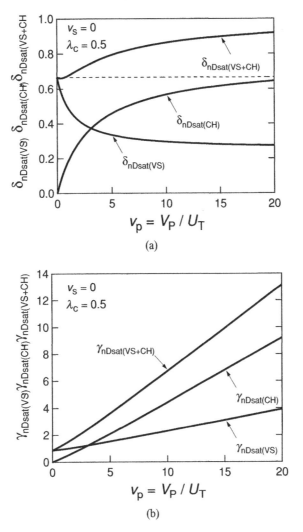

Figure 9.28 (a) Thermal noise excess factors in saturation versus normalized pinch-off voltage v_p. $\delta_{nD(VS)}$ accounts for VS only, $\delta_{nD(CH)}$ accounts for carrier heating only, and $\delta_{nD(VS+CH)}$ accounts for both VS and carrier heating. (b) Thermal noise excess factors in saturation versus normalized pinch-off voltage v_p. $\gamma_{nD\,sat(VS)}$ accounts for VS only, $\gamma_{nD\,sat(CH)}$ accounts for carrier heating only, and $\gamma_{nD\,sat(VS+CH)}$ accounts for both VS and carrier heating

vertical field is also considered. Mobility reduction due to vertical field results in a higher value of E_c, hence a smaller value of u in (9.87a) which considerably attenuates the effect of CLM.

$\delta_{nD\,sat}$ is plotted versus v_p in Figure 9.29(b) with the same levels of approximation than those used in Figure 9.29(a). When both CLM and mobility reduction due to the vertical field are absent, $\delta_{nD\,sat}$ slightly increases with v_p as already shown in Figure 9.27(b). Even when CLM or mobility reduction due to the vertical field is taken into account separately, $\delta_{nD\,sat}$ still increases with v_p. It is the combination of both effects which causes $\delta_{nD\,sat}$ to decrease with v_p. This behavior and the values obtained in Figure 9.27(b) are in agreement with the earlier

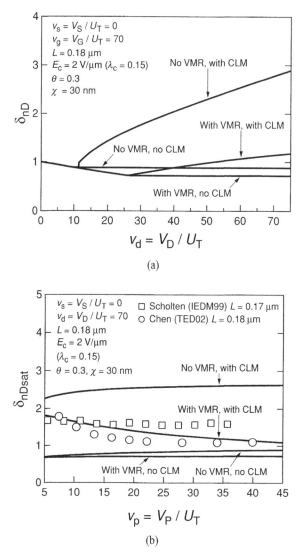

Figure 9.29 (a) Thermal noise parameter δ_{nD} versus normalized drain voltage v_d. (b) Thermal noise parameter $\delta_{nD\,sat}$ in saturation versus normalized pinch-off voltage v_p. The square and circle symbols represent measurements taken from [130] and [117] respectively. Curve labeled no VMR, no CLM accounts for VS and carrier heating only. Curve labeled with VMR, no CLM accounts for VS, carrier heating, and mobility reduction due to the vertical field. Curve labeled no VMR, with CLM accounts for VS, carrier heating, and CLM. Curve labeled with VMR, with CLM accounts for all four effects, namely VS, carrier heating, mobility reduction due to the vertical field, and CLM

results obtained by Scholten [130] and Chen [117] which are represented by the symbols in Figure 9.27(b).

Finally, $\gamma_{nD\,sat}$ is plotted versus v_p in Figure 9.30 again with the same approximations as above. The plots indicate that the effect of vertical field greatly modifies $\gamma_{nD\,sat}$. From the high values obtained earlier when including VS and carrier heating, the effect of mobility reduction due to the vertical field brings $\gamma_{nD\,sat}$ back to values close to 2. This can be explained by

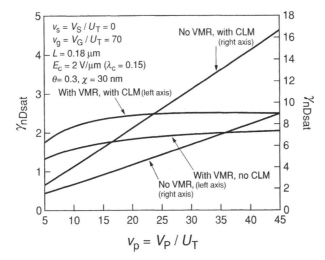

Figure 9.30 Thermal noise excess factor $\gamma_{nD\,sat}$ in saturation versus normalized pinch-off voltage v_p. Curve labeled no VMR, no CLM accounts for VS and carrier heating only. Curve labeled with VMR, no CLM accounts for VS, carrier heating, and mobility reduction due to the vertical field. Curve labeled no VMR, with CLM accounts for VS, carrier heating, and CLM. Curve labeled with VMR, with CLM accounts for all four effects, namely VS, carrier heating, mobility reduction due to the vertical field, and CLM

considering that $\gamma_{nD\,sat}$ decreases as the product $\lambda_c q_s$ decreases and vertical mobility reduction directly reduces λ_c. CLM only slightly increases $\gamma_{nD\,sat}$ to a value of about 2.5.

9.4.4 Summary

If only the VS effect is considered, then the noise conductance G_{nD} becomes smaller compared to the long-channel value. The reason is that the higher noise due to the increase of the inversion charge in the channel required to sustain the drain current with a reduced or even limited velocity is strongly attenuated by the reduction of the transfer function from the local noise source in the channel to the drain terminal caused by the mobility degradation due to VS. Since G_{ds0} is not affected by this mobility reduction, the noise parameter δ_{nD} also gets reduced. However, the gate transconductance is strongly affected by VS and hence the resulting $\gamma_{nD\,sat}$ increases above unity.

VS cannot be considered without carrier heating. The latter has an opposite effect than VS on both δ_{nD} and γ_{nD}, overcompensating the reduction observed in $\delta_{nD\,sat}$ and further increasing $\gamma_{nD\,sat}$. For channel length of the order of 0.1 μm, $\delta_{nD\,sat}$ approximately goes back to values slightly larger than the long-channel value 2/3, whereas $\gamma_{nD\,sat}$ can become larger than 1, typically equal to about 8–10 for an overdrive voltage $V_G - V_{T0}$ of about 0.5 V.

Mobility degradation due to the vertical field causes both G_{nD} and δ_{nD} to decrease slightly because it affects them only through the λ_c parameter. But it affects $\gamma_{nD\,sat}$ very strongly because of the increase in $\lambda_c q_s$, bringing it back to values close to 2. CLM tends to increase all the noise parameters, especially at higher drain (or lower gate) voltages. In summary, the effect of VS and carrier heating try to partly balance each other and also the effect of mobility degradation due to the vertical field and CLM show opposite trends.

We can conclude from the above discussion that because of the presence of counteracting effects, it was possible in the past to present compact thermal noise models *without accounting for all the effects simultaneously and accurately*. As the above factors affect γ_{nD} and δ_{nD} differently, it is therefore very important to distinguish between the thermal noise parameter δ_{nD} and the thermal noise excess factor γ_{nD}. This careful definition of δ_{nD} and γ_{nD} might eventually explain the discrepancies observed between values measured by Scholten [130] and Abidi [131].

10 The Extrinsic Model

The previous chapters were exclusively devoted to the analysis of the intrinsic part of the transistor, defined as the region comprised between the oxide and substrate on top and bottom, and the source and drain junctions on each side. Although the fundamental behavior is indeed dictated by the intrinsic part, the extrinsic part also plays an increasingly important role when either reducing the dimensions and/or increasing the operating frequency. This chapter looks at the different components that constitute this extrinsic part. It starts with the access resistances, which include the source and drain resistances as well as the gate and bulk resistances. They are all presented in Section 10.2. The regions beyond each end of the intrinsic channel include the important overlap capacitances and part of the source and drain access resistances. They are discussed in Section 10.3 with a particular emphasis on their bias dependence. The source and drain junctions are presented in Section 10.4. Finally, Section 10.5 presents the additional noise due to the extrinsic components.

10.1 EXTRINSIC PART OF THE DEVICE

Most of the previous chapters were focused on the so-called *intrinsic part* of the MOS transistor. It is defined by the inside part of the dashed rectangle shown in Figure 10.1 delimited by the source and drain junctions on each side, by the gate oxide and gate electrode on the top and by the substrate on the bottom. This is obviously the most important part of the MOS transistor, since it represents the active part of the device offering the transconductance and enabling amplification. To access the source and drain intrinsic terminals (nodes si and di in Figure 10.1(a)) requires the source and drain extensions (SDE), as well as the source and drain diffusions which are covered with a silicide and contacted by a via. All these parts add some parasitic access resistances which are modeled by the source and drain resistances R_S and R_D. The latter are made of several parts that will be further discussed in Section 10.2.1. The gate is made of polysilicon which is usually covered by silicide in order to lower the gate resistance. Although this resistance is small (in the order of a few $\Omega/\square$), it might be important to account for it particularly for RF *IC* design, where even small series resistances can count. The access to the gate can also be modeled by a simple gate resistance R_G. The modeling of the substrate

Charge-Based MOS Transistor Modeling: The EKV Model for Low-Power and RF IC Design   C. Enz and E. Vittoz
© 2006 John Wiley & Sons, Ltd.

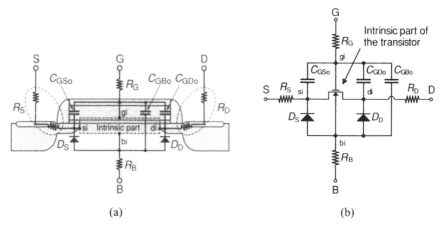

Figure 10.1 (a) Definition of the extrinsic part of the MOS transistor and the extrinsic components including the series access resistances, the overlap parasitic capacitances, and the junction parasitic capacitances. (b) Simple equivalent circuit of the extrinsic part corresponding to (a)

access is a bit more difficult since it strongly depends on the device layout. Modeling it by a simple substrate series resistance R_B is usually sufficient in most cases. More accurate models used for RF IC design will be discussed in more details in Section 11.4.2.

In addition to the four series access resistances R_S, R_D, R_G, and R_B, there are also additional parasitic capacitances. The overlap capacitances between gate and source C_{GSo} and between gate and drain C_{GDo} are due to the overlap of the gate and gate oxide over the SDE. These overlap arise after forming the SDE, by lateral diffusion of the SDE dopants under the gate. These overlap capacitances are made of several parts some of which are bias dependent and will be discussed further in Section 10.3.1. There is also a gate-to-bulk overlap capacitance C_{GBo} which is due to the extension of the gate electrode above the field oxide and on top of the substrate.

In addition, the source and drain junctions and their extensions are modeled by the diodes D_S between the bulk and the source and D_D between the bulk and the drain. As explained in Section 10.4, in dynamic operation they are modeled by two junction capacitances C_{BSj} and C_{BDj}. The latter are obviously bias dependent and are also made of several parts.

Although it is always possible to model the device in great detail taking into account every little series resistances and capacitances, this results in an accurate but usually very complex equivalent circuit. Furthermore, all the components of this equivalent circuit can most of the time not be extracted from experiments in an accurate way, or some not at all. It is therefore important to find the right trade-off between the accuracy required by the circuit designers, which always depends on the circuit application, and the complexity of the equivalent circuit used for the design and the simulations. Also note that most of the parasitic components are distributed resistances and capacitances, which are then modeled by lumped elements. The equivalent circuit shown in Figure 10.1(b) modeling the extrinsic part of the MOS transistor is usually accurate enough for most of the circuit design applications. One exception might be RF circuits, where an even more elaborate equivalent circuit might be required, particularly for the substrate network. This will be discussed further in Chapter 11.

The next sections will discuss each extrinsic component in more detail, particularly its scaling and eventual bias dependence.

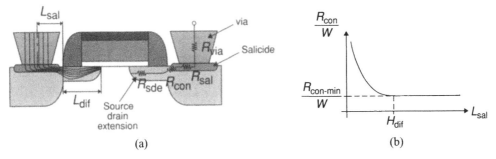

(a)   (b)

Figure 10.2 (a) Components of the source and drain access resistances (on the right) and approximative current flow (on the left). (b) Contact resistance per unit width versus the diffusion width showing that above a certain value H_{dif}, the resistance does not scale as $1/L_{sal}$ because most of the current flows within the salicide instead of going from the bottom of the salicide to the diffusion

10.2 ACCESS RESISTANCES

10.2.1 Source and Drain Resistances

As shown in Figure 10.2(a), the source and drain access resistances are made of several parts including the resistance due to the via R_{via}, the resistance of the salicide R_{sal}, the contact resistance between the salicide and the junction diffusion R_{con}, and the resistance of the SDE R_{sde}. The source (drain) resistance is then given by the series connection of all these components:

$$R_{S(D)} = R_{sde} + R_{con} + R_{sal} + R_{via} \cong R_{sde} + R_{con}. \tag{10.1}$$

The SDE and salicide resistances are scaling as

$$R_{sde} = \frac{L_{dif}}{W} R_{sde\square}, \tag{10.2a}$$

$$R_{sal} = \frac{L_{sal}}{W} R_{sal\square}, \tag{10.2b}$$

where L_{dif} is the length of the SDE and L_{sal} is the half width of the salicide region as shown in Figure 10.2(a). $R_{sde\square}$ and $R_{sal\square}$ are the sheet resistances of the SDE and the salicide, respectively, which have typical values in the $k\Omega$ range for $R_{sde\square}$ and in the Ω range for $R_{sal\square}$. The total via resistance R_{via} depends on the number of via per source or drain diffusion with a typical resistance of a few Ω per via. As can be seen from the above numbers, the total resistance is usually dominated by the contact and the SDE resistances.

Note that the contact resistance R_{con} per unit of finger width does not scale with the salicide length L_{sal} above a certain minimum value defined as H_{dif} as shown in Figure 10.2(b). This is due to the fact that most of the current flows within the salicide instead of going to the diffusion because of the latter higher resistivity as illustrated in Figure 10.2(a). Therefore, increasing the salicide length above H_{dif} does not reduce the contact resistance even though the bottom contact area between salicide and silicon is increased.

As shown in Figure 10.3, the SDE resistance R_{sde} is made of two parts: R_{sde-ov}, which is in the overlap region below the gate and R_{sde-sp} which is outside the gate overlap region below

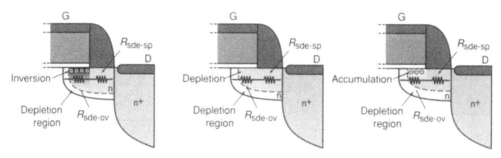

Figure 10.3 SDE resistance splits into the bias-independent part $R_{\text{sde-sp}}$ situated below the spacer and the bias-dependent part $R_{\text{sde-ov}}(V_G, V_{S(D)})$ located in the overlap region

the spacer:

$$R_{\text{sde}} = R_{\text{sde-sp}} + R_{\text{sde-ov}}(V_G, V_{S(D)}). \tag{10.3}$$

As Figure 10.3 illustrates, the resistance $R_{\text{sde-ov}}$ depends on the inversion state of this gate overlap region and hence depends on $V_G - V_S$ on the source side and on $V_G - V_D$ on the drain side. On the other hand, resistance $R_{\text{sde-sp}}$ can be considered as bias independent. Since the gate-to-bulk and drain-to-bulk voltages are usually positive and the source-to-bulk is zero or positive, the overlap regions on the source and drain sides are most of the time biased in accumulation. Increasing the gate-to-bulk voltage will attract even more electrons on the surface and hence reduce the overlap resistances. Note that it is important to account for this bias dependence in order to accurately predict the harmonic distortion [132].

Since the SDE region length L_{dif} is almost constant, the total source and drain resistances R_S and R_D scale only with the finger width and the number of fingers according to

$$R_{S(D)} \cong R_{\text{con}} + R_{\text{sde}}(V_G, V_{S(D)}) \cong \frac{0.5 R_{\text{dsw}}(V_G, V_{S(D)})}{W}, \tag{10.4}$$

where R_{dsw} is the total source and drain resistance per unit width. R_{dsw} is typically in the $k\Omega\ \mu m$ range.

The bias dependence of the source and drain access resistances is illustrated in Figure 10.4, where the total source and drain access resistance is plotted versus the gate-to-source voltage for two different oxide thicknesses [133].

Because of the voltage drop across the source and drain series resistances, the voltages at the intrinsic nodes are smaller than the applied external voltages. Since the current is determined by the intrinsic voltages, the transconductances from the external terminals are smaller than the intrinsic transconductances. This can be easily verified using the small-signal equivalent circuit shown in Figure 10.5. The effective transconductances are given by

$$G_{\text{meff}} \triangleq \left.\frac{\partial I_D}{\partial V_G}\right|_{V_S, V_D} = \frac{G_m}{D} \tag{10.5a}$$

$$G_{\text{mseff}} \triangleq -\left.\frac{\partial I_D}{\partial V_S}\right|_{V_G, V_D} = \frac{G_{\text{ms}} + G_{\text{ds}}}{D} \tag{10.5b}$$

$$G_{\text{mdeff}} \triangleq \left.\frac{\partial I_D}{\partial V_D}\right|_{V_G, V_S} = \frac{G_{\text{md}} + G_{\text{ds}}}{D}, \tag{10.5c}$$

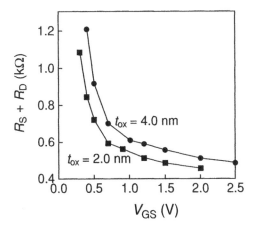

Figure 10.4 Bias dependence of the source and drain resistance (Reproduced by permission of IEEE from [133])

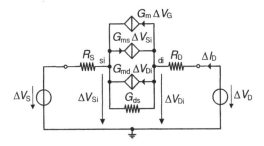

Figure 10.5 Small-signal schematic for calculating the degradation of transconductance due to the source and drain resistances

where

$$D \triangleq 1 + (G_{ms} + G_{ds})R_S + (G_{md} + G_{ds})R_D \qquad (10.6)$$
$$\cong 1 + G_{ms} R_S \quad \text{(in saturation)}.$$

As indicated by (10.5), the intrinsic transconductances are reduced by this factor D which is approximately equal to $1 + G_{ms} R_S$ in saturation.

Similar considerations can be drawn for the drain current which is lowered by the presence of the source and drain series resistances.

10.2.2 The Gate Resistance

The gate resistance starts to play a role typically when it gets equal or larger than the inverse of the gate transconductance. It will not only affect the transistor operation at high frequency, but can also have an effect at low frequency. Indeed, R_G becoming larger than $1/G_m$ will contribute to noise at both low and high frequency, and will also change the frequency behavior at high frequency. It is therefore important to account for it when designing very low-noise circuits

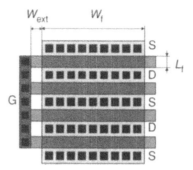

Figure 10.6   Layout of a typical large multifinger device

operating at low frequency, for example sensors front-ends, and at high frequency when designing for example low-noise amplifiers. The transistors used in such circuits are usually made very large and are laid out as multifinger devices as shown in Figure 10.6 for a number of fingers $N_f = 4$. The gate resistance is made of several parts: the resistance R_{Gtop} corresponding to the part that is on top of the channel (in darker gray in Figure 10.6), resistance R_{Gext} corresponding to the part that is outside the channel region (in lighter gray in Figure 10.6), resistance R_{Gvia} corresponding to the vias between metal 1 and the silicided polysilicon and resistance R_{Gcon} corresponding to the contact resistance between the silicide and the polysilicon [134]:

$$R_G = R_{Gtop} + R_{Gext} + R_{Gvia} + R_{Gcon}. \tag{10.7}$$

The part of the gate resistance that is on top and across of the channel R_{Gtop} is modeled by

$$R_{Gtop} = \frac{1}{3} \frac{W_f}{N_f L_f} R_{G\square}, \tag{10.8}$$

where $R_{G\square}$ is the gate silicide sheet resistance, W_f is the finger length (corresponding to the channel width of a single finger) and L_f is the finger width corresponding to the drawn gate length. The factor $1/3$ appearing in (10.8) accounts for the distributed nature of R_{Gtop} as illustrated in Figure 10.7 in order to correctly predict the maximum oscillation frequency [135]. Note that the distributed gate resistance along the channel can be neglected since the finger is usually much longer than wide ($W_f \gg L_f$).

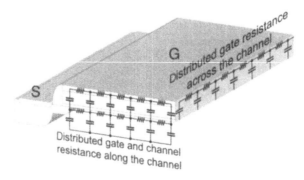

Figure 10.7   Distributed gate and channel resistances

The resistance of the part outside the channel region R_{Gext} depends very much on the geometry and where the gate contacts are placed. In case the gate is contacted along the vertical metal line as shown in Figure 10.6, R_{Gext} is simply given by

$$R_{Gext} = \frac{W_{ext}}{N_f L_f} R_{G\square}.$$ (10.9)

The via resistance R_{Gvia} depends on the number of via N_{via} according to

$$R_{Gvia} = \frac{R_{via}}{N_{via}},$$ (10.10)

where R_{via} is the resistance of a single via.

The silicide-to-polysilicon contact resistance is defined by

$$R_{Gcon} = \frac{\rho_{con}}{N_f W_f L_f},$$ (10.11)

where $1/\rho_{con}$ is the silicide-to-polysilicon specific conductance (in $A/V\ m^2$).

Note that the total gate resistance given by (10.7) can be significantly decreased by connecting the gate on both sides. As shown in Figure 10.8, if the gate resistance is contacted only on

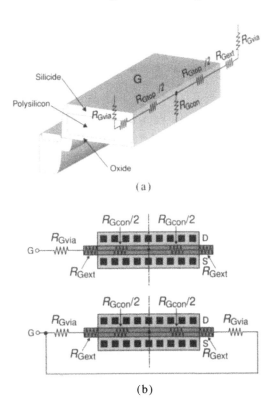

Figure 10.8 (a) Different parts of the gate resistance; (b) contacting the gate at both ends decreases the gate resistance by approximately a factor 4

one side, the total gate resistance is given by (10.7). On the other hand, if the gate is contacted on both sides and the metal is assumed to have a negligible resistance compared to the other components, we have

$$R_G \cong \frac{R_{Gtop}}{4} + \frac{R_{Gext}}{2} + \frac{R_{Gvia}}{2} + R_{Gcon}, \qquad (10.12)$$

which is about four times smaller than the one side contact case corresponding to (10.7). If the layout constraints allow, it is therefore recommended to contact the gate on both sides in order to minimize the gate resistance for the given geometry.

In technologies typically older than $0.18\ \mu m$, the gate current could be completely neglected. Hence, the dc voltage drop across the gate resistance could also be neglected and therefore the gate resistance had no effect on the dc transistor operation. This is no longer the case for ultradeep submicron technologies, where the gate oxide is so thin that a dc tunneling current starts to flow through this oxide. In this case, there is also a small voltage drop across the gate resistance that can also affect the dc operation of the transistor.

10.3 OVERLAP REGIONS

10.3.1 Overlap Capacitances

The different capacitances forming the extrinsic gate-to-source and gate-to-drain parasitic capacitances are shown in Figure 10.9. They are made mainly of three capacitances: the overlap capacitance C_{ov}, the inner fringing-field capacitance C_{if}, and the outer fringing-field capacitance C_{of}

$$C_{GS(D)o} = C_{ov}(V_G, V_{S,(D)}) + C_{if}(V_G, V_{S,(D)}) + C_{of}. \qquad (10.13)$$

Note that $C_{ov}(V_G,\ V_{S,(D)})$ and $C_{if}(V_G,\ V_{S,(D)})$ are strongly bias dependent, whereas C_{of} can be considered as bias independent.

A simple way to model the bias-dependent overlap capacitance C_{ov} is to define an effective overlap length $L_{ov\text{-}eff}$ corresponding to the part of the total overlap length L_{ov} that is constituting

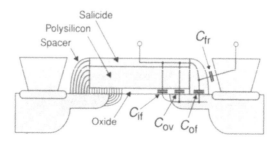

Figure 10.9   Different parts of the total overlap capacitances: the inner fringing-field capacitance C_{if}, the overlap capacitance C_{ov}, and the outer fringing-field capacitance C_{of}. Note that there is also a fringing-field capacitance C_{fr} between the gate electrode and the via

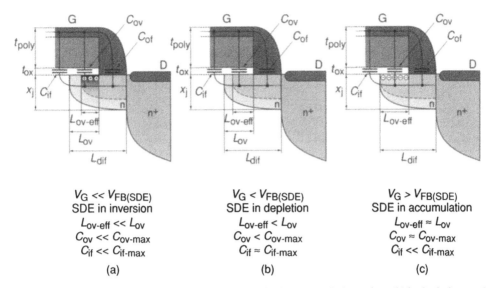

$$V_G \ll V_{FB(SDE)}$$
SDE in inversion
$$L_{ov\text{-}eff} \ll L_{ov}$$
$$C_{ov} \ll C_{ov\text{-}max}$$
$$C_{if} \ll C_{if\text{-}max}$$
(a)

$$V_G < V_{FB(SDE)}$$
SDE in depletion
$$L_{ov\text{-}eff} < L_{ov}$$
$$C_{ov} < C_{ov\text{-}max}$$
$$C_{if} \approx C_{if\text{-}max}$$
(b)

$$V_G > V_{FB(SDE)}$$
SDE in accumulation
$$L_{ov\text{-}eff} \approx L_{ov}$$
$$C_{ov} \approx C_{ov\text{-}max}$$
$$C_{if} \ll C_{if\text{-}max}$$
(c)

Figure 10.10   Bias dependence of C_{if} and C_{ov} when the SDE is (a) in inversion, (b) in depletion, and (c) in accumulation

the bottom plate of C_{ov} as shown in Figure 10.10 [56]:

$$C_{ov}(V_G) \triangleq W L_{ov\text{-}eff}(V_G) C_{ox}. \tag{10.14}$$

Depending on the gate voltage, the overlap region can be either in accumulation when the gate voltage is larger than the flat-band voltage of the SDE region $V_{FB(SDE)}$ (which is close to zero volt [56]) (Figure. 10.10(c)), or in depletion when $V_G < V_{FB(SDE)}$ (Figure. 10.10(b)), or even in inversion when $V_G \ll V_{FB(SDE)}$ (Figure. 10.10(a)).

As shown in Figure 10.10(c), the overlap capacitance is maximum in accumulation ($V_G > V_{FB(SDE)}$) for which the effective overlap length $L_{ov\text{-}eff}$ is about equal to the total overlap length L_{ov}:

$$C_{ov}(V_G > V_{FB(SDE)}) \cong C_{ov\text{-}max} \triangleq W L_{ov} C_{ox}. \tag{10.15}$$

For $V_G < V_{FB(SDE)}$, the overlap capacitance is smaller than $C_{ov\text{-}max}$ and can be modeled empirically by

$$C_{ov} = \begin{cases} C_{ov\text{-}max} & \text{for } V_G \geq V_{FB(SDE)} \\ \dfrac{C_{ov\text{-}max}}{1 + \frac{|V_G|}{V_{Gov}}} & \text{for } V_G < V_{FB(SDE)}, \end{cases} \tag{10.16}$$

where $C_{ov\text{-}max}$ is given by (10.15) and V_{Gov} is a fitting parameter that can be extracted from measurement as explained in [56].

The inner fringing-field capacitance is also bias dependent. When the gate voltage is lower than the channel flat-band voltage, the device is in accumulation and the layer of free holes in the channel region is electrically disconnected from the n$^+$ SDE regions and shields the fringing capacitances reducing them to zero. When V_G increases, the device enters the depletion regime where C_{if} reaches its maximum. As V_G increases further, the device enters in strong inversion and an inversion layer is formed. The inner fringing-field capacitance is again shielded by the inversion layer and decreases down to zero. This behavior is modeled by [56]

$$C_{if} = C_{if\text{-max}} \exp\left[-\left(\frac{V_G - V_{FB} - \Phi_F/2}{3\Phi_F/2}\right)^2\right],$$ (10.17)

where $C_{if\text{-max}}$ is given by

$$C_{if\text{-max}} \triangleq W \frac{\epsilon_{si}}{3\pi} \ln\left[1 + \frac{x_j}{t_{ox}} \sin\left(\frac{\pi}{2}\frac{\epsilon_{ox}}{\epsilon_{si}}\right)\right],$$ (10.18)

where x_j is the depth of the SDE (not the junction).

Finally, the outer fringing-field capacitance can be considered as bias independent and is approximated by [56]

$$C_{of} = W \frac{2\epsilon_{ox}}{\pi} \ln\left(1 + \frac{t_{poly}}{t_{ox}}\right).$$ (10.19)

The above model for $C_{GS(D)o}$ is plotted in Figure 10.11(a) versus V_G for $V_D = V_S = 0$ and for two different values of the SDE region doping N_{sde}. It shows that the overlap capacitance C_{ov} dominates the extrinsic capacitance $C_{GS(D)o}$. Also note the effect of the inner fringing-field component which introduces a little bump in the characteristic. Figure 10.11(b) shows the total gate-to-source (gate-to-drain) capacitance including both the intrinsic and extrinsic parts. It shows that the extrinsic component dominates in the weak and depletion regions, whereas the intrinsic dominates in the moderate and strong inversion regions.

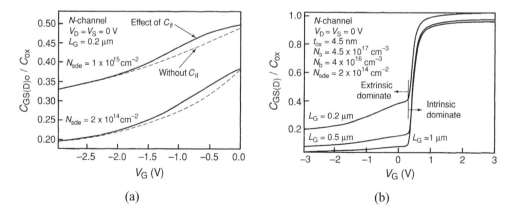

(a)   (b)

Figure 10.11   (a) Effect of the inner fringing-field capacitance given by (10.17) on the total overlap capacitance gate voltage bias dependence. (b) Bias dependence of the total gate-to-source (gate-to-drain) capacitance for different channel lengths (Reproduced by permission of Elsevier Ltd. from [56])

There is also an overlap capacitance C_{GBo} between the gate and the substrate. It is due to the gate extending outside the channel, Since it is along the channel, it is proportional to the gate length

$$C_{GBo} = CGBO\,L_G, \tag{10.20}$$

where $CGBO$ is a capacitance per unit length and L_G is the gate drawn length.

10.3.2 Overlap Gate Leakage Current

In addition to the gate leakage current between the gate and the channel region as discussed in Section 8.6, the source and drain overlap regions also contribute to the gate leakage current and may dominate in some bias conditions [112].

10.4 SOURCE AND DRAIN JUNCTIONS

10.4.1 Source and Drain Diodes Large-Signal Model

As shown in Figure 10.1, the source and drain junctions are simply modeled by two diodes D_S and D_D connected between bulk and source and bulk and drain, respectively. They are characterized by the classical current–voltage relation

$$I_j = -I_s \left(\exp \frac{-V_R}{\eta U_T} - 1 \right), \tag{10.21}$$

where V_R is the reverse voltage applied across the junction, i.e., $V_R = V_{SB}$ on the source side and $V_R = V_{DB}$ on the drain side and I_s is given by

$$I_s = q\,A_D n_i^2 \left(\frac{D_p}{L_p N_{diff}} + \frac{D_n}{L_n N_b} \right), \tag{10.22}$$

where D_n, D_p and L_n, L_p are the diffusion coefficients and diffusion length of electrons and holes, respectively. N_b and N_{diff} are the doping concentrations in the P-type substrate and in the N-type source and drain junctions, respectively. The ideality factor η is ideally equal to unity when the current is dominated by the diffusion current and gets larger than 1 when accounting for recombination and high injection. In reverse mode, equation (10.21) indicates that the current saturates to I_s as soon as $V_R > 5\eta U_T$. But this is without accounting for the generation current due to generation of electron–hole pairs in the depletion region. Actually, in reverse bias, the current is dominated by this generation current which is given by

$$I_{gen} = \frac{q\,A_D n_i d}{\tau_g}, \tag{10.23}$$

where A_D is the diode cross-sectional diode area, τ_g is the generation time constant of the carrier in the depletion region, and d is the depletion width which depends on the reverse voltage V_R according to

$$d = \sqrt{\frac{2\epsilon_{si}}{q} \frac{N_b + N_{diff}}{N_b \, N_{diff}}} \, \sqrt{V_R + \Phi_B}, \qquad (10.24)$$

where Φ_B is the built-in potential

$$\Phi_B = U_T \ln \frac{N_b \, N_{diff}}{n_i^2}. \qquad (10.25)$$

In most cases the junctions are N$^+$-P type with $N_{diff} \gg N_b$ and (10.24) simplifies to

$$d \cong \sqrt{\frac{2\epsilon_{si}}{q \, N_b}} \, \sqrt{V_R + \Phi_B}. \qquad (10.26)$$

The current flowing in the reverse-biased source (drain) junction is then given by

$$I_{S(D)B} \cong I_s + I_{gen} \qquad (10.27)$$

and depends on the bias voltages mainly through the generation current I_{gen}.

If the reverse bias voltage is increased further, so does the electric field in the depletion region until it reaches a critical value E_{jc} corresponding to the avalanche breakdown voltage V_{br}

$$V_{br} = \frac{\epsilon_{si} E_{jc}^2}{2q} \left(\frac{1}{N_b} + \frac{1}{N_{diff.}} \right). \qquad (10.28)$$

Equation (10.28) shows that the breakdown voltage decreases when increasing the doping on either side of the junction. As shown in Figure 10.12(a), as soon as V_R gets slightly larger than V_{br}, the reverse current starts to increase sharply.

The small-signal equivalent circuit of the diodes is shown in Figure 10.12(c), where the junction capacitances are described in Section 10.4.2 and the differential conductances in Section 10.4.3.

10.4.2 Source and Drain Junction Capacitances

The source and drain junction capacitances C_{BSj} and C_{BDj} of Figure 10.12(c) model the variations of the depletion charge due to a change of the depletion width. A junction capacitance can be simply modeled as a parallel plate capacitor with silicon as dielectric and separated by a distance d

$$C_j = \frac{\epsilon_{si}}{d}, \qquad (10.29)$$

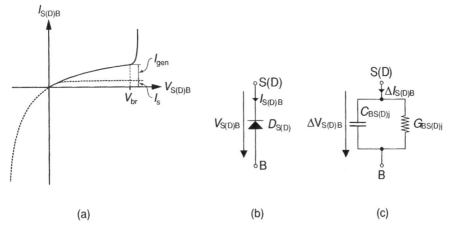

Figure 10.12 Modeling of the source and drain junctions: (a) current–voltage characteristic; (b) large-signal model; (c) small-signal equivalent circuit

where C_j is actually a capacitance per unit area. Note that even though (10.29) is usually derived assuming an abrupt junction (or step profile), it is actually valid for any doping profile. On the other hand, the voltage dependence given in (10.24) assumes abrupt junctions. The junction capacitance can be rewritten as

$$C_j = \frac{C_{j0}}{\sqrt{1 + \frac{V_R}{\Phi_B}}},\tag{10.30}$$

where C_{j0} is the value of the capacitance (per unit area) at equilibrium (i.e., for $V_R = 0$)

$$C_{j0} \triangleq \sqrt{\frac{\epsilon_{si} q}{2\Phi_B} \frac{N_b\, N_{diff}}{N_b + N_{diff}}} \cong \sqrt{\frac{\epsilon_{si}\, q\, N_b}{2\Phi_B}}.\tag{10.31}$$

Note that the above equation holds only for $V_R > -\Phi_B$, which is usually the case since the junction have to be reverse biased in order to maintain the diode leakage current small compared to the drain current.

In real diodes, the doping profile is not abrupt as assumed in the derivation of (10.30). For practical diodes the one-half exponent in (10.30) is replaced by the grading coefficient m, resulting in the following expression for C_j

$$C_j = \frac{C_{j0}}{\left(1 + \frac{V_R}{\Phi_B}\right)^m},\tag{10.32}$$

where m typically ranges between 0.2 and 0.5.

Since the doping levels are very different on top and on bottom of the junctions and in the SDE regions, the junction capacitances have to be split into several parts:

$$C_{BS(D)j} = A_{S(D)}\, C_{jbw} + (P_{S(D)} - W_{eff})\, C_{jsw} + W\, C_{jswg},\tag{10.33}$$

where C_{jbw} refers to the bottom-wall capacitance per unit area, C_{jsw} to the side-wall capacitance per unit length of the perimeter that is on the isolated sides, and C_{jswg} to the side-wall capacitance per unit length of the part of the perimeter that is along the gate and the SDE region. A_S (A_D) is the total source (drain) diffusion area, P_S (P_D) is the total source (drain) diffusion perimeter, and W is the total transistor width.

Note that for devices that are inside a well, an additional junction capacitance $C_{BB'j}$ between the well and the substrate has to be considered. The latter is decomposed into

$$C_{BB'j} = A_W C_{jbww} + P_W C_{jsww}, \tag{10.34}$$

where C_{jbww} refers to the bottom-wall capacitance per unit area and C_{jsww} to the side-wall capacitance per unit length of the well. A_W corresponds to the total well area, whereas P_W is the total well perimeter.

10.4.3 Source and Drain Junction Conductances

The source and drain junctions small-signal schematic should be completed with two differential conductances G_{BSj} and G_{BDj} that are connected in parallel with the junction capacitances as shown in Figure 10.12(c). The conductances are obtained by differentiating the expression of the leakage current (10.27) resulting in

$$G_{BS(D)j} = \frac{A_D n_i}{\tau_g} \frac{N_b + N_{diff}}{N_b N_{diff}} C_j \cong \frac{A_D n_i}{\tau_g N_b} C_j. \tag{10.35}$$

These conductances can often be neglected since they are usually much smaller than the intrinsic (trans)conductances and output conductance G_{ds}. However, they may become non-negligible at very low channel current, for which the intrinsic (trans)conductances become very small, or if the effect of G_{ds} has been canceled as is the case in a cascode configuration.

10.5 EXTRINSIC NOISE SOURCES

The different noise sources appearing at low frequency in a MOS transistor are represented in the small-signal schematic of Figure 10.13.[1] The overall noise is usually dominated by the intrinsic part of the device representing the active part and corresponding in Figure 10.13 to the noise source I_{nD} [114,134]. It comprises both the flicker and thermal noise due to the channel, which were already presented in Sections 6.2 and 6.3. In addition, all the access resistances, namely the source and drain resistances R_S and R_D but also the gate and the substrate resistances R_G and R_B are also noisy and contribute to the thermal and to some extend also to the flicker noise (see Section 6.3.3). They are represented in Figure 10.13 by the noise current sources I_{nRS}, I_{nRD}, I_{nRG}, and I_{nRB} respectively. If their contributions to the flicker noise is neglected,

---

[1] This small-signal equivalent circuit does not include the noise appearing at high frequency. The latter are discussed in Chapter 13.

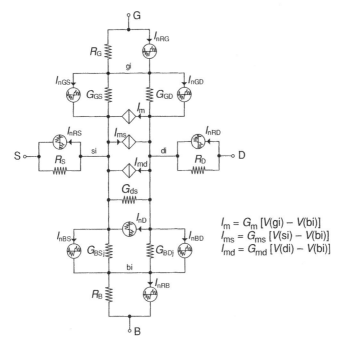

$$I_m = G_m [V(gi) - V(bi)]$$
$$I_{ms} = G_{ms} [V(si) - V(bi)]$$
$$I_{md} = G_{md} [V(di) - V(bi)]$$

Figure 10.13 Low-frequency small-signal equivalent circuit with the main noise sources, including the sources coming from the extrinsic part of the transistor

they show only thermal noise and have power spectral densities (PSD) given by

$$S_{I_{nRS}^2} = \frac{4kT}{R_S}, \tag{10.36a}$$

$$S_{I_{nRD}^2} = \frac{4kT}{R_D}, \tag{10.36b}$$

$$S_{I_{nRG}^2} = \frac{4kT}{R_G}, \tag{10.36c}$$

$$S_{I_{nRB}^2} = \frac{4kT}{R_B}. \tag{10.36d}$$

The leakage currents I_{SB} and I_{DB} of the source-to-bulk and drain-to-bulk junctions also contribute as shot noise. They are represented in Figure 10.13 by the noise current sources I_{nBS} and I_{nBD} which have PSD given by

$$S_{I_{nBS}^2} = 2q I_{SB}, \tag{10.37a}$$

$$S_{I_{nBD}^2} = 2q I_{DB}, \tag{10.37b}$$

where q is the elementary charge.

As discussed in Section 8.6, in deep submicron technologies the gate oxide is so thin that a tunneling current is flowing in the gate. This current is split between the source and the drain, giving rise to a current I_{GS} flowing from the intrinsic gate (gi) to the intrinsic source (si) and another current I_{GD} flowing from the intrinsic gate (gi) to the intrinsic drain (di). Both of these currents show shot noise. They are represented in Figure 10.13 by the noise current sources I_{nGS} and I_{nGD} having PSD

$$S_{I_{nGS}^2} = 2q\,I_{GS}, \qquad (10.38a)$$
$$S_{I_{nGD}^2} = 2q\,I_{GD}. \qquad (10.38b)$$

The conductances G_{GS} and G_{GD} in Figure 10.13 represent the small-signal differential conductances corresponding to these leakage currents.

Part III

The High-Frequency Model

The aggressive scaling of CMOS technologies which is going on since more than 25 years has allowed to increase the number of transistors per chips and hence extend the functionality and in the same time dramatically push the speed performance. Although these tremendous speed improvements have been mainly driven by the requirements of VLSI digital chips, they can also be exploited for analog RF circuits. Today, ultradeep submicron (UDSM) technologies have catched-up or even surpassed the transit frequencies achieved by bipolar transistors. This clearly opens the door to full CMOS highly integrated solutions for wireless applications. After several years of intensive research that has demonstrated the feasibility of using CMOS technologies for RF applications, real products using CMOS also for the RF portion of a chip are now emerging. Several examples of single-chip systems, including the radio transceiver together with the baseband digital processor and fully integrated in CMOS, are on the market today. Nevertheless, the design of RF ICs for real products remains a challenge due to the strong constraints on power consumption and noise that leave little margins for the RF IC designer. It is therefore crucial to be able to accurately predict the performance of CMOS RF circuits in order to improve design efficiency and reduce time-to-market. This requires MOS transistor models that are accurate over a wide range of bias, from dc to RF and for a large set of geometries. Part III presents an overview of the high-frequency aspects of MOS transistor modeling for RF IC design. Chapter 11 presents the equivalent circuit at RF, whereas Chapter 12 focuses on the small-signal circuit. RF thermal noise is finally presented in Chapter 13.

11 Equivalent Circuit at RF

This chapter first presents the structure and layout of a typical RF MOS transistor. It then briefly looks at what is really changing when moving to higher frequency. Several figures of merit widely used to evaluate and compare different devices and technologies are defined in Section 11.3. They include the transit frequency, the maximum frequency of oscillation, and the minimum noise figure. Section 11.3.4 points out that the moderate inversion offers a good trade-off between the power consumption, the low-voltage operation, the noise, and the linearity, all being of major importance for designing RF circuits. Section 11.4 then presents the equivalent large-signal circuit valid at RF. It highlights the importance of a correct modeling of the substrate.

11.1 RF MOS TRANSISTOR STRUCTURE AND LAYOUT

RF MOS transistors are usually designed as large devices in order to achieve the desired transconductance required to meet the RF requirements. As shown in Figure 11.1, they are usually laid out as multifinger devices, because in deep submicron CMOS processes, the maximum finger length (corresponding to the unit transistor width W_f) is limited. This is due to the so-called "narrow-line effect" increasing the silicided polysilicon sheet resistance as the finger width (corresponding to the transistor gate length L_f) decreases due to grain boundary problems [136]. Typical devices have up to 10 or more fingers. The total transistor effective width is then simply $W = N_f W_f$.

11.2 WHAT CHANGES AT RF?

When increasing the operating frequency for a given transistor, the characteristics such as the gain or the transconductance (or more precisely the transadmittance) start to degrade. The sources of this degradation must be distinguished between those coming from the intrinsic part of the device (the channel region) and those related to the extrinsic part of the transistor (i.e., all the parasitic components discussed in Chapter 10). The frequency limit of the intrinsic part

Charge-Based MOS Transistor Modeling: The EKV Model for Low-Power and RF IC Design   C. Enz and E. Vittoz
© 2006 John Wiley & Sons, Ltd.

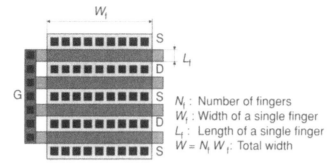

Figure 11.1   Layout of a typical RF MOS transistor

is set by the frequency ω_{qs} delimiting QS and NQS operation given by (5.32) which is repeated here for convenience

$$\omega_{qs} = \omega_{spec}\,\Omega_{qs}(q_s, q_d),\tag{11.1}$$

where ω_{spec} is defined in (5.33) as $\omega_{spec} \triangleq \mu U_T/L_f^2$ and $\Omega_{qs}(q_s, q_d)$ is the normalized QS frequency, which is bias dependent according to (5.32). In strong inversion and saturation, (5.32) reduces to (5.34) which is repeated below

$$\Omega_{qs} \cong \frac{15}{2}q_s = \frac{15}{2}\sqrt{i_f} = \frac{15}{4}\frac{V_P - V_S}{U_T}.\tag{11.2}$$

In order not to have any degradations due to NQS operation, the QS frequency ω_{qs} has to be higher than the operating frequency (typically by a factor 5–7). This condition is achieved by increasing Ω_{qs} either by choosing a sufficiently high bias for a given channel length or by increasing ω_{spec} by reducing the channel length at a given bias or both. Note that, as stated by (5.33), the QS frequency at a given operating point is inversely proportional to the square of the channel length, as long as there is no velocity saturation.

The limitations due to the extrinsic part are strongly related to the layout, but in general the frequency limitations are mainly due to the extrinsic capacitances and particularly the capacitance at the drain, namely the drain-to-bulk junction capacitance C_{BDj} and the gate-to-drain overlap capacitance C_{GDo}. The latter also affects the signal coupling between the gate and the drain.

Some of the limitations described above are characterized by several figures of merit which evaluate the ability of a device to operate at RF. They are discussed in Section 11.3.

11.3 TRANSISTOR FIGURES OF MERIT

11.3.1 Transit Frequency

A very common way to characterize the high-frequency performance of an active device is to look at the frequency at which the extrapolated current gain h_{21} of a small-signal

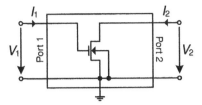

Figure 11.2 Small-signal common-source amplifier as a two-port network

common-source amplifier stage falls to unity. The current gain h_{21} of such a two-port shown in Figure 11.2 is given by

$$h_{21} \triangleq \left. \frac{I_2}{I_1} \right|_{V_2=0} = \frac{Y_{21}}{Y_{11}} = \frac{G_m - j\omega(C_m + C_{GD})}{j\omega C_G} \simeq \frac{G_m}{j\omega C_G} = \frac{\omega_t}{j\omega}, \qquad (11.3)$$

where I_1 corresponds to the small-signal current entering the gate terminal (Port 1) and I_2 corresponding to the small-signal current entering the drain terminal (Port 2).

Frequency ω_t is the unity gain transit frequency given by

$$\omega_t = \frac{G_m}{C_G} = \frac{G_m}{C_{Gi} + C_{Go}} = \omega_{spec} \frac{g_m}{c_{Gi} + c_{Go}}, \qquad (11.4)$$

where C_{Gi} is the total intrinsic capacitance at the gate defined by

$$C_{Gi} \triangleq C_{OX} c_{Gi} = C_{GSi} + C_{GDi} + C_{GBi}, \qquad (11.5)$$

where $C_{OX} \triangleq W L_f C_{ox}$. According to (5.51a) and (5.52), the total gate capacitance C_{Gi} simplifies to

$$C_{Gi} = \frac{C_{OX}}{n}(n - 1 + c_{GSi} + c_{GDi}), \qquad (11.6)$$

where c_{GSi} and c_{GDi} are the normalized intrinsic gate-to-source and gate-to-drain capacitances given by (5.50). In strong inversion and saturation, according to (5.53) $c_{GSi} \cong 2/3$ and $c_{GDi} \cong 0$, resulting in

$$C_{Gi} \cong C_{OX}\left(1 - \frac{1}{3n}\right). \qquad (11.7)$$

C_{Go} is the total overlap capacitance at the gate (see Figure 10.1):

$$C_{Go} \triangleq C_{OX} c_{Go} = C_{GSo} + C_{GDo} + C_{GBo}. \qquad (11.8)$$

Neglecting the fringing-field components of the gate-to-source and gate-to-drain overlap capacitances, they can be approximated by

$$C_{GSo} = C_{GDo} \cong W L_{ov} C_{ox}. \qquad (11.9)$$

Usually $C_{GBo} < C_{GSo} + C_{GDo}$ and C_{Go} can be roughly approximated by

$$C_{Go} \cong 2WL_{ov}C_{ox}. \tag{11.10}$$

The transit frequency in strong inversion and saturation is then approximately

$$\omega_t \cong \frac{\omega_{spec}}{1 - \frac{1}{3n} + \frac{2L_{ov}}{L_f}} \frac{n(V_P - V_S)}{U_T}. \tag{11.11}$$

Equation (11.11) shows that ω_t is actually a fraction of the QS frequency ω_{qs} given by

$$\frac{\omega_t}{\omega_{qs}} \cong \frac{4}{15} \frac{n}{1 - \frac{1}{3n} + \frac{2L_{ov}}{L_f}}. \tag{11.12}$$

For a minimum length device, the overlap length can be a significant fraction of the gate length. Assuming for example that $L_{ov}/L_f = 0.2$, $n = 1.2$, ω_t is about 3.5 times smaller than ω_{qs}.

This transit frequency can be fairly high (typically above 100 GHz) and sometimes cannot be measured directly. It can nevertheless be extracted from a lower frequency measurement of h_{21} according to

$$\omega_t = \Im\{h_{21}\}\omega_{spot}, \tag{11.13}$$

where ω_{spot} is a sufficiently low frequency (typically 1 GHz) at which the current gain h_{21} shows a -20 dB/dec slope. An example of measured current gain h_{21} calculated from the de-embedded Y-parameters of an RF N-channel MOS transistor is plotted in Figure 11.3(a) for a specific bias. The curve labeled "analytic (full)" corresponds to the gain h_{21} calculated from (11.3) directly with the Y-parameters, whereas curve labeled "analytic (simple)" is obtained from the approximation given in (11.3). It shows that the analytic expressions and the simulations are very close to the measured results.

Since G_m and C_G are both bias dependent, f_t is also. This is illustrated in Figure 11.4 which plots the transit frequency versus the inversion factor for two devices in saturation having two different channel lengths and for three different drain bias voltages. Figure 11.4(a) has a lin–log scale and clearly indicates that f_t reaches a maximum called the peak f_t $f_{t\text{-peak}}$. This maximum corresponds to the situation where the gate voltage starts to become large enough for the transistor to leave saturation and enter in the linear region. When entering the linear region, the drain transconductance G_{md} starts to increase and hence the gate transconductance $G_m = (G_{ms} - G_{md})/n$ starts to decrease since the source transconductance G_{ms} remains constant. Also, the intrinsic gate-to-drain capacitance starts to increase from nearly zero to a value close to that of the gate-to-source capacitance. The combined effect of G_m decrease and C_{GDi} increase results in a sharp f_t drop. Note that it is very important to correctly model the bias dependence also of the overlap capacitances and particularly C_{GDo} to accurately model f_t in this region.

Figure 11.5 shows how the peak f_t scales with the transistor length. From (11.4), ω_t is proportional to ω_{spec} which is inversely proportional to L_f^2. ω_t should therefore scale as $1/L_f^2$ which is about the case in the regions above 0.25 μm. Below that, $f_{t\text{-peak}}$ tends to increase slower than at the $1/L_f^2$ rate. This is due to the effect of velocity saturation. Indeed, at high bias and

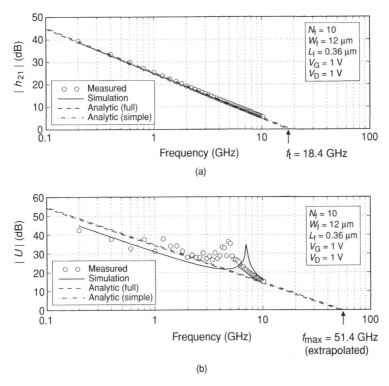

Figure 11.3 Measured, simulated, and analytic results for the current gain h_{21}: (a) for the extraction of the transit frequency f_t and the unilateral power gain U; (b) for the extraction of the maximum frequency of oscillation

for short-channel devices, the carriers enter velocity saturation. As explained in Section 9.1.3, when the carrier velocity is saturated, the transconductance in saturation does not depend on the channel length anymore as stated by (9.62). The gate transconductance is then given by

$$G_{m\ sat} \cong W C_{ox} v_{sat}, \tag{11.14}$$

resulting in a transit frequency given by

$$\omega_t \cong \frac{v_{sat}}{L_f(1 - \frac{1}{3n} + \frac{2L_{ov}}{L_f})}, \tag{11.15}$$

which scales only as $1/L_f$. This explains the -1 slope followed by the points that are below 0.25 µm in Figure 11.5.

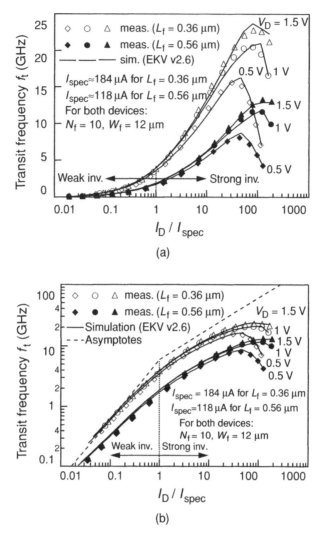

Figure 11.4 Transit frequency versus inversion factor for two channel lengths and three drain bias voltages: (a) lin–log scale [49]; (b) log–log scale [52] (Reproduced by permission of IEEE from [49] and [52])

Figure 11.5 also shows that sub 0.1 µm devices can reach transit frequencies higher than 100 GHz.

11.3.2 Maximum Frequency of Oscillation f_{max}

The transit frequency is only a very simple and partial way to characterize the ability of a device to operate at RF. Another figure of merit that also accounts for the gate resistance R_G and the drain-to-bulk capacitance C_{GD} can be defined from the unilateral power gain U corresponding to the maximum available gain of a two-port (corresponding to the transducer

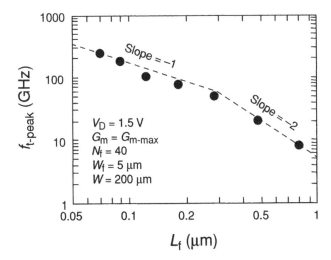

Figure 11.5 Peak transit frequency versus gate length (Reproduced by permission of IEEE from [137])

gain with matched source and load impedances, i.e., $Y_G = Y_{11}^*$ and $Y_L = Y_{22}^*$) with its feedback transadmittance neutralized (i.e., $Y_{12} = 0$) [138]. The advantage of using the unilateral power gain is that it can be defined even if the two-port is unstable in matched condition. The unilateral power gain can be expressed from the Y-parameters as [138]

$$U = \frac{|Y_{21}|^2}{4(G_{11}G_{22} - G_{12}G_{21})},\tag{11.16}$$

where $G_{kl} \triangleq \Re\{Y_{kl}\}$ with $k, l \in \{1, 2\}$. Deriving the Y-parameters from the simple QS small-signal model presented in Figure 5.14 leads to

$$U \cong \frac{G_m^2}{4R_GC_G(G_{DS}C_G + G_mC_{GD})\omega^2} \cong \frac{G_m}{4R_GC_GC_{GD}\omega^2} = \left(\frac{\omega_{max}}{\omega}\right)^2,\tag{11.17}$$

where ω_{max} is the frequency at which the extrapolated unilateral gain reaches unity. It is given by

$$\omega_{max} \cong \frac{G_m}{2\sqrt{R_GC_G(G_{DS}C_G + G_mC_{GD})}} \cong \frac{1}{2}\sqrt{\frac{G_m}{R_GC_GC_{GD}}} \cong \frac{1}{2}\sqrt{\frac{\omega_t}{R_GC_{GD}}}.\tag{11.18}$$

Equation (11.18) shows that the smaller the R_GC_{GD} product the higher ω_{max}. Therefore, the R_GC_{GD} product is sometimes also used as a figure of merit.

The unilateral power gain is plotted in Figure 11.3(b) versus frequency for the same device and bias point used for calculating the current gain and the transit frequency. Unlike the current gain, the unilateral power gain shows some resonance after which it decreases faster than −20 dB/dec. This higher slope region is not shown in Figure 11.3(b) because the measurements were performed only up to 10 GHz. The value of ω_{max} extrapolated from the −20 dB/dec slope can therefore be significantly higher than the actual value of ω at which U becomes unity. It is therefore important to specify how f_{max} has been obtained from the measured data, either by the point at which U is equal to unity if the measurements go at sufficiently high frequency or by extrapolation with a -20 dB/dec slope in case f_{max} cannot be measured directly. Note that the latter method is more advantageous and usually preferred since it gives higher (but erroneous!) values of f_{max}.

11.3.3 Minimum Noise Figure

Having a sufficiently high f_t and f_{max} and hence a high gain is not the only requirement for RF active devices. They should also have as little noise as possible. This feature is measured by the noise factor F or noise figure $NF \triangleq 10 \log F$. The noise factor is defined as the ratio of the total noise power measured at some point along the amplification chain (usually at the output) to the noise produced by the input generator only and measured at that same point. We will come back to these definitions in more details in Section 13.1. The noise factor depends on the generator admittance and becomes minimum for a particular value of this generator admittance. This situation corresponds to noise matching. The minimum value of the noise factor F_{min} (or minimum noise figure NF_{min}) represents what the device can ultimately achieve in terms of minimum thermal noise contribution and is therefore often used as a figure of merit. It is not that easy to find a simple analytical expression for the minimum noise factor of an RF MOS transistor that is accurate. Nevertheless, some approximations discussed in more details in Section 13.3.2 lead to the following very simple expression

$$F_{min} \cong 1 + \frac{\omega}{\omega_t}, \tag{11.19}$$

which accounts for all the dominant noise contributions. Equation (11.19) shows that the noise factor is a function of frequency and is directly linked to the transit frequency. It actually starts to degrade proportionally to frequency when the operating frequency gets higher than the transit frequency. This is illustrated in Figure 11.6(a) which plots the available power gain and

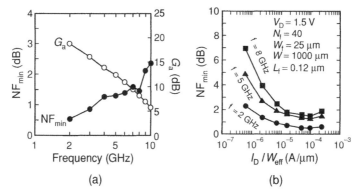

(a)   (b)

Figure 11.6 (a) Minimum noise figure NF_{min} and available power gain G_a versus frequency at a given operating point; (b) minimum noise figure NF_{min} versus drain bias current density for three different operating frequencies (Reproduced by permission of IEEE from [137])

the minimum noise figure verus frequency for a given operating point. This plot also illustrates the fact that the minimum noise figure is obviously minimum when the gain is maximum.

Since the minimum noise figure depends on the transit frequency, it also depends on the bias. Figure 11.6(b) shows the minimum noise figure versus the drain bias current density (current per transistor width) for three different operating frequencies. It clearly shows that there is an optimum bias point where NF_{min} is minimum. Note that this optimum bias occurs about at the same current density for the three different frequencies.

11.3.4 Moderate and Weak Inversion for RF Circuits

The high transit frequency of ultradeep submicron (UDSM) CMOS processes can be traded with power consumption to implement RF circuits operating in the gigahertz frequency range. This can be done by moving the operating point from strong inversion to moderate or even weak inversion, in order to spend just the required power to achieve the desired performance. There are several advantages to bias the transistor in moderate or weak inversion. The first advantage is the increase of the current efficiency (measured by the G_m/I_D ratio) which results in a further reduction of the power consumption. Secondly, the decrease of the bias voltages results in lower electrical fields within the device. This avoids velocity saturation and hot electron effects. Having no velocity saturation results in f_t scaling as $1/L_f^2$ compared to only $1/L_f$ when velocity saturation is present. This means that scaling is more effective for devices biased in the weak and moderate inversion region than in strong inversion. Thirdly, having no hot electron effects avoids the increase of the noise excess factor.

Finally, the reduction of the bias voltages better accommodates the use of low supply voltages that are imposed by the scaling of UDSM technologies.

On the other hand, moving toward weak inversion changes the $I_D - V_G$ characteristic from a quasi-quadratic to an exponential function, which clearly degrades the device linearity. Moderate inversion therefore represents a good trade-off between power consumption, noise, and linearity.

Part of the power is just used to fight against the extrinsic components such as the overlap and junction capacitances. There might be a concern that the time constants in moderate and weak inversion might be completely dominated by these extrinsic components and therefore counterbalance the advantage of the current efficiency increase. A way to investigate this issue is by looking at the total transit time τ_t defined as $\tau_t \triangleq 1/(2\pi f_t)$, which can be decomposed into $\tau_t \triangleq \tau_i + \tau_e$, where $\tau_i = C_{Gi}/G_m$ corresponds to the transit time of the intrinsic part with C_{Gi} being the total gate intrinsic capacitance $C_{Gi} = C_{GSi} + C_{GDi} + C_{GBi}$. The time constant τ_i ultimately represents the lowest time constant the device can achieve for a given operating point. The time constant τ_e corresponds to the additional delay introduced by the extrinsic part of the device due to the overlap capacitances and the series resistances:

$$\tau_e = \frac{C_{Go}}{G_m} + R_D C_{GD} + nR_S \left(C_{GB} + C_{GD} + \frac{n-1}{n} C_{GS} \right), \qquad (11.20)$$

where $C_{Go} = C_{GSo} + C_{GDo} + C_{GBo}$ is the total gate overlap capacitance. Usually the contributions of the source and drain series resistances can be neglected and hence $\tau_e \cong C_{Go}/G_m$.

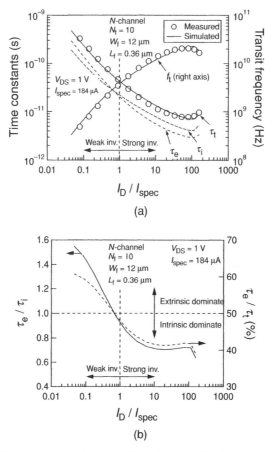

Figure 11.7 Transit frequency versus inversion factor in saturation for two channel lengths and three drain bias voltages (Reproduced by permission of Springer from [48])

The transit times τ_i, τ_e, and τ_t are plotted together with the transit frequency f_t in Figure 11.7(a) versus the inversion factor. The ratio between the extrinsic and the intrinsic transit times is plotted in Figure 11.7(b), which shows that extrinsic parasitics account for about 40% of the total transit time in strong inversion and about 50% in moderate inversion. This means that the ratio of parasitic to intrinsic time constants does not degrade dramatically when moving the operating point from strong to moderate inversion. This is another good reason for moderate inversion to be considered for RF operation with deep submicron devices in order to meet the low-voltage and low-power requirements.

11.4 EQUIVALENT CIRCUIT AT RF

11.4.1 Equivalent Circuit at RF

A cross section of a single-finger MOS transistor is presented in Figure 11.8(a). Although it is always possible to have a detailed equivalent circuit that accounts for all the physical elements

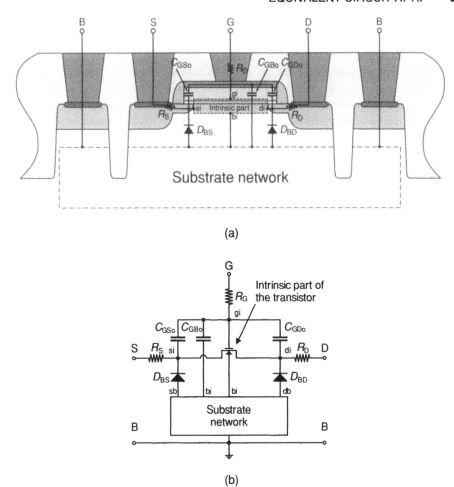

Figure 11.8 (a) Single finger RF MOS transistor cross section with box representing the substrate network connecting the intrinsic bulk node *bi* and nodes *sb* and *db* to the actual bulk terminal *B*. (b) Equivalent RF circuit with substrate network box

that are part of the RF MOS transistor, it is often too complex to be implemented as a compact model or a subcircuit for circuit simulation purpose. Moreover, many of the component values would be difficult or even impossible to extract and the subcircuit would contain too many internal nodes which would significantly reduce the simulation efficiency. Like it is often the case in modeling, a trade-off has to be found between accuracy and efficiency. A good compromise is obtained when simplifying the complete detailed equivalent circuit to the one presented in Figure 11.8(b). This equivalent circuit is made of the intrinsic part of the MOS transistor, corresponding to the active part of the device and represented in Figure 11.8(b) by the MOS transistor symbol. All the other elements are only parasitic components corresponding to the extrinsic part of the device. They are made essentially of capacitances and resistances that play an increasingly important role as the operating frequency rises. Both the intrinsic model and the extrinsic components have already been described in details in previous chapters. The substrate network box represents the part of the substrate that connects the intrinsic substrate terminal bi, the bottom terminals of the D_{BS} and D_{BD} diodes as well as the bottom terminal of the gate-to-bulk overlap capacitance C_{GBo} to the actual substrate terminal B. The latter will be discussed in more details in the next section.

Note that the equivalent circuit of Figure 11.8(b) does not include the parasitic components related to the test structure, such as the pad capacitances, the lead series resistances, and inductances. The latter will have to be carefully de-embedded from the measurements to bring the reference planes close to the useful device. For example, all the measurements presented afterward have been cautiously de-embedded using a two-step procedure [139, 140].

11.4.2 Intradevice Substrate Coupling and Substrate Resistive Networks

At high frequency, the impedances of the junction capacitances become small enough for the RF signal at the drains to couple to the nearby source diffusions and to the bulk contact through the junction capacitances and the substrate as illustrated in Figure 11.9(a). The doping levels of UDSM CMOS processes are sufficiently high so that the substrate can be considered as purely

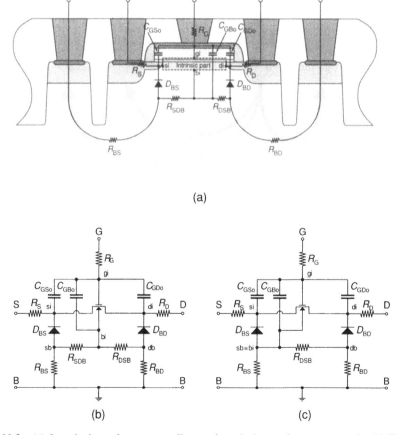

Figure 11.9 (a) Intradevice substrate coupling and equivalent substrate network. (b) Equivalent circuit with resistive substrate network [136, 141]. (c) Equivalent circuit with simplified Π equivalent resistive substrate network [48, 49, 52]

resistive and hence this coupling can be modeled by a simple resistive network. Depending on the technology and on the frequency range to be covered, this network can reduce to a simple resistance or may need to be more complex. Many different substrate resistive networks have been proposed in the literature. A good compromise is to use the Π resistive circuit made of resistances R_{SDB}, R_{DSB}, R_{BS}, and R_{BD} as shown in Figure 11.9(b) [141]. Resistances R_{SDB} and R_{DSB} represent all the coupling occurring from drains to sources, whereas R_{BS} and R_{BD} correspond to the coupling from source and drain to bulk. The partitioning of the total resistance $R_{SDB} + R_{DSB}$ between R_{SDB} and R_{DSB} by choosing the location of the intrinsic substrate node bi is not straightforward. On the other hand, the total resistance $R_{SDB} + R_{DSB}$ is usually small compared to R_{BS} and R_{BD} and simulations have shown that connecting the intrinsic node bi either to the left or to the right of these resistances has very little influence on the Y-parameters. Therefore, the intrinsic substrate node bi can be connected to the source side, and series resistances R_{SDB} and R_{DSB} can be replaced by a single resistance [48,49,52]. This is done in Figure 11.9(c) where only resistance R_{DSB} has been kept while R_{SDB} has been set to zero. This is advantageous for circuit simulation since it simplifies the circuit by saving one component and one node, but it makes the circuit slightly unsymmetrical. It is however a good trade-off which from experience has shown to be sufficient for most RF circuit simulations [48,49,52].

Figure 11.10 shows the cross sections of multifinger RF MOS transistors where only the most important substrate resistances have been drawn. In order to match the equivalent circuit shown in Figure 11.9(c), the equivalent capacitances and resistances have to be calculated from the individual capacitances and resistances shown in Figure11.10. Since all the source (drain) diffusions are connected together via metal layers (assumed to have negligible resistances compared to the substrate resistances), the junction capacitances C_{BSj} and C_{BDj} can reasonably be approximated as the parallel connection of all individual source and drain junction capacitances:

$$C_{BSj} = \sum_{k=1}^{N_s} C_{BSjk}, \tag{11.21a}$$

$$C_{BDj} = \sum_{k=1}^{N_d} C_{BDjk}, \tag{11.21b}$$

where N_s and N_d are the number of source and drain diffusions, respectively and $N_f = N_s + N_d$ is the total number of fingers. The same applies for the substrate drain-to-source, source-to-bulk, and respectively drain-to-bulk resistances:

$$\frac{1}{R_{DSB}} = \sum_{k=1}^{N_f} \frac{1}{R_{DSBk}}, \tag{11.22a}$$

$$\frac{1}{R_{BS}} = \sum_{k=1}^{N_s} \frac{1}{R_{BSk}}, \tag{11.22b}$$

$$\frac{1}{R_{BD}} = \sum_{k=1}^{N_d} \frac{1}{R_{BDk}}. \tag{11.22c}$$

By symmetry, all the individual source (drain) junction capacitances are equal to the one of a single source (drain) diffusion $C_{BSjk} \cong C_{BSjf}$ ($C_{BDjk} \cong C_{BDjf}$). The same is valid for the

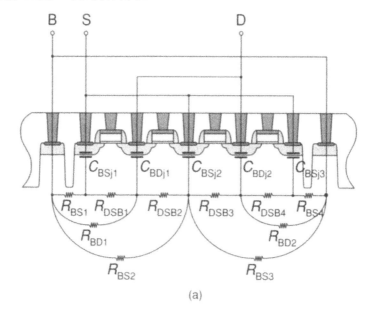

(a)

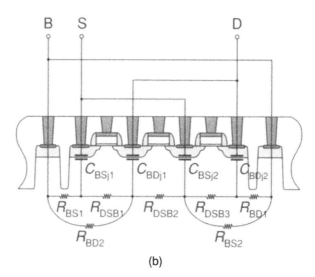

(b)

Figure 11.10   Substrate resistances for (a) even and (b) odd number of fingers

individual drain-to-source substrate resistances, leading to

$$C_{BSj} \cong N_s \, C_{BSjf}, \tag{11.23a}$$

$$C_{BDj} \cong N_d \, C_{BDjf}, \tag{11.23b}$$

$$R_{DSB} \cong \frac{L_f}{N_f \, W_f} \, R_{DSB\text{-}sh}, \tag{11.23c}$$

where C_{BSjf} and C_{BDjf} are the junction capacitances of a single source and drain diffusion and $R_{DSB\text{-}sh}$ is the drain-to-source substrate sheet resistance.

The calculation of the source-to-bulk and drain-to-bulk substrate resistances needs to distinguish between even and odd number of fingers as shown in Figure 11.10. For the even number of fingers transistor (c.f. Figure 11.10(a)), $R_{BS1} \ll R_{BS2}$ and $R_{BS4} \ll R_{BS3}$ since the outer source diffusions are closer to the bulk contact than the inner source diffusions and by symmetry $R_{BS1} \cong R_{BS4}$ and $R_{BD1} \cong R_{BD2}$, resulting in

$$\frac{1}{R_{BS}} \cong \frac{1}{R_{BS1}} + \frac{1}{R_{BS4}} \cong \frac{2}{R_{BS1}}, \tag{11.24a}$$

$$\frac{1}{R_{BD}} \cong \frac{1}{R_{BD1}} + \frac{1}{R_{BD2}} \cong \frac{2}{R_{BD1}}, \tag{11.24b}$$

for N_f even.

For an odd number of fingers (c.f. Figure 11.10(b)), $R_{BS1} \ll R_{BS2}$, $R_{BD1} \ll R_{BD2}$ and $R_{BS1} \cong R_{BD1}$, which results in

$$R_{BS} \cong R_{BD} \cong R_{BS1}. \tag{11.25}$$

From (11.24) and (11.25), resistances R_{BS} and R_{BD} are basically dominated by the source (drain) diffusions which are the closest to the substrate contact. Their scaling strongly depends on the geometry of the bulk contact. For example, if there are only bulk contacts at each end of the device as shown in Figure 11.11(a), R_{BS} and R_{BD} are determined mainly by the source and drain diffusions that are the closest to the substrate contact, resulting in a scaling with the finger width

$$\frac{1}{R_{BS}} \cong \frac{2W_f}{r_{BS\text{-}end}}, \tag{11.26a}$$

$$\frac{1}{R_{BD}} \cong \frac{2W_f}{r_{BD\text{-}end}}, \tag{11.26b}$$

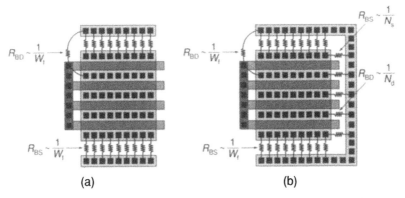

(a)   (b)

Figure 11.11   (a) Intradevice substrate coupling and equivalent substrate network; (b) equivalent circuit with resistive substrate network

for N_f even and

$$\frac{1}{R_{BS}} \cong \frac{W_f}{r_{BS\text{-}end}}, \tag{11.27a}$$

$$\frac{1}{R_{BD}} \cong \frac{1}{R_{BS}}, \tag{11.27b}$$

for N_f odd, where $r_{BS\text{-}end}$ and $r_{BD\text{-}end}$ are the source-to-bulk and drain-to-bulk substrate resistances for a unit width.

The scaling law becomes much more complex in the more realistic case where the substrate contact partly surrounds the diffusions ("horseshoe" substrate contact of Figure 11.11(b)). In this case, part scales with the finger width and part depends on the length of the lateral substrate contact which is proportional to the number of fingers

$$\frac{1}{R_{BS}} \cong \frac{2W_f}{r_{BS\text{-}end}} + \frac{N_s}{r_{lat}}, \tag{11.28a}$$

$$\frac{1}{R_{BD}} \cong \frac{2W_f}{r_{BD\text{-}end}} + \frac{N_d}{r_{lat}}, \tag{11.28b}$$

for N_f even and

$$\frac{1}{R_{BS}} \cong \frac{W_f}{r_{BS\text{-}end}} + \frac{(N_f+1)/2}{r_{lat}}, \tag{11.29a}$$

$$\frac{1}{R_{BD}} \cong \frac{1}{R_{BS}}, \tag{11.29b}$$

for N_f odd. r_{lat} is the lateral source-to-bulk and drain-to-bulk substrate resistances per source and drain diffusion.

The substrate resistances R_{BS}, R_{BD}, and R_{DSB} are in principle also bias dependent due to changes of the depletion width around the diffusions which affect the length of the resistive path. As stated by (11.23c), for a large number of fingers, R_{DSB} becomes much smaller than R_{BS} and R_{BD} so that it can be ignored. Resistances R_{BS} and R_{BD} can then be considered as being connected in parallel and can be replaced by a single substrate resistance R_B as shown in Figure 11.12(d). This substrate resistance shows only a weak bias dependence [142].

Other substrate networks have been published in the literature [143, 144]. Some of them are reproduced in Figure 11.12. Those presented in Figures 11.12(a) and 11.12(b) have already been discussed above. The one presented in Figure 11.12(c) [143] was derived for an epitaxial process. The two top horizontal resistances model the coupling within the epitaxial layer, whereas the three vertical ones model the coupling to the substrate. The last one presented in Figure 11.12(d) [144] is valid for RF transistors having many fingers. Indeed, since R_{DSB} scales as $1/N_f$, for a large number of fingers R_{DSB} becomes much smaller than R_{BS} and R_{BD} and can therefore be neglected. The Π network reduces to a simple resistance corresponding to parallel connection of R_{BS} and R_{BD}.

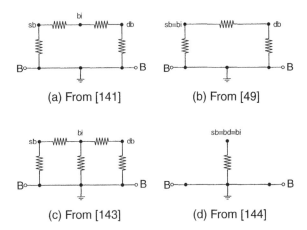

(a) From [141]

(b) From [49]

(c) From [143]

(d) From [144]

Figure 11.12   Several resistive substrate networks: (a) from [141], (b) from [49], (c) from [143], and (d) from [144]

11.4.3 Practical Implementation Issues

The MOS compact models available in circuit simulators such as Spice have four terminals but usually do not include the gate resistance and the substrate network. In order to have access to the internal nodes of the RF MOS transistor and implement the equivalent circuit of Figure 11.9(c) in a Spice simulator, most of the time a subcircuit approach is used. Note that not all the extrinsic components that are already available in the compact model (i.e., source and drain resistances, overlap capacitances, and junction diodes) can be used. For example, the source and drain series resistors in most compact models are only "soft" resistances embedded in the expression used to calculate the drain current. They account for the dc voltage drop across the source and drain resistances and its effect on the static drain current, but they do not add any poles. They have therefore to be added outside of the compact model as "real" resistors. Also, the source-to-bulk and drain-to-bulk diodes of the compact model have their anodes connected to the same node. Depending on the substrate network, their anodes have to be connected to two separate nodes (as shown in Figure 11.9(c)). In this case, the diodes internal to the compact model have to be turned off (by setting some appropriate values of the diode parameters) and two external diodes D_{BS} and D_{BD} have to be added in the subcircuit as shown in Figure 11.9(c). The overlap capacitances C_{GSo}, C_{GDo}, and C_{GBo} are usually also part of most compact models, but not all provide good bias-dependent models. This bias-dependence has imperatively to be accounted for in order to obtain a RF MOST model that is valid over a large bias range. Note that before even looking at the RF operation it is important to have a good dc model, since all the small-signal parameters are derived from it.

12 The Small-Signal Model at RF

After deriving the large-signal equivalent circuit in the previous chapter, this chapter focuses on the small-signal equivalent circuit at RF. The Y-parameters are derived in Section 12.2 directly from the quasi-static (QS) RF small-signal circuit. They are then compared with measurements highlighting the effect of the substrate network on the output admittance Y_{22}. The extension of the quasi-static model to include non-quasi-static (NQS) effects is also presented. Finally, the large-signal operation is briefly discussed and it is concluded that distortion mainly arises from the static $I - V$ characteristic, the contributions coming from the nonlinearity of the bias-dependent capacitances, and access resistances being negligible.

12.1 THE EQUIVALENT SMALL-SIGNAL CIRCUIT AT RF

The QS small-signal equivalent circuit including the substrate network corresponding to Figure 11.9(c) is shown in Figure 12.1(a) for operation in the linear region and in Figure 12.1(b) for saturation. Note that the capacitances include both the intrinsic and extrinsic capacitances:

$$C_{GS} = C_{GSi} + C_{GSo}, \tag{12.1a}$$

$$C_{GD} = C_{GDi} + C_{GDo}, \tag{12.1b}$$

$$C_{GB} = C_{GBi} + C_{GBo}, \tag{12.1c}$$

$$C_{BS} = C_{BSi} + C_{BSj}, \tag{12.1d}$$

$$C_{BD} = C_{BDi} + C_{BDj}, \tag{12.1e}$$

where the intrinsic capacitances are given by (5.50a) and (5.51a), the overlap capacitances by (10.13), and (10.20), and the junction capacitances by (10.33).

Charge-Based MOS Transistor Modeling: The EKV Model for Low-Power and RF IC Design   C. Enz and E. Vittoz
© 2006 John Wiley & Sons, Ltd.

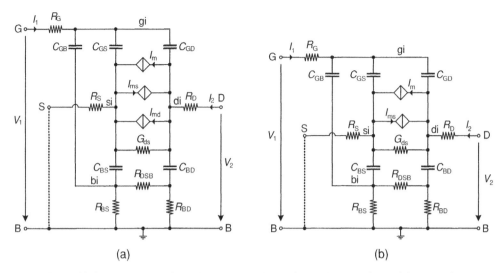

(a)   (b)

Figure 12.1 Equivalent RF small-signal circuit: (a) in the linear region; (b) in saturation

The voltage-controlled current sources (VCCS) are defined by

$$I_m \triangleq Y_m [V(gi) - V(bi)], \tag{12.2a}$$

$$I_{ms} \triangleq Y_{ms} [V(si) - V(bi)], \tag{12.2b}$$

$$I_{md} \triangleq Y_{md} [V(di) - V(bi)], \tag{12.2c}$$

where $V(k)$ with $k \in \{gi, si, di, bi\}$ stands for the potential at node k. The transadmittances Y_m, Y_{ms} and Y_{md} are given by (5.58), (5.56), and (5.57), which are repeated here for convenience:

$$Y_m = G_m (1 - j\omega\tau_{qs}) = G_m - j\omega C_m, \tag{12.3a}$$

$$Y_{ms} = G_{ms} (1 - j\omega\tau_{qs}) = G_{ms} - j\omega C_{ms}, \tag{12.3b}$$

$$Y_{md} = G_{md} (1 - j\omega\tau_{qs}) = G_{md} - j\omega C_{md}. \tag{12.3c}$$

Remember that the gate transadmittance, transconductance, and transcapacitance are related to the source and drain transadmittances, transconductances, and transcapacitances according to (5.37), (5.9), and (5.59):

$$Y_m = \frac{Y_{ms} - Y_{md}}{n}, \tag{12.4a}$$

$$G_m = \frac{G_{ms} - G_{md}}{n}, \tag{12.4b}$$

$$C_m = \frac{C_{ms} - C_{md}}{n}. \tag{12.4c}$$

12.2 Y-PARAMETERS ANALYSIS

The small-signal behavior of RF MOS transistors at high frequency is validated by measuring the S-parameters versus frequency at several operating points of a single transistor connected in common source as shown in Figure 12.1 by the dashed line. The S-parameters are usually measured directly on wafer using probes connecting the pads. Most often the input and output ports are connected by ground-signal-ground or GSG probes. The measured S-parameters therefore also include the effect of the RF pads used to connect the device. The effects of the pads have then to be de-embedded from the measured S-parameters using either a one-, two- or even a three-step procedure [139, 140] in order to move the reference planes from the end of the tips to the gate and drain nodes. The measurements shown below used a two-step de-embedding procedure which is usually sufficient for measurements up to 10 GHz. The two-step de-embedding procedure requires the measurement of the open and short structures. The open structure is simply the same as the RF MOS transistor except that the RF MOS transistor is taken out leaving the gate and drain open. The short structure is the same but now the RF MOS transistor is replaced by a short circuit between the gate and the drain.

For convenience, the de-embedded S-parameters can then be transformed into Y-parameters which are often easier to analyze [138, 145, 146]. The measured Y-parameters can then be compared either to the analytical or eventually the simulated Y-parameters corresponding to the equivalent circuit shown in Figure 11.9(c).

The equivalent small-signal circuit in saturation of Figure 12.1(b) will be validated by first deriving the corresponding analytical Y-parameters and comparing them to the de-embedded measurements. Since the capacitances are all approximately proportional to the total gate width $W = N_f W_f$ and since the source and drain terminal resistances are inversely proportional to W (see (10.4)), the time constants due to the terminal resistances depend only on the gate length L_f, the overlap length L_{ov}, or the diffusion width H_{dif}. The latter dimensions are usually taken as minimum to achieve the highest cutoff frequency. Therefore the poles due to the terminal resistances are at a much higher frequency than typically the transit frequency, so that they can be basically neglected when calculating the Y-parameters and the related quantities. Neglecting the substrate resistances also in the small-signal circuit of Figure 12.1(b) (i.e., assuming that they are zero) allows to derive the following analytical expressions for the Y-parameters in saturation:

$$Y_{11} \cong \frac{j\omega C_G}{1 + j\omega R_G C_G}, \tag{12.5a}$$

$$Y_{12} \cong \frac{-j\omega C_{GD}}{1 + j\omega R_G C_G}, \tag{12.5b}$$

$$Y_{21} \cong \frac{G_m - j\omega(C_{GD} + C_m)}{1 + j\omega R_G C_G}, \tag{12.5c}$$

$$Y_{22} \cong \frac{G_{ds} + \omega^2 R_G C_{GD} C_m + j\omega(C_{GD} + C_{BD})}{1 + j\omega R_G C_G}, \tag{12.5d}$$

where it has been assumed that $G_m R_G \ll 1$ and $G_{ds} \ll G_m$. Capacitance C_G is the total capacitance at the gate

$$C_G \triangleq C_{GS} + C_{GD} + C_{GB}, \tag{12.6}$$

which includes both the intrinsic and extrinsic capacitances. Equation (12.5) can be further simplified assuming that $\omega R_G C_G \ll 1$:

$$Y_{11} \cong \omega^2 R_G C_G^2 + j\,\omega C_G, \tag{12.7a}$$

$$Y_{12} \cong -\omega^2 R_G C_{GD} C_G - j\,\omega C_{GD}, \tag{12.7b}$$

$$Y_{21} \cong G_m - \omega^2 R_G C_G (C_{GD} + C_m) - j\,\omega(C_{GD} + C_m), \tag{12.7c}$$

$$Y_{22} \cong G_{ds} + \omega^2 R_G (C_G C_{BD} + C_G C_{GD} + C_{GD} C_m) \tag{12.7d}$$
$$+ j\,\omega(C_{BD} + C_{GD}).$$

One of the advantage of having the simple analytical expressions for the Y-parameters given by (12.7) is that they can be used for a direct extraction of the RF model parameters from measurements as presented in [147, 148]. For example, C_G can be extracted as $\Im\{Y_{11}\}/\omega$ and C_{GD} as $|\Im\{Y_{12}\}|/\omega$. C_{GS} can then be extracted as

$$C_{GS} = C_G - C_{GD} - C_{GB} \cong C_G - C_{GD}, \tag{12.8}$$

since in strong inversion and in saturation, C_{GB} is usually much smaller than C_{GS}. The gate resistance can be extracted from (12.7) as

$$R_G = \frac{\Re\{Y_{11}\}}{\Im\{Y_{11}\}^2}. \tag{12.9}$$

The extraction of the capacitances in saturation and of the gate resistance in the linear region are illustrated in Figures 12.2(a) and 12.2(b) respectively. The fact that the extracted values of the components are constant over frequency indicates that the equivalent circuit and the simplified analytical expressions are correct in the frequency range considered.

The Y-parameters given by (12.7) are plotted in Figure 12.3 (dashed line) and compared to the measured values (symbols) obtained from a two-step de-embedding process. As can be seen from Figure 12.3, the analytical expressions match the measurements very well even up to 10 GHz except for Y_{22}. This discrepancy is due to the substrate coupling effect which has been

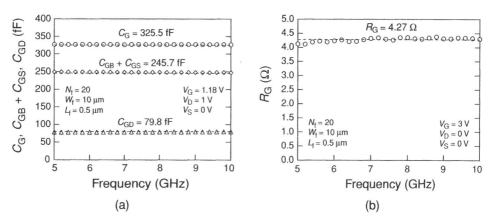

(a)   (b)

Figure 12.2 Direct extraction of some of the components of the small-signal circuit of Figure 12.1(b): (a) gate capacitances C_G, C_{GD}, and $C_{GB} + C_{GS}$; (b) gate resistance R_G (Reproduced by permission of IEEE from [147])

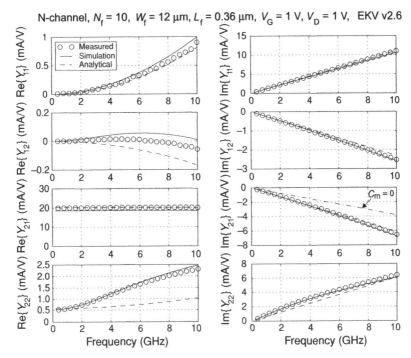

Figure 12.3 Comparison between the measured, simulated, and analytical Y-parameters versus frequency [48,49,52]. The simulated results are obtained from an ac simulation corresponding to the QS small-signal circuit presented in Figure 12.1(b), whereas the analytical results are obtained directly from (12.7) (Reproduced by permission of IEEE and Springer)

ignored in the derivation of (12.7). Including the substrate network leads to very complex expressions of the Y-parameters that are not easy to use. The substrate network has been included in the model used for simulation as described by the complete QS circuit of Figure 11.9(c) using the EKV v2.6 compact model for the intrinsic part. Most parameters specific to the RF part have been extracted using the methodology presented in [147, 148]. The simulations with the complete QS model of Figure 12.1(b) are also presented in Figure 12.3 (straight lines). They show a very good match with measurements including the output admittance Y_{22}. Note that the discrepancies in $\Re\{Y_{12}\}$ are not critical since Y_{12} is dominated by its imaginary part corresponding approximately to ωC_{GD} which is about 10 times larger than $\Re\{Y_{12}\}$. Similar results have been obtained for other operating points and other device geometries using the same scalable model [48,49,52].

Figure 12.3 also shows that there may be a big discrepancy in $\Im\{Y_{21}\}$ if the transcapacitance C_m is neglected in the expression of Y_{21} given by (12.7).

Note that the results of $\Re\{Y_{11}\}$ are particularly sensitive to the de-embedding procedure. Also note that accurately modeling the bias dependence of C_{GDo} is crucial to fit $\Im\{Y_{12}\}$ at high bias.

The extraction of the substrate resistances requires a more complicated procedure. After having extracted R_G and R_D, they are de-embedded from the Y-parameters, resulting in the prime Y-parameters corresponding to the circuit of Figure 12.4(a). The Y'_{22} parameter of Figure 12.4(a) is obtained by grounding the intrinsic gate node gi. The resulting circuit can be further simplified by assuming that $\omega R_S C_{GS} \ll 1$ (i.e., replacing resistance R_S by a short circuit) resulting in the circuit of Figure 12.4(b). The Y'_{22} parameter of the circuit of Figure 12.4(b)

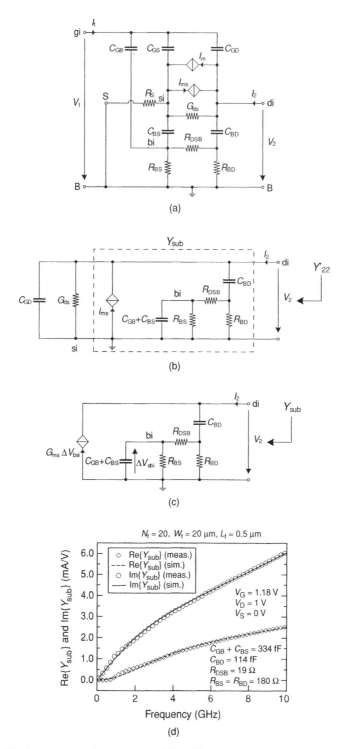

Figure 12.4 Π substrate network extraction [147, 148]. (a) Small-signal circuit obtained after de-embedding of access resistances R_G and R_D. (b) Y_{22} parameter of the circuit shown in (a) after simplification assuming $\omega R_S C_{GS} \ll 1$ (i.e., replacing resistance R_S by a short circuit). (c) Substrate admittance Y_{sub} obtained after de-embedding of C_{GD} and G_{ds} (Reproduced by permission of IEEE from [147])

is then given by

$$Y_{22}' = Y_{sub} + G_{ds} + j\,\omega C_{GD}, \tag{12.10}$$

where Y_{sub} corresponds to the admittance of the circuit shown in Figure 12.4(c) which includes the substrate network. It can be obtained from the Y_{22}' parameter as

$$Y_{sub} = Y_{22}' - G_{ds} - j\,\omega C_{GD}, \tag{12.11}$$

where G_{ds} is extracted from Y_{22}' as

$$R_{ds} \triangleq \frac{1}{G_{ds}} = \frac{1}{\Re\{Y_{22}'\}}\bigg|_{\omega=0} - R_S, \tag{12.12}$$

since $Y_{sub} = 0$ at $\omega = 0$. An example of an extraction of Y_{sub} from measured Y-parameters is shown in Figure 12.4(d). Resistances R_{DSB}, R_{BS}, and R_{BD} can unfortunately not be extracted directly but require some optimization on the circuit shown in Figure 12.4(c). The result of this fitting is shown in Figure 12.4(d) with the values of the components of the Y_{sub} circuit where it has been assumed that $R_{BS} = R_{BD}$.

When nonminimum channel length devices are used (particularly P-channel transistors since their mobility is much lower than that of N-channel devices), they might operate in NQS regime. This is illustrated by Figure 12.5, showing the measured de-embedded Y-parameters

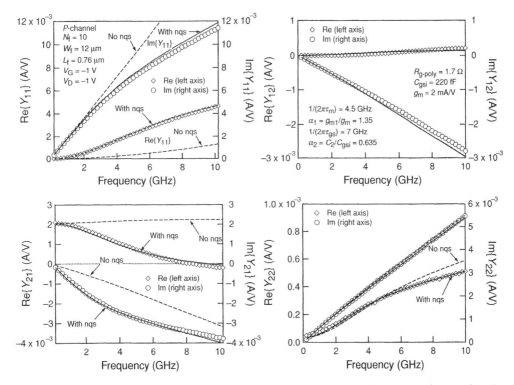

Figure 12.5 Comparison of the Y-parameters versus frequency measured on a nonminimum length P-channel device with the results obtained from the simple QS static model (dashed lines) of Figure 12.1(b) and from the first-order NQS model (Reproduced by permission of IEEE from [49])

of a P-channel RF MOS transistor having a length of 0.75 μm. The Y-parameters of the QS model are plotted in Figure 12.5 as dashed lines. A large discrepancy can be observed, particularly in the Y_{11} and Y_{21} parameters since the device is biased in saturation. As shown in Figure 12.5, the fit can already be significantly improved by using a first-order NQS model. The latter uses the circuit of Figure 12.7, where the intrinsic transadmittances and admittances are replaced by their first-order approximation expressions (5.42) and (5.49), respectively.

A good fit of the Y-parameters over frequency at a particular operating point is not that difficult to obtain. On the other hand, having the simulated Y-parameters fit the measurements over a wide range of bias at a given frequency is much more difficult to achieve. It not only requires an accurate intrinsic compact model but also requires accurate bias-dependent models for the extrinsic components. The bias dependence of the equivalent circuit shown in Figure 11.9(c) has been checked at a frequency of 1 GHz over a wide bias range by sweeping the gate voltage. The measured and simulated Y-parameters are plotted versus the inversion factor i_f in Figure 12.6 for the same device used in Figure 12.3. An excellent fit is obtained over

N-channel, $N_f = 10$, $W_f = 12$ μm, $L_f = 0.36$ μm, $f = 1$ GHz, $V_D = 0.5, 1, 1.5$ V, EKV v2.6

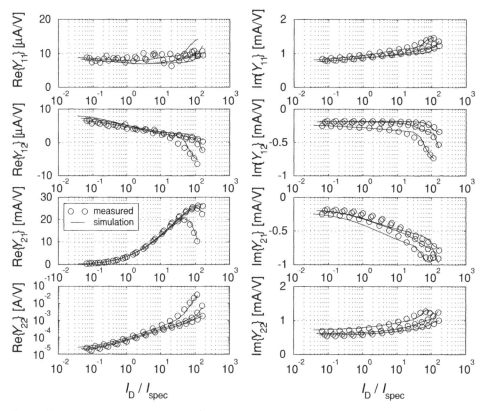

Figure 12.6 Comparison between the measured and simulated Y-parameters versus inversion factor for $V_D = 0.5, 1, 1.5$ V [48, 49, 52]. The simulated results are obtained from an ac simulation performed for each bias point and corresponding to the QS small-signal circuit presented in Figure 12.1(b) (Reproduced by permission of IEEE from [52])

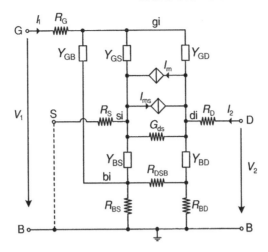

Figure 12.7   Full RF NQS small-signal equivalent circuit in saturation

more than three decades of currents, which also validates the bias dependence of the intrinsic and extrinsic parts.

The full NQS effects can be included by replacing the intrinsic part by the NQS model discussed in Section 5.2 resulting in the circuit shown in Figure 12.7 for operation in saturation. The admittances Y_{GB}, Y_{GS}, Y_{GD}, Y_{BS}; and Y_{BD} of Figure 12.7 include the intrinsic NQS admittances plus the extrinsic components:

$$Y_{GB} \triangleq Y_{GBi} + C_{GBo}, \tag{12.13a}$$

$$Y_{GS} \triangleq Y_{GSi} + C_{GSo}, \tag{12.13b}$$

$$Y_{GD} \triangleq Y_{GDi} + C_{GDo}, \tag{12.13c}$$

$$Y_{BS} \triangleq Y_{BSi} + C_{BSj}, \tag{12.13d}$$

$$Y_{BD} \triangleq Y_{BDi} + C_{BDj}. \tag{12.13e}$$

The small-signal circuit of Figure 12.1(a) should be completed with the different noise sources. This will be done in Chapter 13. The next section will look at the large-signal model at RF.

12.3 THE LARGE-SIGNAL MODEL AT RF

The small-signal circuit discussed above is useful for the design and simulation of RF blocks that operate in small-signal such as low-noise amplifiers (LNA). On the other hand, many other RF circuits such as mixers or voltage-controlled oscillators (VCO) often operate in large signal. The device nonlinearities are generating harmonic frequency components and intermodulation products that have to be accurately predicted. To this purpose, it is important to identify all the possible sources of nonlinearity in the device. Looking to the large-signal equivalent circuit of Figure 11.9, the intrinsic part is obviously nonlinear, but as discussed in Chapter 10, actually all the extrinsic components are nonlinear as well, since they are bias dependent.

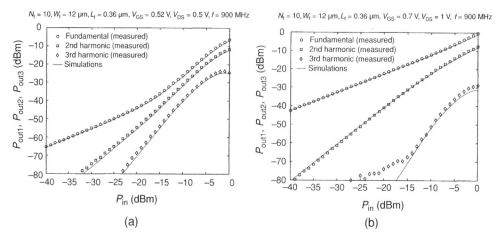

Figure 12.8 Comparison between the measured and simulated output power of the fundamental, second and third harmonics of a common source RF amplifier at 900 MHz versus the power of the input signal at the gate, for a low-bias condition (a) and medium bias condition (b)

The low-frequency large-signal behavior is mainly captured by the static nonlinear $I - V$ characteristics, whereas at RF, the nonlinearities of the capacitances may also contribute. The large-signal equivalent circuit given in Figure 11.9 has been evaluated and compared to measurements performed at 900 MHz. The simulation have been performed with a harmonic balanced simulator on a common-source device. Note that the subcircuit parameters have been extracted from dc and Y-parameter measurements and no additional fitting was required. Figure 12.8 shows the fundamental, second, and third harmonics versus the input power for two different bias points. The match between simulations and measurement is very good for both the low and medium bias condition. The main contributor to the nonlinearities remains the intrinsic nonlinear $I - V$ characteristic. Indeed keeping all the capacitances (extrinsic and

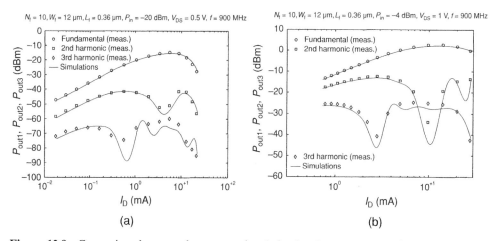

Figure 12.9 Comparison between the measured and simulated output power of the fundamental, second and third harmonics of a common-source RF amplifier at 900 MHz versus the bias current for two different input signal power and bias conditions (Reproduced by permission of IEEE from [52])

intrinsic) and the access resistances constant does not significantly change the results presented in Figure 12.8.

Figure 12.9 presents the fundamental, second and third harmonics as a function of the bias drain current for a given input power and frequency. The match between measurement and simulation is again very good. The model even captures the different nulls appearing in the second and third harmonics of Figure 12.9. Again, the main nonlinear contribution comes from the $I - V$ characteristic, the capacitances and access resistances playing only a secondary role.

The above measurements evaluated only the equivalent circuit looking at the power spectrum without consideration for the phase. More advanced RF nonlinear measurement techniques including both amplitude and phase can be performed [149].

13 The Noise Model at RF

The chapter starts with the theory of noisy two-port networks where it is shown that two noise sources (for example, a series voltage source and a shunt current source at the input) as well as their correlation admittance are required to fully characterize the noisy two-port. The important definition of noise figure is introduced and it is shown that this noise figure can be minimized by setting the source admittance to an appropriate value. It is shown that the noisy two-port can be characterized by a total of four noise parameters, including the minimum noise figure, the input-referred noise resistance, and the real and imaginary parts of the optimum source admittance that minimize the noise figure. The noise model discussed in the previous chapters was exclusively looking at the noise produced at the drain. At high frequency, the capacitive coupling between the channel and the gate on one hand and between the channel and bulk on the other hand induce noise currents to flow in the gate and bulk terminals in addition to the noise produced at the source and drain. A complete non-quasi-static thermal noise model is then presented in Section 13.2, where the induced gate and substrate noise and their correlation to the drain noise are derived. The chapter ends with the noise analysis of a common-source amplifier, deriving the four noise parameters with different levels of approximation.

13.1 THE HF NOISE PARAMETERS

13.1.1 The Noisy Two-Port

The noise at the output of an amplifier arises from the noise generated within the amplifier plus the noise already present at the input and amplified by the amplifier. Such a small-signal noisy amplifier can be represented by the noisy two-port network shown in Figure 13.1(a). The output noise current I_{nout} includes both the noise coming from the amplifier and the amplified input noise. Note that we look at the short-circuited output noise current because we will characterize the two-port using the Y-parameters which are defined in short-circuit conditions. The noisy two-port of Figure 13.1(a) can be represented by a noiseless two-port to which two noise sources have to be added (actually one noise source per port) [145, 146]. Figure 13.1(b) shows one possible representation where two current noise sources have been added to the

Charge-Based MOS Transistor Modeling: The EKV Model for Low-Power and RF IC Design   C. Enz and E. Vittoz
© 2006 John Wiley & Sons, Ltd.

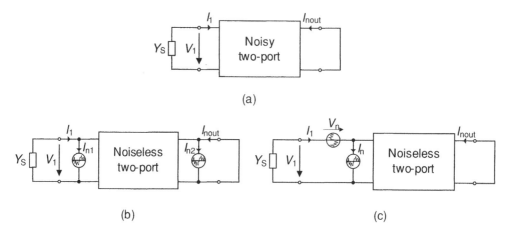

(a)

(b)   (c)

Figure 13.1   Different equivalent representations of a linear noisy two-port [145, 146]. (a) Noisy two-port network. (b) Admittance representation of the noisy two-port network with noiseless two-port in the middle. (c) Chain matrix representation of the noisy two-port network with noiseless two-port in the middle

noiseless two-port, one at the input port (I_{n1}) and one at the output port (I_{n2}). The two-port equations are then modified according to

$$I_1 = Y_{11} V_1 + Y_{12} V_2 + I_{n1}$$
$$I_2 = Y_{21} V_1 + Y_{22} V_2 + I_{n2},$$

(13.1)

where the complex currents and voltages are phasors defined by $I \triangleq \widehat{I} e^{j\omega t}$ and $V \triangleq \widehat{V} e^{j\omega t}$ with $\widehat{I}$ and $\widehat{V}$ being the current and voltage amplitudes. The current noise sources I_{n1} and I_{n2} can be evaluated from the noisy two-port internal noise sources by short-circuiting the input and output ports:

$$I_{n1} = I_1|_{V_1=V_2=0},$$ (13.2a)
$$I_{n2} = I_2|_{V_1=V_2=0}.$$ (13.2b)

This representation is useful for example for including the induced gate noise and the drain noise. Noise current source I_{n1} then represents the induced gate noise, whereas I_{n2} represents the drain noise.

For the calculation of the noise figure, it is more convenient to refer both noise sources directly at the input of the two-port, as shown in Figure 13.1(c), and use the ABCD matrix representation:

$$V_1 = A V_2 - B I_2 + V_n$$
$$I_1 = C V_2 - D I_2 + I_n,$$

(13.3)

where V_n is a noise voltage source that represents all the noise of the device referred to the input when the source impedance is zero (input short-circuited) and I_n is a noise current source that represents all the noise of the device referred to the input when the source admittance is zero (input open circuited):

$$V_n = B \, I_2|_{V_1=V_2=0} = -\frac{1}{Y_{21}} \, I_2|_{V_1=V_2=0}, \tag{13.4a}$$

$$I_n = D \, I_2|_{I_1=V_2=0} = -\frac{Y_{11}}{Y_{21}} \, I_2|_{I_1=V_2=0}. \tag{13.4b}$$

V_n is calculated using (13.4a) by first evaluating the Y-parameters, calculating the output current I_2 with the input and output in short circuit, and dividing this value by $-Y_{21}$. Similarly, I_n is calculated by first evaluating the output current I_2 with the input open and the output in short circuit and multiplying this value by $-Y_{11}/Y_{21}$. Examples of V_n and I_n noise sources are presented in Section 13.3.

The two noise sources V_n and I_n of Figure 13.1(c) are related to the noise sources I_{n1} and I_{n2} of Figure 13.1(b) by

$$V_n = -\frac{I_{n2}}{Y_{21}}, \tag{13.5a}$$

$$I_n = I_{n1} - \frac{Y_{11}}{Y_{21}} I_{n2}. \tag{13.5b}$$

13.1.2 The Correlation Admittance

Since both of the noise sources V_n and I_n (or equivalently I_{n1} and I_{n2}) are due to the same physical noise sources within the two-port, they are usually correlated. To account for the correlation existing between sources I_n and V_n, noise source I_n can be written as

$$I_n = I_{nu} + I_{nc} = I_{nu} + Y_c \, V_n, \tag{13.6}$$

where I_{nu} stands for the uncorrelated part of I_n, and I_{nc} represents the part of I_n that is fully correlated to V_n. The definition of the correlation admittance Y_c is obtained by multiplying (13.6) by V_n^* and averaging. This results in

$$\overline{I_n V_n^*} = \underbrace{\overline{I_{nu} V_n^*}}_{=0} + Y_c \, \overline{|V_n|^2} = Y_c \, \overline{|V_n|^2}, \tag{13.7}$$

where the first term of the right-hand side of (13.7) is zero since by definition I_{nu} is not correlated to V_n. Y_c is then given by

$$Y_c = \frac{\overline{I_n V_n^*}}{\overline{|V_n|^2}}. \tag{13.8}$$

The mean-square value of source I_n is given by

$$\overline{|I_n|^2} = \overline{|I_{nu}|^2} + |Y_c|^2 \, \overline{|V_n|^2} + \underbrace{Y_c^* \, \overline{I_{nu} V_n^*}}_{=0} + \underbrace{Y_c \, \overline{V_n I_{nu}^*}}_{=0}$$

$$= \overline{|I_{nu}|^2} + \underbrace{|Y_c|^2 \, \overline{|V_n|^2}}_{\triangleq \overline{|I_{nc}|^2}}. \tag{13.9}$$

The correlated and uncorrelated parts of I_n can also be written in terms of the correlation factor c

$$\overline{|I_{nu}|^2} = \overline{|I_n|^2} - |Y_c|^2 \, \overline{|V_n|^2} = (1 - |c|^2) \, \overline{|I_n|^2}, \tag{13.10a}$$

$$\overline{|I_{nc}|^2} = |Y_c|^2 \, \overline{|V_n|^2} = |c|^2 \, \overline{|I_n|^2}, \tag{13.10b}$$

with

$$c = \frac{\overline{I_n V_n^*}}{\sqrt{\overline{|I_n|^2} \, \overline{|V_n|^2}}} = Y_c \sqrt{\frac{\overline{|V_n|^2}}{\overline{|I_n|^2}}}. \tag{13.11}$$

The same relations apply for the power spectral densities (PSD) S_v and S_i of noise sources V_n and I_n respectively

$$S_{iu} = S_i - |Y_c|^2 \, S_v = (1 - |c|^2) \, S_i, \tag{13.12a}$$

$$S_{ic} = |Y_c|^2 \, S_v = |c|^2 \, S_i, \tag{13.12b}$$

$$S_i = S_{iu} + S_{ic}, \tag{13.12c}$$

where S_{iu} and S_{ic} are the parts of PSD S_i that are respectively uncorrelated and fully correlated to source V_n. It is convenient to treat the noise sources V_n and I_n as if they were thermal noise sources defined by

$$S_v \triangleq 4kT R_v, \tag{13.13a}$$

$$S_i \triangleq 4kT G_i, \tag{13.13b}$$

$$S_{iu} \triangleq 4kT G_{iu}, \tag{13.13c}$$

$$S_{ic} \triangleq 4kT G_{ic}. \tag{13.13d}$$

The above defined resistance R_v and conductances G_i, G_{iu}, and G_{ic} are then related according to (13.12)

$$G_{iu} = G_i - |Y_c|^2 \, R_v = (1 - |c|^2) \, G_i, \tag{13.14a}$$

$$G_{ic} = |Y_c|^2 \, R_v = |c|^2 \, G_i, \tag{13.14b}$$

$$G_i = G_{iu} + G_{ic}, \tag{13.14c}$$

where the correlation factor c is given by

$$c = Y_c \sqrt{\frac{R_v}{G_i}}, \tag{13.15}$$

or its square magnitude

$$|c|^2 = (G_c^2 + B_c^2) \frac{R_v}{G_i}, \tag{13.16}$$

with $Y_c = G_c + jB_c$.

Note that in general R_v, G_i, G_{iu}, and G_{ic} are frequency dependent since they depend on the real physical noise sources within the two-port through the Y-parameters.

To be fully characterized, the noise of a two-port requires four *noise parameters*, for example, R_v, G_{iu}, G_c, and B_c. As will be shown in the next section, all these noise parameters are required to evaluate the noise factor.

13.1.3 The Noise Factor

The noise factor is defined as the ratio of the total output noise to the output noise due only to the noise already present at the two-port input:

$$F \triangleq \frac{\text{total output noise}}{\text{output noise due to input source}} = 1 + \frac{N_a}{N_s}, \tag{13.17}$$

where N_a is the noise power added by the two-port and N_s is the noise power coming from the source. To calculate the noise factor of a two-port in terms of the equivalent noise sources V_n and I_n and their correlation admittance, we first have to evaluate the total noise at the two-port output. Since the current noise contribution coming from the source I_{nrs} and that coming from V_n and I_n are amplified the same way, we can just calculate the total noise current $I_{n\,tot}$ in short-circuit condition at the input of the two-port. With the help of Figure 13.2, the current source $I_{n\,tot}$ is given by

$$I_{n\,tot} = I_{nrs} + I_n + Y_s\,V_n, \tag{13.18}$$

and the mean-square value is

$$\overline{|I_{n\,tot}|^2} = \overline{|I_{nrs} + I_n + Y_s\,V_n|^2} = \overline{|I_{nrs}|^2} + \overline{|I_n + Y_s\,V_n|^2}, \tag{13.19}$$

which can be rewritten in terms of correlated and uncorrelated noise as

$$\begin{aligned}
\overline{|I_{n\,tot}|^2} &= \overline{|I_{nrs}|^2} + \overline{|I_{nu} + I_{nc} + Y_s\,V_n|^2} \\
&= \overline{|I_{nrs}|^2} + \overline{|I_{nu} + (Y_c + Y_s)\,V_n|^2} \\
&= \overline{|I_{nrs}|^2} + \overline{|I_{nu}|^2} + |Y_c + Y_s|^2\,\overline{|V_n|^2}.
\end{aligned} \tag{13.20}$$

The noise factor is then simply given by

$$F \triangleq \frac{\overline{|I_{n\,tot}|^2}}{\overline{|I_{nrs}|^2}} = 1 + \frac{\overline{|I_{nu}|^2} + |Y_c + Y_s|^2\,\overline{|V_n|^2}}{\overline{|I_{nrs}|^2}}. \tag{13.21}$$

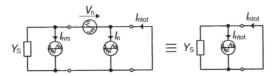

Figure 13.2   Total equivalent noise at the input of a noisy two-port

The above definition of the noise factor uses power or mean square values and is therefore independent of frequency. Another definition uses the above defined PSD instead of the mean-square values:

$$
\begin{aligned}
F &= 1 + \frac{S_{iu} + |Y_c + Y_s|^2 S_v}{4kTG_s} = 1 + \frac{G_{iu} + |Y_c + Y_s|^2 R_v}{G_s} \\
&= 1 + \frac{G_{iu}}{G_s} + \frac{R_v}{G_s}\left[(G_s + G_c)^2 + (B_s + B_c)^2\right],
\end{aligned}
\tag{13.22}
$$

where G_s and B_s are defined by $Y_s \triangleq G_s + jB_s$.

Note that the noise factor defined by (13.22) is in general frequency dependent, since R_v, G_{iu}, Y_c, and Y_s can all depend on frequency. For this reason, it is sometimes also called the *spot noise factor*. The two definitions of the noise factor given by (13.21) and (13.22) are about equivalent for narrow band systems since

$$
\overline{|V_n|^2} \cong S_v \, B = 4kTR_v \, B, \tag{13.23a}
$$
$$
\overline{|I_n|^2} \cong S_i \, B = 4kTG_i \, B, \tag{13.23b}
$$
$$
\overline{|I_{nu}|^2} \cong S_{iu} \, B = 4kTG_{iu} \, B, \tag{13.23c}
$$
$$
\overline{|I_{nc}|^2} \cong S_{ic} \, B = 4kTG_{ic} \, B, \tag{13.23d}
$$

where B is the system noise bandwidth.

13.1.4 Minimum Noise Factor

Once the two-port is characterized by its four noise parameters R_v, G_{iu}, G_c, and B_c, the noise factor given by (13.22) reaches a minimum value called F_{min} (or $NF_{min} \triangleq 10 \log F_{min}$) for a particular value of the source admittance $Y_{opt} \triangleq G_{opt} + jB_{opt}$. The optimum source conductance G_{opt} and susceptance B_{opt} are obtained from the differentiation of (13.22) and can be expressed in terms of the four noise parameters according to

$$
G_{opt} = \sqrt{\frac{G_{iu}}{R_v} + G_c^2} = \sqrt{\frac{G_i}{R_v} - B_c^2} \tag{13.24a}
$$
$$
B_{opt} = -B_c. \tag{13.24b}
$$

The minimum noise factor F_{min} can then be written as

$$
F_{min} = 1 + 2R_v(G_{opt} + G_c) = 1 + 2R_v\left(\sqrt{\frac{G_{iu}}{R_v} + G_c^2} + G_c\right). \tag{13.25}
$$

The actual noise factor can be written in terms of F_{min}, R_v, G_{opt}, B_{opt} and the source conductance G_s and susceptance B_s as

$$
F = F_{min} + \frac{R_v}{G_s}\left[(G_s - G_{opt})^2 + (B_s - B_{opt})^2\right]. \tag{13.26}
$$

Equation (13.26) clearly indicates that for the noise factor to reach its minimum value, F_{min}, requires both conditions $G_s = G_{opt}$ and $B_s = B_{opt}$. This situation is called *noise matching*. In the same way the noise factor can be minimized, the power gain can also be maximized by setting the source admittance to some appropriate value. The latter situation is called *gain matching*. Unfortunately, most of the time noise matching does not coincide with gain matching.

Usually the parameters that are available from measurements are the four de-embedded noise parameters F_{min}, R_v, G_{opt}, and B_{opt}.[1] The latter allow to calculate the noise factor for any source impedance according to (13.26).

The four measured noise parameters F_{min}, R_v, G_{opt}, and B_{opt} can also be used to evaluate the four parameters characterizing the noise sources V_n and I_n of the noisy two-port of Figure 13.1(c) from (13.24a), (13.25), and (13.24b) respectively:

$$G_i = |Y_{opt}|^2 R_v = (G_{opt}^2 + B_{opt}^2) R_v, \qquad (13.27a)$$

$$G_c = \frac{F_{min} - 2R_v G_{opt} - 1}{2R_v}, \qquad (13.27b)$$

$$B_c = -B_{opt}. \qquad (13.27c)$$

Note that R_v is identical in both sets of parameters. Equations (13.27) allow to go from the measured noise parameter to the equivalent noisy two-port representation.

The correlated and uncorrelated parts of G_i can then be calculated as

$$G_{ic} = |c|^2 G_i, \qquad (13.28a)$$

$$G_{iu} = (1 - |c|^2) G_i, \qquad (13.28b)$$

where the correlation coefficient is given by

$$c = Y_c \frac{R_v}{G_i} = \frac{G_c + jB_c}{\sqrt{G_{opt}^2 + B_{opt}^2}} \qquad (13.29)$$

or

$$|c|^2 = \frac{G_c^2 + B_c^2}{G_{opt}^2 + B_{opt}^2}. \qquad (13.30)$$

13.2 THE HIGH-FREQUENCY THERMAL NOISE MODEL

At RF, the flicker noise is not present and the total noise is dominated by the thermal noise component which then sets the fundamental limit to signal resolution. As presented in Section 6.2, this channel noise is commonly modeled as a shunt current source between drain and source as shown in Figure 6.5. This simple model is not sufficient to predict the noise behavior at frequencies that get close or even beyond the channel cutoff frequency. At high frequency, the

[1] Actually most of the time noise measurement systems give the optimum reflection coefficient Γ_{opt} from which the optimum admittance can be calculated using the definition $\Gamma_{opt} = (Y_0 - Y_{opt})/(Y_0 + Y_{opt})$, where $1/Y_0 = 50\ \Omega$ is the characteristic impedance.

capacitive coupling existing between the noisy channel and the gate and bulk terminals induce additional noise currents to flow through those terminals in addition to the drain-to-source noise current [91, 150, 151]. Also, at higher frequency the channel can be looked at as a nonuniform RC transmission line introducing some phase shift to the signals traveling along the channel. All this results into a drain noise current that is no longer equal to the source noise current as it was assumed at lower frequency in Section 6.2. Since the physical source of thermal noise is still due to the channel resistance, the terminal noise currents are obviously correlated [91, 150, 151]. These additional noise currents and their correlation coefficients strongly affect all the four noise parameters [151], namely the input referred noise resistance R_v, the optimum source conductance G_{opt}, the optimum source susceptance B_{opt}, but particularly the minimum noise figure NF_{min}. It is therefore crucial to correctly model all these additional noise currents as well as their related correlation coefficients.

The downscaling of CMOS technology has resulted in a significant increase of the maximum transit frequency f_t, reaching or even exceeding 100 GHz for sub 0.1 µm technologies [152]. For RF applications operating in the gigahertz frequency range, the ratio between the transit frequency (or the channel cutoff frequency) and the operating frequency can then be increased in order to avoid any non-quasi-static effects. This is usually done at the expense of power consumption which has become a very important specification, particularly for portable and battery-operated devices. To save power, it is therefore important not to waste any bandwidth and therefore to operate with the lowest possible cutoff frequency. Also, the downscaling of CMOS technology is combined with a reduction of the supply voltage which results in a decrease of the overdrive voltage. For these two reasons, the operating points are therefore pushed away from the traditional strong inversion region toward moderate and even weak inversion [49, 50, 52]. For RF IC design in deep submicron CMOS processes, it therefore becomes more and more important to be able to correctly predict the terminal noise currents and their correlation coefficients in moderate and weak inversion regimes.

13.2.1 Generalized High-Frequency Noise Model

The thermal noise model presented in Section 6.2 is valid only at low frequency, i.e., for frequencies much lower than the channel cutoff frequency. At higher frequency the capacitive coupling between the channel and the gate and bulk terminals has to be accounted for. From multiport noise theory, it is known that each port requires its own noise source which can be either a voltage or current source. The MOS transistor is a four-terminal device and therefore requires four noise sources as indicated in Figure 13.3. Current noise sources have been chosen since all the following derivations are carried out using Y-parameters. As shown in Figure 13.3(b), the noisy MOS transistor of Figure 13.3(a) can then be replaced by a noiseless transistor and four additional noise current sources ΔI_{nD}, ΔI_{nS}, ΔI_{nG}, and ΔI_{nB} having PSD $S_{\Delta I_{nD}^2}$, $S_{\Delta I_{nS}^2}$, $S_{\Delta I_{nG}^2}$, and $S_{\Delta I_{nB}^2}$, respectively. Since the noise appearing at each terminal is generated from the same physical thermal noise source in the channel, the noise current sources ΔI_{nD}, ΔI_{nS}, ΔI_{nG}, and ΔI_{nB} are correlated. This correlation is accounted for by the cross-power spectral densities (CPSD) $S_{\Delta I_{nk} \Delta I_{nl}^*}$ with $k \neq l \in \{D, S, G, B\}$.

The signs of the terminal currents and the related noise current sources are defined as indicated in Figure 13.3. Note that, although one can choose any definition for the signs of these currents, it is important to be consistent in order to account for the correct correlations existing between them.

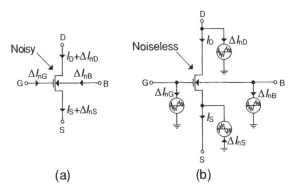

Figure 13.3   (a) HF noisy MOS transistor with terminal currents including noise fluctuations. (b) Equivalent HF noise circuit with noiseless MOS transistor

The PSD and CPSD are derived in Section 13.2.3 using the approach already described in Section 6.1 and applied to the high frequency case in the next section.

13.2.2 The Two-Transistor Approach at High Frequency

The procedure to derive the PSD and CPSD is similar to the two-transistor approach used in Sections 6.1 and 6.2 to derive the PSD of the thermal noise at the drain at low frequency, except that the equivalent small-signal circuit of Figure 6.3(b) has now to be replaced by the general non-quasi-static circuit as shown in Figure 13.4.[2] For long channel, we have shown in Section 6.2 that the PSD of the thermal noise voltage source coming from the infinitesimal portion of the channel resistance comprised between x and $x + \Delta x$ is simply proportional to the corresponding infinitesimal resistance ΔR and is given by

$$S_{\delta V_n^2} = \Delta R^2 S_{\delta I_n^2} = 4kT \, \Delta R = 4kT \, \frac{\Delta x}{W \mu_0 (-Q_i)}. \tag{13.31}$$

To simplify the further notation it is convenient to normalize the noise voltage and current source PSDs to $S_{V_{spec}^2} \triangleq 4kT/G_{spec}$ and $S_{I_{spec}^2} \triangleq 4kT G_{spec}$ respectively. The normalized noise voltage PSD due to the infinitesimal piece of channel ΔR is then given by

$$S_{\delta v_n^2}(\xi) \triangleq \frac{S_{\delta V_n^2}(x)}{S_{V_{spec}^2}} = G_{spec} \, \Delta R = \frac{\Delta \xi}{q_i(\xi)}, \tag{13.32}$$

with $\Delta \xi \triangleq \Delta x / L$.

The effect of noise source δV_n on the different terminal currents is obtained from an analysis of the non-quasi-static small-signal equivalent circuit of Figure 13.4(b) observing that admittances Y_{GBi1} and Y_{GBi2} have no influence since they are both short-circuited. The transfer functions from the local noise source δV_n in the channel to the terminal currents δI_{nD}, δI_{nS},

---

[2] Note that here we will be using the Thévenin equivalent noise voltage source instead of the Norton noise current source to make the derivation. As explained in Section 6.1 the two approaches are equivalent.

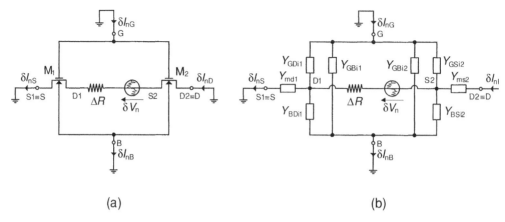

(a)   (b)

Figure 13.4 (a) Single transistor split into two separate transistors to evaluate the noise transadmittances between channel noise source δV_n and the terminal noise currents. (b) Corresponding non-quasi-static small-signal circuit

δI_{nG}, and δI_{nB} are then given by

$$Y_{nD}(\omega, x) \triangleq \frac{\delta I_{nD}}{\delta V_n} = -\frac{Y_{ms2}(Y_{md1} + Y_{GDi1} + Y_{BDi1})}{Y_{md1} + Y_{ms2} + Y_{GDi1} + Y_{BDi1} + Y_{GSi2} + Y_{BSi2}}, \qquad (13.33a)$$

$$Y_{nS}(\omega, x) \triangleq \frac{\delta I_{nS}}{\delta V_n} = -\frac{Y_{md1}(Y_{ms2} + Y_{GSi2} + Y_{BSi2})}{Y_{md1} + Y_{ms2} + Y_{GDi1} + Y_{BDi1} + Y_{GSi2} + Y_{BSi2}}, \qquad (13.33b)$$

$$Y_{nG}(\omega, x) \triangleq \frac{\delta I_{nG}}{\delta V_n} = \frac{Y_{ms2}Y_{GDi1} - Y_{md1}Y_{GSi2} + Y_{GDi1}Y_{BSi2} - Y_{GSi2}Y_{BDi1}}{Y_{md1} + Y_{ms2} + Y_{GDi1} + Y_{BDi1} + Y_{GSi2} + Y_{BSi2}}, \qquad (13.33c)$$

$$Y_{nB}(\omega, x) \triangleq \frac{\delta I_{nB}}{\delta V_n} = \frac{Y_{ms2}Y_{BDi1} - Y_{md1}Y_{BSi2} + Y_{BDi1}Y_{GSi2} - Y_{BSi2}Y_{GDi1}}{Y_{md1} + Y_{ms2} + Y_{GDi1} + Y_{BDi1} + Y_{GSi2} + Y_{BSi2}}. \qquad (13.33d)$$

Equations (13.33) can be further simplified remembering the basic relations between the bulk-to-drain and gate-to-drain admittances, and the bulk-to-source and gate-to-source admittances as stated in Section 5.2 by (5.39) which is repeated here for convenience:

$$Y_{BDi1} = (n-1)Y_{GDi1},$$
$$Y_{BSi2} = (n-1)Y_{GSi2}.$$

This leads to

$$Y_{nD}(\omega, x) = -\frac{Y_{ms2}(Y_{md1} + nY_{GDi1})}{Y_{md1} + Y_{ms2} + n(Y_{GDi1} + Y_{GSi2})}, \qquad (13.34a)$$

$$Y_{nS}(\omega, x) = -\frac{Y_{md1}(Y_{ms2} + nY_{GSi2})}{Y_{md1} + Y_{ms2} + n(Y_{GDi1} + Y_{GSi2})}, \qquad (13.34b)$$

$$Y_{nG}(\omega, x) = \frac{Y_{ms2}Y_{GDi1} - Y_{md1}Y_{GSi2}}{Y_{md1} + Y_{ms2} + n(Y_{GDi1} + Y_{GSi2})}, \qquad (13.34c)$$

$$Y_{nB}(\omega, x) = (n-1)Y_{nG}(\omega, x). \qquad (13.34d)$$

From (13.34d) it can be seen that the transadmittance from the noise source δV_n to the bulk noise current δI_{nB} is $n-1$ times that to the gate noise current δI_{nG}. This results in a PSD of the induced bulk noise current δI_{nB} that is $(n-1)^2$ that of the induced gate noise current δI_{nG}:

$$S_{\delta I_{nB}^2} = (n-1)^2 S_{\delta I_{nG}^2}. \tag{13.35}$$

Assuming a constant slope factor n, (13.35) is also valid for the total induced gate and bulk currents ΔI_{nG} and ΔI_{nB}:

$$S_{\Delta I_{nB}^2} = (n-1)^2 S_{\Delta I_{nG}^2}. \tag{13.36}$$

The noise transadmittances in (13.34) can be normalized by introducing the normalization conductances of each half transistors M_1 and M_2 defined by [78, 153]

$$G_{spec1} \triangleq 2n\mu_0 \frac{W}{x} C_{ox} U_T = \frac{G_{spec}}{\xi}, \tag{13.37a}$$

$$G_{spec2} \triangleq 2n\mu_0 \frac{W}{L-x} C_{ox} U_T = \frac{G_{spec}}{1-\xi}, \tag{13.37b}$$

and the frequency can be normalized using the specific frequency of each half transistors M_1 and M_2 defined as [78, 153]

$$\omega_{spec1} \triangleq \frac{\mu_0 U_T}{x^2} = \frac{\omega_{spec}}{\xi^2}, \tag{13.38a}$$

$$\omega_{spec2} \triangleq \frac{\mu_0 U_T}{(L-x)^2} = \frac{\omega_{spec}}{(1-\xi)^2}, \tag{13.38b}$$

with

$$\omega_{spec} \triangleq \frac{\mu_0 U_T}{L^2} \tag{13.39}$$

being the specific frequency of the full-length transistor. Introducing the above normalizations in (13.34) results in [78, 153]

$$y_{nD}(\Omega, \xi) \triangleq \frac{Y_{nD}}{G_{spec}} \tag{13.40a}$$

$$= -\frac{y_{ms2}(\Omega(1-\xi)^2)[y_{md1}(\Omega\xi^2) + n y_{GDi1}(\Omega\xi^2)]}{(1-\xi)y_{md1}(\Omega\xi^2) + \xi y_{ms2}(\Omega(1-\xi)^2) + n[(1-\xi)y_{GDi1}(\Omega\xi^2) + \xi y_{GSi2}(\Omega(1-\xi)^2)]},$$

$$y_{nS}(\Omega, \xi) \triangleq \frac{Y_{nS}}{G_{spec}} \tag{13.40b}$$

$$= -\frac{y_{md1}(\Omega\xi^2)[y_{ms2}(\Omega(1-\xi)^2) + n y_{GSi2}(\Omega(1-\xi)^2)]}{(1-\xi)y_{md1}(\Omega\xi^2) + \xi y_{ms2}(\Omega(1-\xi)^2) + n[(1-\xi)y_{GDi1}(\Omega\xi^2) + \xi y_{GSi2}(\Omega(1-\xi)^2)]},$$

$$y_{nG}(\Omega, \xi) \triangleq \frac{Y_{nG}}{G_{spec}} \tag{13.40c}$$

$$= \frac{y_{ms2}(\Omega(1-\xi)^2)y_{GDi1}(\Omega\xi^2) - y_{md1}(\Omega\xi^2)y_{GSi2}(\Omega(1-\xi)^2)}{(1-\xi)y_{md1}(\Omega\xi^2) + \xi y_{ms2}(\Omega(1-\xi)^2) + n[(1-\xi)y_{GDi1}(\Omega\xi^2) + \xi y_{GSi2}(\Omega(1-\xi)^2)]},$$

$$y_{nB}(\Omega, \xi) \triangleq \frac{Y_{nB}}{G_{spec}} = (n-1)y_{nG}(\Omega, \xi). \tag{13.40d}$$

where $\Omega \triangleq \omega/\omega_{\text{spec}}$ is the normalized frequency. The above noise transadmittances are then used in the next section to derive the terminal noise current PSD and CPSD.

13.2.3 Generic PSDs Derivation

The PSD and CPSD of each noisy terminal currents due to the channel thermal noise voltage source δV_n are given by

$$S_{\delta I_{nk}^2}(\omega, x) = |Y_{nk}(\omega, x)|^2 S_{\delta V_n^2}(x), \tag{13.41a}$$

$$S_{\delta I_{nk}\delta I_{nl}^*}(\omega, x) = Y_{nk}(\omega, x)Y_{nl}^*(\omega, x) S_{\delta V_n^2}(x), \tag{13.41b}$$

with $k \neq l \in \{D, S, G, B\}$.

Integrating these relations from source to drain and assuming that all noise sources in the channel are uncorrelated, the following normalized relation can be derived [78, 153]:

$$S_{\Delta I_{nk}^2}(\Omega) \triangleq \frac{S_{\Delta I_{nk}^2}(\omega)}{S_{I_{\text{spec}}^2}} = \int_0^1 |y_{nk}(\Omega, \xi)|^2 \frac{d\xi}{q_i(\xi)}, \tag{13.42a}$$

$$S_{\Delta I_{nk}\Delta I_{nl}^*}(\Omega) \triangleq \frac{S_{\Delta I_{nk}\Delta I_{nl}^*}(\omega)}{S_{I_{\text{spec}}^2}} = \int_0^1 y_{nk}(\Omega, \xi)y_{nl}^*(\Omega, \xi) \frac{d\xi}{q_i(\xi)}. \tag{13.42b}$$

At this point it is useful to introduce the following additional variables [78, 153]

$$\chi(\xi) \triangleq q_i(\xi) + \frac{1}{2}, \quad \chi_s \triangleq q_s + \frac{1}{2}, \quad \chi_d \triangleq q_d + \frac{1}{2}, \tag{13.43}$$

which greatly simplify the evaluation of the integrals in (13.42). Using these intermediate variables, integrating the same current in the complete transistor and in transistors M_1 and M_2 leads to

$$i_d = \chi_s^2 - \chi_d^2, \quad i_d\xi = \chi_s^2 - \chi^2, \quad i_d(1 - \xi) = \chi^2 - \chi_d^2. \tag{13.44}$$

Finally, the different normalized PSD and CPSD of (13.42) become [78, 153]

$$S_{\Delta I_{nk}^2}(\Omega) = \frac{1}{i_d} \int_{\chi_d}^{\chi_s} |y_{nk}(\Omega, \chi)|^2 \frac{4\chi}{2\chi - 1} d\chi, \tag{13.45a}$$

$$S_{\Delta I_{nk}\Delta I_{nl}^*}(\Omega) = \frac{1}{i_d} \int_{\chi_d}^{\chi_s} y_{nk}(\Omega, \chi)y_{nl}^*(\Omega, \chi) \frac{4\chi}{2\chi - 1} d\chi, \tag{13.45b}$$

where variable ξ used in the expressions of the normalized transadmittances given by (13.40) must also be expressed as a function of χ using the following relation:

$$\xi = \frac{\chi_s^2 - \chi^2}{i_d}. \tag{13.46}$$

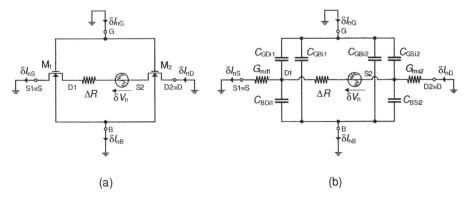

Figure 13.5 (a) Single transistor split into two separate transistors to evaluate the noise transadmittances between channel noise source δV_n and the terminal noise currents. (b) Corresponding quasi-static small-signal circuit

It is unfortunately not possible to get closed-form algebraic expressions for the integrals (13.45a) and (13.45b). However, a simple first-order approximation will be derived in the next section.

13.2.4 First-Order Approximation

A first-order approximation can be obtained by replacing each transistor M_1 and M_2 by their quasi-static small-signal model as shown in Figure 13.5(b), which clearly illustrates the capacitive coupling existing between the channel and the gate and bulk terminals inducing noise currents to flow. The small-signal Y-parameters used in (13.34) then reduce to $Y_{md1} = G_{md1}$, $Y_{ms2} = G_{ms2}$, $Y_{GDi1} = j\omega\, C_{GDi1}$, and $Y_{GSi2} = j\omega\, C_{GSi2}$.

Within the limits of this crude approximation, the PSD of the drain noise current ΔI_{nD} can then be calculated using (13.45a) as [78, 153]

$$S_{\Delta i^2_{nD}} \cong S_{\Delta i^2_{nS}} \cong \frac{4\chi_s^2 - 3\chi_s + 4\chi_s\chi_d - 3\chi_d + 4\chi_d^2}{6(\chi_s + \chi_d)}. \qquad (13.47)$$

Note first that in this first-order approximation, the PSD of the drain noise current is equal to the PSD of the source noise current. Therefore, it can be represented by a single current noise source connected between the drain and the source of the noiseless transistor as in the low-frequency case shown in Figure 6.5(b). In addition, after replacing χ_s and χ_d with their definitions (13.43), it appears that the first-order non-quasi-static drain current noise PSD (13.47) is identical to the low-frequency expression given in (6.19) [49, 52]. For the drain and source noise currents, this first-order model hence reduces to the simple low-frequency case.

The PSD of the induced gate noise current ΔI_{nG} can be evaluated from (13.45a) as [78, 153]

$$S_{\Delta i^2_{nG}} = \frac{S_{\Delta i^2_{nB}}}{(n-1)^2}$$

$$\cong \Omega^2 \frac{16\chi_s^4 + 16\chi_d^4 + 80\chi_s^3\chi_d + 80\chi_s\chi_d^3 + 168\chi_s^2\chi_d^2 - 15\chi_s^3 - 15\chi_d^3 - 75\chi_s^2\chi_d - 75\chi_s\chi_d^2}{540n^2(\chi_s + \chi_d)^5},$$

$$(13.48)$$

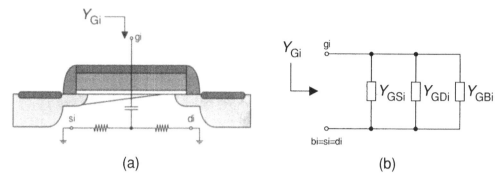

Figure 13.6 Admittance seen at the gate of the intrinsic transistor

which is *proportional to the square of the frequency*. This is simply due to the capacitive coupling existing between the channel and the gate. The bias dependence is a little bit more complex than the PSD of the drain current noise. As mentioned above, the PSD of the induced bulk noise current is simply $(n-1)^2$ that of the induced gate noise current given by (13.48).

For $V_D = V_S$, the transistor is basically a passive resistor[3] capacitively coupled to the gate and the bulk. Hence, the Nyquist theorem applies and the induced gate noise for $V_D = V_S$ should therefore be equal to the noise of the conductance $G_{Gi} \triangleq \Re\{Y_{Gi}\}$ seen at the gate. This is no longer strictly true for $V_D \neq V_S$, but a thermal noise parameter at the gate δ_{nG} can be defined as

$$\delta_{nG} \triangleq \frac{G_{nG}}{G_{Gi}} = \frac{S_{\Delta i_{nG}^2}}{g_{Gi}}, \tag{13.49}$$

where G_{nG} is the thermal noise conductance at the gate defined by

$$S_{\Delta I_{nG}^2} \triangleq 4kT\, G_{nG}(\omega) \propto \omega^2. \tag{13.50}$$

Similar to the definition of the thermal noise parameter at the drain, δ_{nG} measures the deviation of the actual PSD of the induced gate noise current with respect to that of the conductance G_{Gi}. With the quasi-static model, Y_{Gi} is simply given by

$$Y_{Gi} = j\,\omega\,C_{Gi} = j\,\omega\,(C_{GSi} + C_{GDi} + C_{GBi}), \tag{13.51}$$

which has no real component. Therefore a non-quasi-static approach is needed to evaluate the real part of Y_{Gi}. As shown in Figure 13.6, the gate sees the oxide capacitance in series with part of the channel, which is responsible for the real part of Y_{Gi}. The Y_{Gi} admittance is simply equal to the parallel connection of the intrinsic gate-to-source, gate-to-drain, and gate-to-bulk admittances:

$$Y_{Gi} = Y_{GSi} + Y_{GDi} + Y_{GBi}. \tag{13.52}$$

---

[3] Actually a resistor is always passive, but in this case, the term passive is used to emphasise the fact that for $V_D = V_S$ the transistor can be looked at as a passive RC network and the Nyquist and Bode theorems apply.

From the first-order non-quasi-static model described in Section 5.2, assuming $\omega \, \tau_{qs} \ll 1$, the latter are given by

$$Y_{GSi} \cong \frac{j \, \omega \, C_{GSi}}{1 + j \, \omega \, \tau_{qs}/2} \cong j \, \omega \, C_{GSi} \, (1 - j \, \omega \, \frac{\tau_{qs}}{2}), \tag{13.53a}$$

$$Y_{GDi} \cong \frac{j \, \omega \, C_{GDi}}{1 + j \, \omega \, \tau_{qs}/2} \cong j \, \omega \, C_{GDi} \, (1 - j \, \omega \, \frac{\tau_{qs}}{2}), \tag{13.53b}$$

$$Y_{GBi} = \frac{n-1}{n}(j \, \omega \, C_{OX} - Y_{GSi} - Y_{GDi}), \tag{13.53c}$$

where τ_{qs} is the channel time constant which is related to the quasi-static frequency ω_{qs} defined by (5.32) according to

$$\frac{\tau_{qs}}{\tau_{spec}} = \frac{\omega_{spec}}{\omega_{qs}} = \frac{1}{\Omega_{qs}} = \frac{1}{30} \frac{4q_s^2 + 4q_d^2 + 12q_sq_d + 10q_s + 10q_d + 5}{(q_s + q_d + 1)^3}. \tag{13.54}$$

Notice that τ_{qs}/τ_{spec} does not depend on the transistor geometry, but is bias dependent according to (13.54). Ω_{qs} corresponds to the limit between the quasi-static and non-quasi-static model. This means that for normalized frequencies close or even larger than Ω_{qs}, non-quasi-static effects have to be accounted for.

Introducing (13.53) into (13.52) results in

$$\begin{aligned} Y_{Gi} &\cong \omega^2 \frac{\tau_{qs}(C_{GSi} + C_{GDi})}{2n} + j \, \omega \, \frac{(n-1)C_{OX} + C_{GSi} + C_{GDi}}{n} \\ &= \omega^2 \frac{\tau_{qs}C_{OX}(c_{GSi} + c_{GDi})}{2n} + j \, \omega \, C_{OX} \frac{n-1+c_{GSi} + c_{GDi}}{n}, \end{aligned} \tag{13.55}$$

where $C_{OX} \triangleq W L_f \, C_{ox} = N_f \, W_f \, L_f \, C_{ox}$ and c_{GSi} and c_{GDi} are the normalized gate-to-source and gate-to-drain intrinsic capacitances, which depend on the normalized source and drain charge densities according to (5.50):

$$c_{GSi} \triangleq \frac{C_{GSi}}{C_{OX}} = \frac{2}{3}q_s \frac{q_s + 2q_d + 3/2}{(q_s + q_d + 1)^2}, \tag{13.56a}$$

$$c_{GDi} \triangleq \frac{C_{GDi}}{C_{OX}} = \frac{2}{3}q_d \frac{q_d + 2q_s + 3/2}{(q_s + q_d + 1)^2}. \tag{13.56b}$$

The expression of the gate noise parameter δ_{nG} will not be detailed here, but it can be evaluated from (13.49), (13.48), and (13.55). It is plotted in Figure 13.7(a) versus the inversion factor for different values of the i_r/i_f ratio. As illustrated in Figure 13.7(a), δ_{nG} remains close to unity in all bias conditions since it is kept between 1 in weak inversion and 4/3 in strong inversion and saturation:

$$\delta_{nG} = \begin{cases} 1 & \text{in weak inversion} \\ 4/3 & \text{in strong inversion and saturation.} \end{cases} \tag{13.57}$$

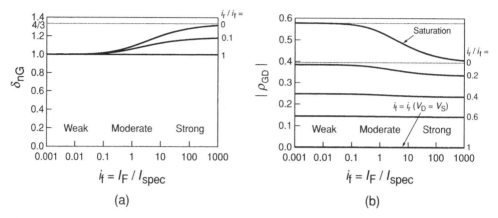

Figure 13.7 (a) Noise factor δ_{nG} and (b) magnitude of the drain-gate correlation coefficient ρ_{GD} versus the inversion factor i_f for different i_r/i_f ratios (from $i_r = i_f$ to saturation where $i_r = 0$)

In strong inversion and saturation, $c_{GDi} \cong 0$ and τ_{qs} is given by

$$\tau_{qs} = \frac{C_m}{G_m} \cong \frac{4}{15} \frac{C_{OX}}{G_m} = \frac{2}{5} \frac{C_{GSi}}{G_m}. \tag{13.58}$$

The real part of the input admittance is then given by

$$G_{Gi} \cong \frac{(\omega C_{GSi})^2}{5nG_m} = \frac{(\omega C_{GSi})^2}{5G_{ms}} = \frac{5}{4} \frac{G_m}{n}(\omega\tau_{qs})^2, \tag{13.59}$$

The gate noise conductance in strong inversion and saturation then simplifies to

$$G_{nG}(\omega) \cong \delta_{nG} \frac{(\omega C_{GSi})^2}{5G_{ms}} = \delta_{nG} \frac{5}{4} \frac{G_m}{n}(\omega\tau_{qs})^2 = \frac{5}{3} \frac{G_m}{n}(\omega\tau_{qs})^2, \tag{13.60}$$

since in strong inversion and saturation the gate thermal noise parameter $\delta_{nG} \cong 4/3$.

Equation (13.60) is in accordance with earlier results for strong inversion and saturation found for example in [151]. It can be further simplified by replacing G_{ms} by $G_{spec}q_s$ with $G_{spec} = I_{spec}/U_T = 2n\mu_0 C_{ox}W/LU_T$, resulting in

$$G_{nG} \cong \frac{8}{135} \frac{WL^3\omega^2}{nq_s} \frac{C_{ox}}{\mu_0 U_T} = \frac{16}{135} \frac{WL^3\omega^2 C_{ox}}{\mu_0(V_G - V_{T0} - nV_S)}. \tag{13.61}$$

According to (13.61), the induced gate noise in strong inversion and saturation is proportional to the cube of the gate length and inversely proportional to the overdrive voltage. For a given transistor width and for a given overdrive voltage, the induced gate noise quickly decreases when reducing the gate length.

It should be noted here that induced gate noise is actually always present for $\omega > 0$. Its effect should be compared to the effect of the other noise sources and particularly to the effect

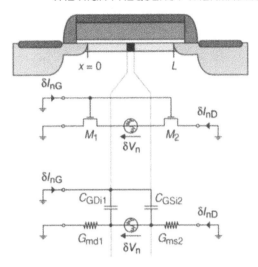

Figure 13.8 Two-transistor model for $V_D = V_S$ used for evaluation of the correlation between the induced gate noise current and the drain current noise

of the dominant drain noise. This comparison can be done only after having accounted for the correlation existing between the induced gate noise and the drain noise, which is evaluated below.

The CPSD between the gate and the drain noise currents is obtained from (13.45b) [78, 153]:

$$S_{\Delta i_{nG} \Delta i_{nD}^*} \cong S_{\Delta i_{nG} \Delta i_{nS}^*} \cong \frac{j\Omega}{18n} \frac{(\chi_s - \chi_d)(\chi_s^2 + 4\chi_s \chi_d + \chi_d^2)}{(\chi_s + \chi_d)^3}. \tag{13.62}$$

As expected $s_{\Delta i_{nG} \Delta i_{nD}^*}$ is not null, meaning that the gate noise is partly correlated with the drain noise. At frequency low enough for this first-order approximation to be valid, the correlation factor ρ_{GD}, as defined by equation (13.71) below, is independent of the frequency Ω. As illustrated in Figure 13.7(b), it is always null for $V_D = V_S$. This can be explained with the help of Figure 13.8. The gate current can be expressed in terms of the channel thermal noise voltage source as

$$\delta I_{nG} = Y_{nG}\, \delta V_n, \tag{13.63}$$

where the transadmittance Y_{nG} is given by

$$Y_{nG} = \frac{G_{md1} G_{ms2}}{G_{md1} + G_{ms2}} \frac{j\,\omega\left(\frac{C_{GDi1}}{G_{md1}} - \frac{C_{GSi2}}{G_{ms2}}\right)}{1 + j\,\omega\frac{C_{GDi1} + C_{GSi2}}{G_{md1} + G_{ms2}}}. \tag{13.64}$$

Notice the minus sign in the numerator of (13.64). For $V_D = V_S$, the channel is uniform from source to drain and hence

$$G_{md1} = \frac{G_{ms}}{\xi}, \tag{13.65a}$$

$$G_{ms2} = \frac{G_{ms}}{1 - \xi}, \tag{13.65b}$$

$$C_{GDi1} = \xi C_{GDSi}, \tag{13.65c}$$

$$C_{GSi2} = (1 - \xi)C_{GDSi}, \tag{13.65d}$$

with $G_{ms} = G_{md1} + G_{ms2}$ and $C_{GDSi} = C_{GDi1} + C_{GSi2}$. Replacing (13.65) into (13.64) results in

$$Y_{nG} = \frac{-j\,\omega C_{GDSi}(1 - 2\xi)}{1 + j\,\omega \frac{C_{GDSi}}{G_{ms}}(1 - \xi)\xi}. \tag{13.66}$$

From (13.66), it is interesting to note that $Y_{nG}(0.5) = 0$ and $Y_{nG}(1 - \xi) = -Y_{nG}(\xi)$, so that the sum $Y_{nG}(\xi) + Y_{nG}(1 - \xi)$ is zero. This is simply the result of the source and drain full symmetry occurring when the transistor is biased with $V_D = V_S$. When summing all the contributions along the channel to obtain the total induced gate noise current, the contribution at ξ is fully correlated with the one at $1 - \xi$ and since the one at the middle of the channel is zero, the part of the total induced current that could be correlated with the drain current is canceled out, leaving only the noncorrelated part. This explains why the correlation coefficient ρ_{GD} is zero for $V_D = V_S$.

Since Ω, χ_s, χ_d, and n in (13.62) are all real, the first-order approximation of the CPSD $S_{\Delta i_{nG} \Delta i_{nD}^*}$ and the correlation coefficient ρ_{GD} is purely imaginary, which is simply due to the capacitive coupling. It is plotted versus the inversion factor in Figure 13.7(b). In saturation ρ_{GD} is given by[4]

$$\rho_{GD} = \frac{+j\sqrt{5}(2q_s^2 + 6q_s + 3)}{\sqrt{(4q_s + 3)(32q_s^3 + 114q_s^2 + 132q_s + 45)}} \qquad \text{(saturation)}, \tag{13.67}$$

which has the following asymptotes:

$$\rho_{GD} = \begin{cases} +j/\sqrt{3} \approx +j\,0.6 & \text{in weak inversion and saturation} \\ +j\sqrt{5/32} \approx +j\,0.4 & \text{in strong inversion and saturation.} \end{cases} \tag{13.68}$$

Note that (13.68) is in agreement with the early result found for strong inversion by Van der Ziel [91] and the more recent result in weak inversion presented in [153].

---

[4] Note that with the sign definition of the noise current source ΔI_{nG} given in Figure 13.3(b) (i.e., current flowing from gate to ground), the imaginary part of the correlation coefficient ρ_{GD} is positive. In reference [151], the current is taken positive flowing from source to gate, leading to a negative value of the imaginary part of ρ_{GD}.

Finally, since the PSD of the drain and source noise currents are identical, the CPSD between the drain and the source noise currents is equal to the PSD of the drain noise current [78, 153]:

$$S_{\Delta i_{nD} \Delta i_{nS}^*} \cong S_{\Delta i_{nD}^2} \cong \frac{4\chi_s^2 - 3\chi_s + 4\chi_s\chi_d - 3\chi_d + 4\chi_d^2}{6(\chi_s + \chi_d)}. \tag{13.69}$$

Although this first-order model is usually sufficient for most circuit design purposes, it is interesting to investigate the full non-quasi-static model. This is done in the next section.

13.2.5 Higher Order Effects

The behavior at frequencies higher than Ω_{qs} has been explored by numerically integrating equations (13.45a) and (13.45b) and using the complete non-quasi-static expressions of the transadmittances.

To get readable results, only the deviation from the first-order behavior described above is plotted. A new set of parameters are therefore defined by [78, 153]

$$\kappa_k \triangleq S_{\Delta i_{nk}^2} \bigg/ \left\{ S_{\Delta i_{nk}^2} \right\}_{\text{(first order)}} \tag{13.70}$$

with $\kappa_B = \kappa_G$. The three independent coefficients κ_D, κ_S, and κ_G are plotted in Figure 13.9(a) versus the normalized frequency $\Theta \triangleq \Omega/\Omega_{qs}$, in the linear region and saturation, at different levels of inversion. The magnitude and phase of the correlation coefficients defined by

$$\rho_{kl} \triangleq \frac{S_{\Delta i_{nk} \Delta i_{nl}^*}}{\sqrt{S_{\Delta i_{nk}^2} S_{\Delta i_{nl}^2}}} \tag{13.71}$$

are plotted versus $\Theta \triangleq \Omega/\Omega_{qs}$ in Figures 13.9(b) and 13.9(c) respectively. The lines represent the results obtained from the numerical integration of (13.45a) and (13.45b), whereas the symbols correspond to the results obtained from a 16-segment simulation using the approach described in [114, 134]. Each segment has been simulated using the intrinsic part of the EKV compact model instead of the MOS Model 11 used in [114, 134].

It can be seen from the bottom part of Figure 13.9(a) that at high frequency, the induced gate noise $s_{\Delta I_{nG}^2}$ becomes smaller than expected from the first-order approximation. It still increases, but only proportionally to $\sqrt{\Omega}$ instead of Ω^2 (since $\kappa_G \propto \Omega^{-3/2}$). Note that the actual curves do not vary with the inversion factor and that even the linear region and saturation behaviors are very close to each other.

The top part of Figure 13.9(a) shows that the drain and source noise PSDs tend to slowly increase with frequency, about at a $\sqrt{\Omega}$ rate. Note that although this frequency behavior might seem surprising at a first glance, a similar behavior has already been observed by L.-J. Pu and Y. Tsividis in [154]. In the linear region, the curve is again independent of the state of inversion. In saturation mode, the behavior is more complex. In the strong inversion region ($IC \gg 1$), both κ_D and κ_S tend toward the linear region curve, whereas in weak inversion ($IC \ll 1$), the drain noise is kept almost constant while the source noise still increases in a similar fashion as the linear region curve.

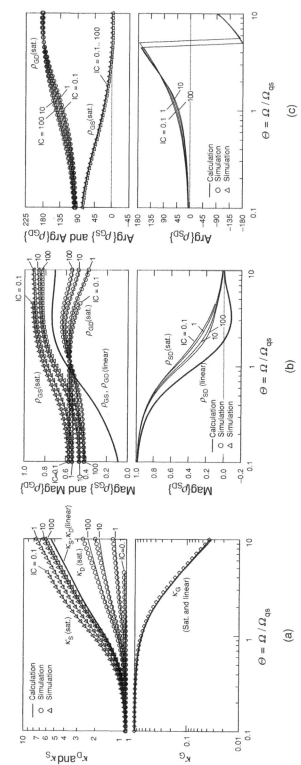

Figure 13.9 High-frequency effects on the noise PSDs and correlation coefficients referred to the first-order model [153]. The lines correspond to the result of the numerical integration of (13.45a) and (13.45b). The symbols correspond to the results obtained from a 16-segment simulation using the approach described in [114, 134] (Reproduced by permission of IEE from [153])

The magnitude of ρ_{GD} and ρ_{GS} are plotted on the top part of Figure 13.9(b). They follow a complex pattern that is hardly predictable in saturation, even qualitatively. In the linear regime, both correlation coefficients ρ_{GD} and ρ_{GS} are null at low frequency, since as explained above, the two identical channel portions placed symmetrically relatively to the center point contribute to the gate noise with the same magnitude but with an opposite correlation sign. At higher frequency, though, the drain and source currents are no longer forced equal. In the linear region, both correlation coefficients ρ_{GD} and ρ_{GS} remain equal but increase towards an asymptotic value of about 0.71. At the same time, the total correlation between the drain and source current, plotted in the bottom part of Figure 13.9(b), vanishes to zero since each elementary channel section can only contribute to the noise current of the nearby terminal and is completely damped before reaching the remote one.

The phase of the correlation coefficients is plotted in Figure 13.9(c), which shows that the low-frequency capacitive correlation of the induced gate noise (i.e., ρ_{GD}) tends toward a purely real value at higher frequency. This phenomenon is related to the additional transconductance phase shifts of the non-quasi-static regime, but is rather hard to explain directly. The source–drain correlation phase increases quickly when the distributed effects become significant.

Note that the calculation obtained from (13.45a) and (13.45b) match very well to the results obtained from the simulation.

Finally, Figure 13.10 shows that the mobility reduction due to the vertical field and velocity saturation have only a small effect on the HF noise, even in strong inversion ($IC = 100$).

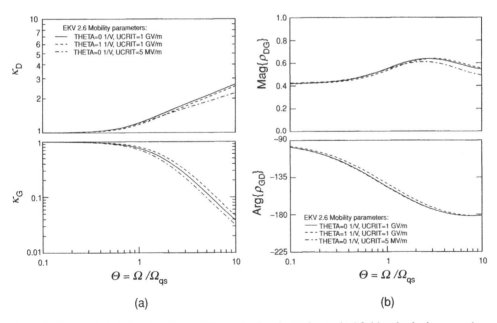

(a)   (b)

Figure 13.10 Effects induced by the mobility reduction due to the vertical field and velocity saturation in strong inversion ($IC = 100$) on κ_D, κ_G, and ρ_{GD} using the EKV2.6 model parameters THETA and UCRIT. The parameter set (THETA $= 0$ V^{-1}, UCRIT $= 1$ GV/m), (THETA $= 1$ V^{-1}, UCRIT $= 1$ GV/m), (THETA $= 0$ V^{-1}, UCRIT $= 5$ MV/m) correspond respectively to: no mobility reduction and no velocity saturation, mobility reduction without velocity saturation and velocity saturation without mobility reduction (Reproduced by permission of IEE from [153])

13.3 HF NOISE PARAMETERS OF A COMMON-SOURCE AMPLIFIER

13.3.1 Simple Equivalent Circuit Including Induced Gate Noise and Drain Noise

Figure 13.11 shows a common-source amplifier and its simplified equivalent small-signal circuit, where capacitance C_{GS} includes both the intrinsic and extrinsic parts (mainly the overlap in this case). The latter circuit is made extremely simple in order to allow for hand calculation. It is assumed that the transistor is biased in strong inversion and saturation. The equivalent circuit of Figure 13.11 includes the drain current noise source ΔI_{nD} and the induced gate noise current source ΔI_{nG} having PSDs

$$S_{\Delta I_{nD}^2} = 4kT G_{nD}, \tag{13.72a}$$

$$S_{\Delta I_{nG}^2} = 4kT G_{nG}(\omega), \tag{13.72b}$$

respectively, with

$$G_{nD} = \gamma_{nD} G_m, \tag{13.73a}$$

$$G_{nG}(\omega) = \delta_{nG} \frac{(\omega C_{GS})^2}{5n G_m} = \beta_{nG} \frac{(\omega C_{GS})^2}{G_m}, \tag{13.73b}$$

with

$$\beta_{nG} = \frac{\delta_{nG}}{5n} = \frac{4}{15n}, \tag{13.74}$$

since according to (13.57), $\delta_{nG} = 4/3$ in strong inversion and saturation.

We will now calculate the noise parameters R_v, G_i, Y_c of the equivalent noisy two-port network of Figure 13.11. Remember that the two noise sources ΔI_{nD} and ΔI_{nG} are correlated with a correlation coefficient ρ_{GD} given by (13.68), which is purely imaginary

$$\rho_{GD} = +j c_g, \tag{13.75}$$

with $c_g = \sqrt{5/32} \cong 0.4$ in strong inversion and saturation. This makes the analysis a bit more complicated than if both sources were uncorrelated.

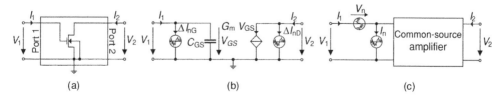

(a)   (b)   (c)

Figure 13.11   (a) Common source amplifier, (b) simple HF equivalent circuit including the induced gate noise, and (c) equivalent two-port representation

We first have to calculate the noise sources V_n and I_n from their definitions given in (13.4) which require the Y-parameters Y_{11} and Y_{21}, which are easily calculated as

$$Y_{11} = j\omega C_{GS}, \tag{13.76a}$$
$$Y_{21} = G_m. \tag{13.76b}$$

The output current I_2 when the input and output are short-circuited is given by

$$I_2|_{V_1=V_2=0} = \Delta I_{nD}, \tag{13.77}$$

from which we obtain V_n as

$$V_n = -\frac{1}{Y_{21}} I_2\big|_{V_1=V_2=0} = -\frac{\Delta I_{nD}}{G_m}. \tag{13.78}$$

The output current I_2 when the input is open and the output is short-circuited is

$$I_2\big|_{I_1=V_2=0} = \Delta I_{nD} - \frac{G_m}{j\omega C_{GS}} \Delta I_{nG}, \tag{13.79}$$

from which we get I_n as

$$I_n = -\frac{Y_{11}}{Y_{21}} I_2\big|_{I_1=V_2=0} = \Delta I_{nG} - \frac{j\omega C_{GS}}{G_m} \Delta I_{nD}. \tag{13.80}$$

The mean-square value of V_n is then given by

$$\overline{|V_n|^2} = \frac{\overline{|\Delta I_{nD}|^2}}{G_m^2}, \tag{13.81}$$

whereas the mean-square value of I_n is given by

$$\overline{|I_n|^2} = \overline{|\Delta I_{nG}|^2} + \left(\frac{\omega C_{GS}}{G_m}\right)^2 \overline{|\Delta I_{nD}|^2} + \frac{j\omega C_{GS}}{G_m} (\overline{\Delta I_{nG} \Delta I_{nD}^*} - \overline{\Delta I_{nG}^* \Delta I_{nD}}). \tag{13.82}$$

The mean-square values $\overline{|\Delta I_{nD}|^2}$ and $\overline{|\Delta I_{nG}|^2}$ can be expressed in terms of the PSDs $S_{\Delta I_{nD}^2}$ and $S_{\Delta I_{nG}^2}$ as

$$\overline{|\Delta I_{nD}|^2} = S_{\Delta I_{nD}^2} B = 4kTBG_{nD}, \tag{13.83a}$$
$$\overline{|\Delta I_{nG}|^2} = S_{\Delta I_{nG}^2} B = 4kTBG_{nG}(\omega), \tag{13.83b}$$

where (13.72) have been used.

For narrow-band systems, the definition of the correlation coefficient given by (13.71) can be extended to the mean-square values as

$$\rho_{kl} \triangleq \frac{S_{\Delta i_{nk} \Delta i_{nl}^*}}{\sqrt{S_{\Delta i_{nk}^2} S_{\Delta i_{nl}^2}}} = \frac{\overline{\Delta I_{nk} \Delta I_{nl}^*}}{\sqrt{|\Delta I_{nk}|^2 |\Delta I_{nl}|^2}}. \tag{13.84}$$

The mean-square values $\overline{\Delta I_{nG} \Delta I_{nD}^*}$ and $\overline{\Delta I_{nG}^* \Delta I_{nD}}$ can then be expressed in terms of mean-square values $\overline{|\Delta I_{nD}|^2}$ and $\overline{|\Delta I_{nG}|^2}$ and the correlation coefficient ρ_{GD} according to (13.84):

$$\overline{\Delta I_{nG} \Delta I_{nD}^*} = \rho_{GD} \sqrt{\overline{|\Delta I_{nG}|^2 |\Delta I_{nD}|^2}} = +jc_g 4kT B \sqrt{G_{nG} G_{nD}}, \tag{13.85a}$$

$$\overline{\Delta I_{nG}^* \Delta I_{nD}} = \rho_{GD}^* \sqrt{\overline{|\Delta I_{nG}|^2 |\Delta I_{nD}|^2}} = -jc_g 4kT B \sqrt{G_{nG} G_{nD}}. \tag{13.85b}$$

The noise parameters R_v and G_i are then given by

$$R_v = \frac{\overline{|V_n|^2}}{4kT B} = \frac{G_{nD}}{G_m^2}, \tag{13.86a}$$

$$G_i = \frac{\overline{|I_n|^2}}{4kT B} = G_{nG} + \left(\frac{\omega C_{GS}}{G_m}\right)^2 G_{nD} - \frac{2c_g \omega C_{GS}}{G_m} \sqrt{G_{nG} G_{nD}}. \tag{13.86b}$$

According to (13.8), for calculating Y_c, we need $\overline{I_n V_n^*}$ which is given by

$$\overline{I_n V_n^*} = \frac{1}{G_m} \left(\frac{j\omega C_{GS}}{G_m} \overline{|\Delta I_{nD}|^2} - \overline{\Delta I_{nG} \Delta I_{nD}^*}\right) \tag{13.87}$$

$$= \frac{4kT B}{G_m} \left(\frac{j\omega C_{GS}}{G_m} G_{nD} - jc_g \sqrt{G_{nG} G_{nD}}\right),$$

from which we get

$$Y_c = \frac{\overline{I_n V_n^*}}{\overline{|V_n|^2}} = j \left(\omega C_{GS} - c_g G_m \sqrt{\frac{G_{nG}}{G_{nD}}}\right). \tag{13.88}$$

From (13.88), we see that $G_c = 0$ and $Y_c = j B_c$.

Introducing the expressions of G_{nD} and G_{nG} given by (13.73) finally results in

$$R_v = \frac{\gamma_{nD}}{G_m}, \tag{13.89a}$$

$$G_i = (\omega C_{GS})^2 \frac{\gamma_{nD}}{G_m} \left(1 + \frac{\beta_{nG}}{\gamma_{nD}} - 2c_g \sqrt{\frac{\beta_{nG}}{\gamma_{nD}}}\right), \tag{13.89b}$$

$$G_c = 0, \tag{13.89c}$$

$$B_c = \omega C_{GS} \left(1 - c_g \sqrt{\frac{\beta_{nG}}{\gamma_{nD}}}\right). \tag{13.89d}$$

The uncorrelated and correlated parts of G_i are then obtained using (13.14):

$$G_{iu} = (\omega C_{GS})^2 \frac{\beta_{nG}}{G_m}(1 - c_g^2), \tag{13.90a}$$

$$G_{ic} = (\omega C_{GS})^2 \frac{\gamma_{nD}}{G_m}\left(1 + c_g^2 \frac{\beta_{nG}}{\gamma_{nD}} - 2c_g\sqrt{\frac{\beta_{nG}}{\gamma_{nD}}}\right). \tag{13.90b}$$

From the R_v, G_i, Y_c parameters, we can derive the four noise parameters R_v, G_{opt}, B_{opt}, and F_{min} as a function of the circuit parameters using (13.24) and (13.25), resulting in

$$G_{opt} = \omega C_{GS}\sqrt{\frac{\beta_{nG}}{\gamma_{nD}}(1 - c_g^2)}, \tag{13.91a}$$

$$B_{opt} = -\omega C_{GS}\left(1 - c_g\sqrt{\frac{\beta_{nG}}{\gamma_{nD}}}\right), \tag{13.91b}$$

$$F_{min} = 1 + 2\omega C_{GS}\frac{\gamma_{nD}}{G_m}\sqrt{\frac{\beta_{nG}}{\gamma_{nD}}(1 - c_g^2)}$$

$$\cong 1 + 2\gamma_{nD}\frac{\omega}{\omega_t}\sqrt{\frac{\beta_{nG}}{\gamma_{nD}}(1 - c_g^2)}, \tag{13.91c}$$

where the G_m/C_{GS} ratio has been approximated by the transit frequency $\omega_t \cong G_m/C_{GS}$. Equations (13.91) reveal that, due to the induced gate noise, the noise matching condition is slightly different than the gain matching condition which would require $B_s = -\omega C_{GS}$. Also, the minimum noise factor is strongly depending on the induced gate noise through parameters β_{nG} and c_g. If induced gate noise was ignored (by setting $\beta_{nG} = 0$), the minimum noise factor would be equal to unity. This surprising result can be explained as follows: if the induced gate noise is ignored, there is only the drain noise left and the optimum source conductance is null whereas the optimum source susceptance is $-\omega C_{GS}$. This noise matching situation corresponds to having an inductor with a susceptance value being $-\omega C_{GS}$ and no internal conductance. The input circuit is then an inductance in series with the transistor gate-to-source capacitance. This source inductance will then resonate with the input transistor capacitance at the operating frequency providing an infinite voltage gain at the input. The input referred noise is then nulled, resulting in a unity noise factor.

Equation (13.91c) indicates that the minimum noise factor increases with frequency for a given bias. For a given operating frequency, it can be decreased by increasing the transistor transit frequency. This can be achieved by increasing the transistor bias or by reducing the transistor length (or both). Technology scaling therefore leads to an improved noise factor at a given frequency and bias.

It is also interesting to note that the correlation between the drain noise and the induced gate noise reduces the noise factor compared to the situation where they would be fully uncorrelated.

Note that the above derivation has used the HF noise model which did not include short-channel effects. As shown in Section 9.4, the noise parameter γ_{nD} is affected by effects such as velocity saturation, carrier heating, mobility degradation due to the vertical field, and channel length modulation. The HF noise model developed in Section 13.2 did not include these effects, particularly for the induced gate noise. Now, since the physical noise source for the drain and gate noise is the channel, both parameters γ_{nD} and β_{nG} (or δ_{nG}) should be affected by these effects in a similar way. Hence, their ratio appearing in (13.91) should not be affected in first-order. On the other hand, the γ_{nD} term appearing in the minimum noise factor of (13.91c) should be replaced by the one derived in Section 9.4 to include the short-channel effects.

The noise parameters (13.91) can be further simplified by replacing γ_{nD} and β_{nG} by their expression valid in strong inversion and saturation leading to

$$G_{opt} \cong 0.45\,\omega\,C_{GS}, \qquad (13.92a)$$
$$B_{opt} \cong -0.8\,\omega\,C_{GS}, \qquad (13.92b)$$
$$F_{min} \cong 1 + 0.77\frac{\omega}{\omega_t}, \qquad (13.92c)$$

where $n = 1.3$ has been used. Equations (13.92) give a first-order relation for the noise parameters of a common-source amplifier.

We have seen above that the induced gate noise plays a fundamental role in the noise parameters of a common-source amplifier. We have also seen that due to the induced gate noise, the minimum noise factor becomes frequency dependent. The noise parameters and in particular the minimum noise factor are obtained only in the special condition of noise matching, which are not always possible to satisfy due to additional design constraints. The effective noise factor is then higher than the minimum noise factor and can be expressed in terms of the noise parameters and the actual source conductance and susceptance by (13.26). For a given operating frequency, it is interesting to know above which value of the source impedance the contribution of the induced gate noise to the effective noise factor starts to dominate over the contribution of the drain noise. To derive this condition, we will assume that the noise matching condition is only fulfilled for the imaginary part of the source admittance B_s. In this situation, the actual noise factor is obtained by setting $B_s = B_{opt}$ in (13.26) resulting in

$$F = F_{min} + \frac{R_v}{G_s}(G_s - G_{opt})^2. \qquad (13.93)$$

In this calculation, we are actually more interested in the ratio of the noise added by the amplifier to the noise coming from the source N_a/N_s which is simply given by $F - 1$. Using the noise parameters given in (13.91) results in

$$F - 1 = \gamma_{nD}\frac{G_s}{G_m} + \beta_{nG}(1 - c_g^2)\left(\frac{\omega}{\omega_t}\right)^2\frac{G_m}{G_s}, \qquad (13.94)$$

where the first term is due to the drain noise, whereas the second originates from the induced gate noise. Equation (13.94) is plotted versus G_m/G_s in Figure 13.12 for an operating frequency $\omega/\omega_t = 0.224$ (leading to $G_m/G_{opt} \cong 10$). We can clearly identify the contribution of the drain noise which decreases with respect to G_m/G_s and that of the induced gate noise which increases with respect to G_m/G_s. Both contributions are equal for $G_s = G_{opt}$ for which

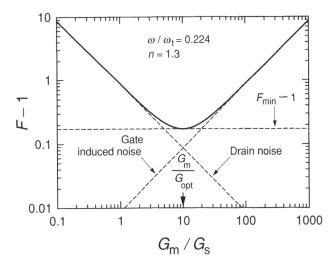

Figure 13.12 Noise added by the common-source amplifier normalized to the source noise versus the source resistance normalized to the transistor transconductance for an operating frequency $\omega/\omega_t = 0.224$

$F = F_{min}$. The drain noise dominates over the induced gate noise for $G_m/G_s \ll G_m/G_{opt}$, whereas the induced gate noise dominates over the drain noise for $G_m/G_s \gg G_m/G_{opt}$. For $G_s = 1/50 \ \Omega^{-1}$ and $\omega/\omega_t = 0.224$, the transconductance needs to be larger than 200 mA/V for the induced gate noise to dominate. This is a rather large transconductance and hence in most practical cases, the G_m/G_s ratio is smaller than G_m/G_{opt} and therefore it is usually the drain noise that actually dominates.

We can also find the frequency ω_{ign} at which the induced gate noise is equal to the drain noise for a given source conductance G_s, by simply equating the first and the second term of (13.94) and solving for ω. This results in

$$\frac{\omega_{ign}}{\omega_t} = \sqrt{\frac{\gamma_{nD}}{\beta_{nG}(1 - c_g^2)}} \frac{G_s}{G_m} \cong \frac{4}{3}\sqrt{\frac{5}{3}} \, n \, \frac{G_s}{G_m} \cong 2.24 \, \frac{G_s}{G_m}, \tag{13.95}$$

where $n = 1.3$ has been used. We see from (13.95) that the frequency ω_{ign} at which the induced gate noise contributes as much as the drain noise is a fraction of the transit frequency ω_t which is inversely proportional to the G_m/G_s ratio.

The above discussion somehow mitigates the importance of the induced gate noise in the case of practical designs such as LNAs because the noise matching conditions are seldom achieved due to other design constraints. Nevertheless, it remains an important contributor to the minimum noise figure, which represents the ultimate noise figure that can be achieved by a given device at a given frequency and for a given bias.

The previous analysis was based on a very simple small-signal equivalent circuit of a common-source amplifier that considered only the drain noise and the induced gate noise. Additional noise sources will be considered in the next section.

13.3.2 Equivalent Circuit Including Induced Gate Noise, Drain Noise, Gate and Substrate Resistances Noise

The above analysis was based on a very simple equivalent circuit that ignored the noise coming from the gate and the substrate resistances. The equivalent circuit accounting for these two additional noise sources is shown in Figure 13.13. Note that, in order to keep the analysis simple, the output conductance, the gate-to-bulk and bulk-to-source capacitances as well as the source and drain access resistances have been ignored. Also, even though the intrinsic gate-to-drain and bulk-to-drain capacitances are zero in strong inversion and saturation, the overlap and junction capacitances would remain. They will also be ignored for the sake of simplicity. Since the circuit is a common-source amplifier, it has its source tied to the ground. On the other hand, because of the substrate resistance, the intrinsic bulk is not at the ground. In this situation, it is better to use the combination of gate and bulk transconductances driven by the small-signal voltages ΔV_{GS} and ΔV_{BS} respectively instead of gate and source transconductances driven by voltages ΔV_{GB} and ΔV_{SB}. The two circuits are equivalent since the small-signal drain current (ignoring the drain noise current) can be written as

$$\Delta I_D = G_m \Delta V_{GB} - G_{ms} \Delta V_{SB} = G_m \Delta V_{GS} + G_{mb} \Delta V_{BS}, \qquad (13.96)$$

where $G_{mb} = G_{ms} - G_m = (n-1)G_m$ is the bulk transconductance. The noise of the substrate resistance R_B is modeled by a current noise source generating a V_{BS} voltage across R_B which is transmitted to the drain by the bulk transconductance G_{mb}.

We will not detail the calculation of the noise parameters R_v, G_i, G_c, and B_c of the equivalent noisy two-port network of Figure 13.13. They can be obtained following the same procedure used in Section 13.3.1, which leads to

$$R_v = \frac{\gamma_{nD}}{G_m} D_n, \qquad (13.97a)$$

$$G_i = \frac{\gamma_{nD}}{G_m} (\omega C_{GS})^2 A_n = \gamma_{nD} G_m \left(\frac{\omega}{\omega_t}\right)^2 A_n, \qquad (13.97b)$$

$$G_c = R_G (\omega C_{GS})^2 \frac{A_n}{D_n}, \qquad (13.97c)$$

$$B_c = \omega C_{GS} \frac{B_n}{D_n}, \qquad (13.97d)$$

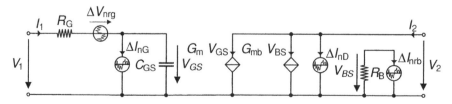

Figure 13.13 More accurate equivalent circuit of the common-source amplifier of Figure 13.11(a) with the gate and substrate resistances in addition

where

$$D_n \triangleq 1 + \alpha_B + \alpha_G + (R_G \omega C_{GS})^2 A_n$$
$$= 1 + \alpha_B + \alpha_G + (\gamma_{nD} \alpha_G)^2 \left(\frac{\omega}{\omega_t}\right)^2 A_n, \tag{13.98a}$$

$$A_n \triangleq 1 + \alpha_B + \frac{\beta_{nG}}{\gamma_{nD}} - 2c_g \sqrt{\frac{\beta_{nG}}{\gamma_{nD}}}, \tag{13.98b}$$

$$B_n \triangleq 1 + \alpha_B - c_g \sqrt{\frac{\beta_{nG}}{\gamma_{nD}}}. \tag{13.98c}$$

α_G and α_B are the thermal noise contributions of the gate and substrate resistances respectively, normalized to the thermal noise of the drain:

$$\alpha_G \triangleq \frac{G_m R_G}{\gamma_{nD}}, \tag{13.99a}$$

$$\alpha_B \triangleq \frac{G_{mb}^2 R_B}{\gamma_{nD} G_m} = \frac{(n-1)^2}{\gamma_{nD}} G_m R_B. \tag{13.99b}$$

The noise parameters G_{opt}, B_{opt}, and F_{min} can then be calculated from R_v, G_i, G_c, and B_c, resulting in

$$G_{opt} = \omega C_{GS} \frac{\sqrt{A_n D_n - B_n^2}}{D_n}, \tag{13.100a}$$

$$B_{opt} = -\omega C_{GS} \frac{B_n}{D_n}, \tag{13.100b}$$

$$F_{min} = 1 + 2\gamma_{nD} \frac{\omega}{\omega_t} \sqrt{A_n D_n - B_n^2 + 2\gamma_{nD}^2 \alpha_G \left(\frac{\omega}{\omega_t}\right)^2 A_n}$$
$$\cong 1 + 2\gamma_{nD} \frac{\omega}{\omega_t} \sqrt{A_n D_n - B_n^2}. \tag{13.100c}$$

In a typical situation, the relative contributions of the gate and substrate resistances are $\alpha_G \approx 5\%$ and $\alpha_B \approx 20\%$ [48, 49, 52]. Using values of parameters γ_{nD}, β_{nG}, and c_g given above for strong inversion with $n = 1.3$ leads to $A_n \approx 1.1$ and $B_n \approx 1$. Assuming further that $\omega/\omega_t = 0.1$ allows to neglect the frequency-dependent part of D_n:

$$D_n \cong 1 + \alpha_B + \alpha_G \cong 1.25. \tag{13.101}$$

The four noise parameters can then be approximated by

$$R_v \cong 1.25 \frac{\gamma_{nD}}{G_m}, \tag{13.102a}$$

$$G_{opt} \cong 0.5 \omega C_{GS}, \tag{13.102b}$$

$$B_{opt} \cong -0.8 \omega C_{GS}, \tag{13.102c}$$

$$F_{min} \cong 1 + \frac{\omega}{\omega_t}. \tag{13.102d}$$

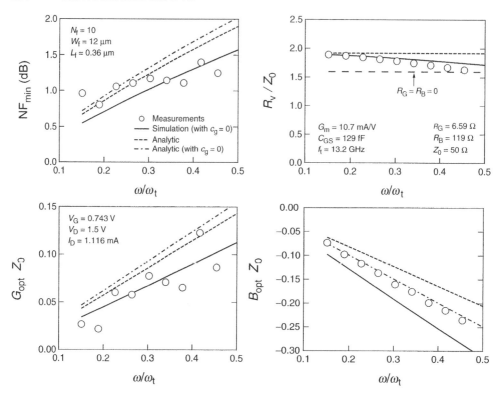

Figure 13.14   Comparison between the measured, simulated, and the analytical results for the four noise parameters

The noise parameters derived above are compared to those measured on an N-channel device with $N_f = 10$, $W_f = 12$ μm, and $L_f = 0.36$ μm biased in strong inversion and saturation. The noise parameters have been carefully de-embedded using the methodology presented in [155]. They are plotted in Figure 13.14 and compared to the results obtained from simulation using the complete subcircuit of Figure 11.9(c) with the additional induced gate noise source added to the subcircuit (but not accounting for the correlation between induced gate noise and drain thermal noise). Also, plotted in dashed lines in Figure 13.14 are the results obtained from (13.102). The dashed-dotted line corresponds to (13.102) without accounting for the gate-to-drain correlation ρ_{GD} by setting c_g to zero. The discrepancies between the analytical and the measured results mainly come from a wrong frequency behavior due to the extremely simple equivalent circuit used for the analytical derivation. Nevertheless, the simple approximation given by (13.102) already gives a reasonable estimation of the minimum noise figure and of the other noise parameters.

References

1. Y. Tsividis, *Operation and Modeling of the MOS Transistor*, 2nd ed. New York: Mc-Graw-Hill, 1999.
2. M. Forrer, R. L. Coultre, A. Beyner, and H. Oguey, *L'aventure de la montre à quartz*. CH-2000 Neuchâtel, Switzerland: Centredoc, 2002 (in French).
3. http://invention.smithsonian.org/centerpieces/quartz/inventors/swissinvent.html.
4. F. Leuenberger and E. Vittoz, "Complementary MOS Low-Power Low-Voltage Integrated Binary Counter," *IEEE Proc.*, vol. 57, no. 9, pp. 1528–1532, Sept. 1969.
5. E. A. Vittoz, B. Gerber, and F. Leuenberger, "Silicon-Gate CMOS Frequency Divider for the Electronic Wrist Watch," *IEEE J. Solid-State Circ.*, vol. 7, no. 2, pp. 100–104, Apr. 1972.
6. E. Vittoz and H. Oguey, "Complementary Dynamic Logic Circuits," *El. Lett.*, vol. 9, no. 4, Jan. 1973.
7. H. Oguey and E. Vittoz, "CODYMOS Frequency Dividers Achieve Low Power Consumption and High Frequency," *El. Lett.*, vol. 9, no. 17, Aug. 1973.
8. M. B. Barron, "Low Level Currents in Insulated Gate Field Effect Transistors," *Solid-State Electron.*, vol. 15, pp. 293–302, 1972.
9. R. M. Swanson and J. D. Meindl, "Ion-Implanted Complementary MOS Transistors in Low-Voltage Circuits," *IEEE J. Solid-State Circ.*, vol. 7, no. 2, pp. 146–153, Apr. 1972.
10. R. R. Troutman and S. T. Chakravarti, "Subthreshold Characteristics of Insulated-Gate Field-Effect Transistors," *IEEE Trans. Circ. Theory*, vol. 20, no. 6, pp. 659–665, Nov. 1973.
11. T. Masuhara, J. Etoh, and M. Nagata, "A Precise MOSFET Model for Low-Voltage Circuits," *IEEE Trans. Electron. Devices*, vol. 21, no. 6, pp. 363–371, June 1974.
12. E. Vittoz and J. Fellrath, "New Analog CMOS IC's Based on Weak Inversion Operation," in *European Solid State Circ. Conf. Dig. Tech. Pap.*, Toulouse, pp. 12–13, Sept. 1976.
13. E. Vittoz and J. Fellrath, "CMOS Analog Integrated Circuits Based on Weak Inversion Operation," *IEEE J. Solid State Circ.*, vol. 12, no. 3, pp. 224–231, June 1977.
14. J. Fellrath and E. Vittoz, "Small Signal Model of MOS Transistors in Weak Inversion," in *Proc. Journées d'Electronique '77*, EPF-Lausanne, pp. 315–324, 1977.
15. P. G. A. Jespers, C. Jusseret, and Y. Leduc, "A Fast Sample and Hold Charge-Sensing Circuit for Photodiode Arrays," *IEEE J. Solid-State Circ.*, vol. 12, no. 3, pp. 232–237, June 1977.
16. H. Wallinga and K. Bult, "Design and Analysis of CMOS Analog Processing Circuits by Means of a Graphical MOST Model," *IEEE J. Solid-State Circ.*, vol. 24, no. 3, pp. 672–680, June 1989.
17. J.-D. Châtelain, *Dispositifs à Semiconducteur*, 2nd ed. ser. Traité d'Électricité. Editions Georgi, 1979, vol. 7.

18. J. J. Ebers and J. L. Moll, "Large-Signal Behavior of Junction Transistors," *Proc. IRE*, vol. 42, no. 12, pp. 1761–1772, Dec. 1954.

19. H. Oguey and S. Cserveny, "Modèle du transistor MOS valable dans un grand domaine de courants," *Bulletin ASE*, vol. 73, no. 3, pp. 113–116, 1982 (in French).

20. H. Oguey and S. Cserveny, "MOS Modelling at Low Current Density," ESAT, Summer Course on Process and Device Modelling, June 1983.

21. C. C. Enz, "High Precision CMOS Micropower Amplifiers," Ph.D. Thesis, Swiss Federal Institute of Technology, Thesis No. 802, 1989.

22. E. Vittoz, "Micropower Techniques," in *Design of MOS VLSI Circuits for Telecommunications*, Y. Tsividis and P. Antognetti, Eds. New Jersey: Prentice-Hall, 1985.

23. E. Vittoz, "Micropower Techniques," in *Design of Analog-Digital VLSI Circuits for Telecommunications and Signal Processing*, J. Franca and Y. Tsividis, Eds. New Jersey: Prentice-Hall, 1994.

24. E. Vittoz, "MOS Transistor," EPFL, Technical Report, 1988.

25. C. C. Enz, F. Krummenacher, and E. A. Vittoz, "An Analytical MOS Transistor Model Valid in All Regions of Operation and Dedicated to Low-Voltage and Low-Current Applications," *Analog Integr. Circ. Signal Process. J. Low-Voltage and Low-Power Des.*, vol. 8, pp. 83–114, July 1995.

26. M. Bagheri and C.Turchetti, "The Need for an Explicit Model Describing MOS Transistors in Moderate Inversion." *El. Lett.*, vol. 21, no. 19, pp. 873–874, 1985.

27. M. A. Maher and C. A. Mead, "A Physical Charge-Conrolled Model for the MOS Transistors," in *Advanced Research in VLSI, Proceeding of the 1987 Stanford Conference*, P. Losleben, Ed. Cambridge, MA: MIT Press, 1987.

28. M. A. Maher and C. Mead, "Fine Points of Transistor Physics," in *Analog VLSI and Neural Systems*. Addison-Wesley, 1989.

29. B. Iñiguez and E. G. Moreno, "A Physically Based C_∞-Continuous Model for Small-Geometry MOSFET's," *IEEE Trans. Electron. Devices*, vol. 42, no. 2, pp. 283–287, Feb. 1995.

30. B. Iñiguez and E. G. Moreno, "C_∞-Continuous Small-Geometry MOSFET Modeling for Analog Applications," in *Proc. 38th Midwest Symp. Circ. and Syst.*, pp. 41–44, Aug. 1995.

31. A. I. A. Cunha, S. M. Acosta, M. C. Schneider, and C. Galup-Montoro, "An Explicit MOSEFT Model for Analog Circuit Simulation," in *Proc. IEEE Int. Symp. Circ. Syst.*, pp. 1592–1595, May 1995.

32. A. I. A. Cunha, M. C. Schneider, and C. Galup-Montoro, "An Explicit Physical Model for Long-Channel MOS Transistor Including Small-Signal Parameters," *Solid-State Electron.*, vol. 38, no. 11, pp. 1945–1952, 1995.

33. A. I. A. Cunha, O. C. Gouveia-Filho, M. C. Schneider, and C. Galup-Montoro, "A Current-Based Model for the MOS Transistor," in *Proc. IEEE Int. Symp. Circ. Syst.*, pp. 1608–1611, June 1995.

34. A. I. A. Cunha, M. C. Schneider, and C. Galup-Montoro, "An MOS Transistor Model for Analog Circuit Design," *IEEE J. Solid-State Circ.*, vol. 33, no. 10, pp. 1510–1519, Oct. 1998.

35. M. Bucher, C. Lallement, C. Enz, F. Theodoloz, and F. Krummenacher, "Scalable Gm/I Based MOSFET Model," in *Proc. Int. Semiconductor Device Res. Symp.*, Charlottesville, Dec. 1997.

36. M. Bucher, J.-M. Sallese, C. Lallement, W. Grabinski, C. C. Enz, and F. Krummenacher, "Extended Charges Modeling for Deep Submicron CMOS," in *Proc. Int. Semiconductor Device Res. Symp.*, Charlottesville, pp. 397–400, Dec. 1999.

37. M. Bucher, "Analytical MOS Transistor Modelling for Analog Circuit Simulation," Ph.D. Thesis, Swiss Federal Institute of Technology, Thesis No. 2114, 1999.

38. J.-M. Sallese and A.-S. Porret, "A Novel Approach to Charge Based Non Quasi Static Model of the MOS Transistor Valid in all Modes of Operation," *Solid-State Electron.*, vol. 44, no. 6, pp. 887–894, June 2000.

39. H. K. Gummel and K. Singhal, "Inversion Charge Modeling," *IEEE Trans. Electron. Devices*, vol. 48, no. 8, pp. 1585–1593, Aug. 2001.

40. H. K. Gummel and K. Singhal, "Intrinsic MOSFET Capacitance Coefficients," *IEEE Trans. Electron. Devices*, vol. 48, no. 10, pp. 2384–2393, Oct. 2001.

41. J.-M. Sallese, M. Bucher, F. Krummenacher, and P. Fazan, "Inversion Charge Linearization in MOSFET Modeling and Rigourous Derivation of the EKV Compact Model," *Solid-State Electron.*, vol. 47, pp. 677–683, 2003.

42. C. Lallement, M. Bucher, and C. C. Enz, "Simple Solution for Modeling the Non-Uniform Substrate Doping," in *Proc. IEEE Int. Symp. Circ. Syst.*, pp. 436–439, May 1996.

43. C. Lallement, M. Bucher, and C. C. Enz, "Modelling and Characterization of Non-Uniform Substrate Doping," *Solid-State Electronics*, vol. 41, no. 12, pp. 1857–1861, Dec. 1997.

44. A.-S. Porret, J.-M. Sallese, and C. C. Enz, "A Compact Non-Quasi-Static Extension of a Charge-Based MOS Model," *IEEE Trans. Electron. Devices*, vol. 48, no. 8, pp. 1647–1654, Aug. 2001.

45. J.-M. Sallese, M. Bucher, and C. Lallement, "Improved Analytical Modeling of Polysilicon Depletion in MOSFETs for Circuit Simulation," *Solid-State Electron.*, vol. 44, no. 6, pp. 905–912, June 2000.

46. M. Bucher, J.-M. Sallese, and C. Lallement, "Accounting for Quantum Effects and Polysilicon Depletion in an Analytical Design-Oriented MOSFET Model," in *Simulation of Semiconductor Processes and Devices*, C. T. D. Tsoukalas, Ed. Springer, 2001, pp. 296–299.

47. C. Lallement, J.-M. Sallese, M. Bucher, W. Grabinski, and P. C. Fazan, "Accounting for Quantum Effects and Polysilicon Depletion From Weak to Strong Inversion in a Charge-Based Design-Oriented MOSFET Model," *IEEE Trans. Electron. Devices*, vol. 50, no. 2, pp. 406–417, Feb. 2003.

48. C. Enz and Y. Cheng, "MOS Transistor Modeling Issues for RF Circuit Design," in *Advances in Analog Circuit Design (AACD'99)*, W. Sansen, J. Huijsing, and R. v. d. Plassche, Eds. Kluwer Book, 1999, with Kind Permission of Springer Science and Business Media.

49. C. Enz and Y. Cheng, "MOS Transistor Modeling for RF IC Design," *IEEE J. Solid-State Circ.*, vol. 35, no. 2, pp. 186–201, © 2000 IEEE.

50. C. Enz, "MOS Transistor Modeling for RF Integrated Circuit Design," in *Proc. IEEE Custom Integr. Circ. Conf.*, pp. 189–196, May 2000.

51. C. Enz, "MOS Transistor Modeling for RF IC Design," in *Proc. Gallium Arsenide Semiconductor Appl. Symp. (GAAS 2000)*, pp. 536–539, Oct. 2000.

52. C. Enz, "An MOS Transistor Model for RF IC Design Valid in All Regions of Operation," *IEEE Trans. Microw. Theory Tech.*, vol. 50, no. 1, pp. 342–359, © 2002 IEEE.

53. A. S. Roy and C. C. Enz, "Compact Modeling of Thermal Noise in the MOS Transistor," in *Proc. 11th Int. Conf. Mixed Des. Integr. Circ. and Syst. (MIXDES)*, Szczecin, Poland, pp. 71–78, June 2004.

54. A. S. Roy and C. C. Enz, "Compact Modeling of Thermal Noise in the MOS Transistor," *IEEE Trans. Electron. Devices*, vol. 52, no. 4, pp. 611–614, Apr. 2005.

55. A. S. Roy and C. C. Enz, "An Analytical Thermal Noise Model of the MOS Transistor Valid in All Modes of Operation," in *Int. Conf. Noise and Fluctuations*, Sept. 2005.

56. F. Prégaldiny, C. Lallement, and D. Mathiot, "A Simple Efficient Model of Parasitic Capacitances of Deep-Submicron LDD MOSFETs," *Solid-State Electron.*, vol. 46, no. 12, pp. 2191–2198, copyright 2002.

57. M. Bucher, C. C. Enz, F. Krummenacher, J.-M. Sallese, C. Lallement, and A.-S. Porret, "The EKV 3.0 Compact MOS Transistor Model: Accounting for Deep-Submicron Aspects," in *Workshop Compact Model. Int. Conf. Model. Simul. Microsyst.*, Puerto Rico, pp. 670–673, Apr. 2002.

58. M. Bucher, J.-M. Sallese, F. Krummenacher, D. Kazazis, C. Lallement, W. Grabinski, and C. Enz, "EKV 3.0: An Analog Design-Oriented MOS Transistor Model," in *Proc. 9th Int. Conf. Mixed Des. Integr. Circ. Syst. (MIXDES)*, Wroclaw, Poland, June 20–22, 2002.

59. G. Machado, C. C. Enz, and M. Bucher, "Estimating Key Parameters in the EKV MOST Model for Analogue Design and Simulation," in *Proc. IEEE Int. Symp. Circuits Syst.*, pp. 1588–1591, May 1995.

60. M. Bucher, C. Lallement, and C. C. Enz, "An Efficient Parameter Extraction Methodology for the EKV MOST Model," in *Proc. IEEE Int. Conf. Microelectron. Test Struct.*, pp. 145–150, Mar. 1996.

61. W. Grabinski, "EKV v2.6 Parameter Extraction Tutorial," in *ICCAP Users' Web Conf.*, Dec. 2001.

62. W. Grabinski, "EKV v2.6 Parameter Extraction Tutorial," in *ICCAP Users' Conf.*, Berlin, 2002.

63. http://legwww.epfl.ch/ekv/.

64. J.-M. Sallese, F. Krummenacher, F. Prégaldiny, C. Lallement, A. S. Roy, and C. Enz, "A Design Oriented Charge-Based Current Model for Symmetric DG MOSFET and its Correlation with the EKV Formalism," *Solid-State Electron.*, vol. 49, no. 3, pp. 485–489, Mar. 2005.

65. A. S. Roy, J.-M. Sallese, and C. Enz, "A Closed-Form Charge-Based Expression for Drain Current in Symmetric and Asymmetric Double Gate MOSFET," in *European Solid-State Dev. Res. Conf. Dig. Tech. Papers*, Grenoble, Sept. 2005.

66. A. S. Grove and D. J. Fitzgerald, "Surface Effects on p–n Junctions: Characteristics of Surface-Charge Regions Under Non-Equilibrium Conditions," *Solid-State Electron.*, vol. 9, pp. 783–806, 1966.

67. S. M. Sze, *Semiconductor Devices: Physics and Technology*, 2nd ed. John Wiley & Sons, 1981.

68. E. Vittoz, "MOS Transistors Operated in the Lateral Bipolar Mode and Their Applications in CMOS Technology," *IEEE J. Solid-State Circ.*, vol. 18, no. 6, pp. 273–279, June 1983.

69. A.-S. Porret, T. Melly, C. C. Enz, and E. A. Vittoz, "Design of High-Q Varactors for Low-Power Wireless Applications Using a Standard CMOS Process," *IEEE J. Solid-State Circ.*, vol. 35, no. 3, pp. 337–345, Mar. 2000.

70. H. C. Pao and C. T. Sah, "Effect of Diffusion Current On Characteristics of Metal-Oxide (Insulator)-Semiconductor Transistors," *Solid-State Electron.*, vol. 9, pp. 927–937, 1966.

71. J. R. Brews, "A Charge-Sheet Model of the MOSFET," *Solid-State Electron.*, vol. 21, pp. 345–355, 1978.

72. J. A. v. Nielen and O. W. Memelink, "The Influence of the Substrate upon the DC Characteristics of Silicon MOS Transistors," *Philips Res. Repts*, vol. 22, pp. 55–71, 1967.

73. E. Vittoz, C. Enz, and F. Krummenacher, "A Basic Property of MOS Transistors and its Circuit Implications," in *Workshop on Compact Model. Int. Conf. Model. Simul. Microsyst.*, vol. 2. Computational Publications, San Fransisco, pp. 246–249, Feb. 2003.

74. M. Degrauwe, O. Leuthold, E. Vittoz, H. Oguey, and A. Descombes, "CMOS Voltage References Using Lateral Bipolars," *IEEE J. Solid-State Circ.*, vol. 20, pp. 1151–1157, Dec. 1985.

75. E. Vittoz and X. Arreguit, "Linear Networks Based on Transistors," *El. Lett.*, vol. 29, Feb. 1993.

76. E. Vittoz, "Pseudo-Resistive Networks and Their Applications to Analog Collective Computation," in *Proc. MicroNeuro 97*, Dresden, pp. 163–173, 1997.

77. K. Bult and G. J. G. M. Geelen, "An Inherently Linear and Compact MOST-Only Current Division Technique," *IEEE J. Solid-State Circ.*, vol. 27, pp. 1730–1735, Dec. 1992.

78. A.-S. Porret, "Design of a Low-Power and Low-Voltage UHF Transceiver Integrated in a CMOS Process," Ph.D. Dissertation, EPFL, Thesis No. 2542, 2002.

79. A. S. Roy, J. M. Vasi, and M. B. Patil, "A New Approach to Model Nonquasi-static (NQS) Effects for MOSFETs. Part I: Large-Signal Analysis," *IEEE Trans. Electron. Devices*, vol. 50, no. 12, pp. 2393–2400, Dec. 2003.

80. A. J. Scholten, L. F. Tiemeijer, P. W. H. d. Vreede, and D. B. M. Klaassen, "A Large-Signal Non-Quasi-Static MOS Model for RF Circuit Simulation," in *Proc. Int. Electron. Device Meet.*, pp. 163–166, Dec. 1999.

81. F. M. Klaassen and J. Prins, "Thermal Noise of MOS Transistors," *Philips Res. Repts.*, vol. 22, pp. 505–514, Oct. 1967.

82. F. M. Klaassen, "Characterization of Low 1/f Noise in MOS Transistors," *IEEE Trans. Electron. Devices*, vol. 18, no. 10, pp. 887–891, Oct. 1971.

83. R. v. Langevelde, J. Paasschens, A. J. Scholten, R. Havens, L. Tiemeijer, and D. B. M. Klaassen, "New Compact Model for Induced gate Current Noise," in *Proc. Int. Electron. Device Meet.*, pp. 867–870, Dec. 2003.

84. A. Cappy and W. Heinrich, "High-Frequency FET Noise Performance: A New Approach," *IEEE Trans. Electron. Devices*, vol. 36, no. 2, pp. 403–409, Feb. 1989.

85. W. Shockley, J. A. Copeland, and R. P. James, "The Impedance Field Method of Noise Calculation in Active Semiconductor Devices," in *Quantum Theory of Atoms, Molecules and the Solid-State*, O. Lwdin, Ed. New York: Academic Press, 1966, p. 537.

86. K. M. v. Vliet, A. Friedmann, R. J. J. Zijlstra, A. Gisolf, and A. v. d. Ziel, "Noise in Single Injection Diose. I. A Survey of Methods," *J. Appl. Phys.*, vol. 46, no. 4, pp. 1804–1813, Apr. 1974.

87. K. M. v. Vliet, "The Transfer-Impedance Method for Noise in Field-Effect Transistors," *Solid-State Electron.*, vol. 22, no. 3, pp. 233–236, Mar. 1979.

88. J. Fellrath, "Shot Noise Behaviour of Subthreshold MOS Transistors," *Revue de Physique Appliquée*, vol. 13, pp. 719–723, Dec. 1978.

89. G. Reimbold and P. Gentil, "White Noise of MOS Transistors Operating in Weak Inversion," *IEEE Trans. Electron. Devices*, vol. ED-29, no. 11, pp. 1722–1725, Nov. 1982.

90. R. Sarpeshkar, T. Delbrück, and C. A. Mead, "White Noise in MOS Transistors and Resistors," *IEEE Circ. Devices Mag.*, vol. 9, no. 6, pp. 23–29, Nov. 1993.

91. A. v. d. Ziel, *Noise in Solid State Devices and Circuits*. John Wiley, 1986.

92. C. C. Enz, E. A. Vittoz, and F. Krummenacher, "A CMOS Chopper Amplifier," *IEEE J. Solid-State Circ.*, vol. 22, no. 3, pp. 335–342, June 1987.

93. C. C. Enz and G. C. Temes, "Circuit Techniques for Reducing the Effects of Op-Amp Imperfections: Autozeroing, Correlated Double Sampling and Chopper Stabilization," *Proc. IEEE*, vol. 84, no. 11, pp. 1584–1614, Nov. 1996.

94. A. L. M. Worther, "Semiconductor Surface Physics," R. H. Kingston, Ed. Philadelphia: University of Pennsylvania Press, 1957, p. 27.

95. G. Ghibaudo, "Low Frequency Noise and Fluctuations in Advanced CMOS Devices," in *Symposium on Fluctuations and Noise – Noise in Devices and Circuits*, J. Deen, Z. Celik-Butler, and M. E. Levinstein, Eds., vol. 5113. Santa Fe: SPIE, June 2003, pp. 16–28.

96. M. Valenza, A. Hoffmann, D. Sodini, A. Laigle, F. Martinez, and D. Rigaud, "Overview of the Impact of Downscaling Technology on 1/f Noise in MOSFETs to 90nm," *IEE Proc.-Circ. Devices Syst.*, vol. 151, no. 2, pp. 102–110, Apr. 2004.

97. S. C. Sun and J. D. Plummer, "Electron Mobility in Inversion and Accumulation Layers on Thermally Oxidized Silicon Surfaces," *IEEE Trans. Electron. Devices*, vol. 27, pp. 1497–1508, 1980.

98. K. K. Hung, P. K. Ko, C. Hu, and Y. C. Cheng, "A Physics-Based MOSFET Noise Model for Circuit Simulators," *IEEE Trans. Electron. Devices*, vol. 37, no. 5, pp. 1323–1333, May 1990.

99. G. Reimbold, "Modified 1/f Trapping Noise Theroy and Experiments in MOS Transistors Biased from Weak to Strong Inversion – Influence of Interface States," *IEEE Trans. Electron. Devices*, vol. 31, no. 9, pp. 1190–1198, Sept. 1984.

100. R. Kolarova, T. Skotnicki, and J. A. Chroboczek, "Low Frequency Noise in Thin Gate Oxide MOSFETs," *Microelectron. Reliability*, vol. 41, pp. 579–585, 2001.

101. F. N. Hooge, "1/f Noise," *Physica*, vol. 83B, pp. 14–23, 1976.

102. X. Li and L. K. J. Vandamme, "1/f Noise in Series Resistance of LDD MOSTs," *Solid-State Electron.*, vol. 35, no. 10, pp. 1471–1475, Oct. 1992.

103. X. Li and L. K. J. Vandamme, "An Explanantion of 1/f Noise in LDD MOSFETs from the Ohmic Region to Saturation," *Solid-State Electron.*, vol. 36, no. 11, pp. 1515–1521, Nov. 1993.

104. E. A. Vittoz, "The Design of High-Performance Analog Circuits on Digital CMOS Chips," *IEEE J. Solid-State Circ.*, vol. 20, no. 3, pp. 657–665, June 1985.

105. M. J. M. Pelgrom, A. C. J. Duinmaijier, and A. P. G. Welbers, "Matching Properties of MOS Transistors," *IEEE J. Solid-State Circ.*, vol. 24, no. 5, pp. 1433–1440, Oct. 1989.

106. K. R. Lakshmikumar, R. A. Hadaway, and M. A. Copeland, "Characterization and Modeling of Mismatch in MOS Transistors for Precision Analog Design," *IEEE J. Solid-State Circ.*, vol. 21, no. 6, pp. 1057–1066, Dec. 1986.

107. J. A. Croon, M. Rosmeulen, S. Decoutere, W. Sansen, and H. E. Maes, "An Easy-to-Use Mismatch Model for the MOS Transistor," *IEEE J. Solid-State Circ.*, vol. 37, no. 8, pp. 1056–1064, Aug. 2002.

108. F. Stern, "Quantum Properties of Surface Space-Charge Layers," *Solid-State Electron.*, pp. 499–514, 1974, crit. Rev. Solid State Sci.

109. M. Lenzlinger and E. H. Snow, "Fowler-Nordheim Tunnelling in Thermally Grown SiO_2," *J. Appl. Phys.*, vol. 40, pp. 278–283, Jan. 1969.

110. E. Harari, L. Schmitz, B. Troutman, and S. Wang, "A 256-Bit Nonvolatile Static RAM," in *IEEE Int. Solid-State Circ. Conf. Dig.*, pp. 108–109, Feb. 1978.

111. W. S. Johnson, G. Perlegos, A. Renninger, G. Kuhn, and T. R. Ranganath, "A 16 kB Electrically Erasable Nonvolatile Memory," in *IEEE Int. Solid-State Circ. Conf. Dig.*, pp. 152–153, Feb. 1980.

112. R. v. Langevelde, A. J. Scholten, R. Duffy, F. N. Cubaynes, M. J. Knitel, and D. B. M. Klaassen, "Gate Current: Modeling, ΔL Extraction and Impact on RF Performance," in *Proc. Int. Electron. Device Meet.*, pp. 13.2.1–13.2.4, Dec. 2001.

113. R. v. Langevelde, "MOS Model 11," *Nat. Lab.Unclassified Report*, vol. NL-UR 2001/813, Apr. 2001.

114. A. J. Scholten, L. F. Tiemeijer, R. v. Langevelde, R. J. Havens, V. C. Venezia, A. T. A. Z. v. Duijn-hoven, B. Neinhüs, C. Jungemann, and D. B. M. Klaassen, "Compact Modeling of Drain and Gate Current Noise for RF CMOS," in *Proc. Int. Electron. Device Meet.*, pp. 155–158, Dec. 2002.

115. P. K. Ko, "Approaches to Scaling," in *Advanced MOS Device Physics*, ser. VLSI Electronics Microstructure Science, N. G. Einspruch and G. S. Gildenblat, Eds. Academic Press, 1989, vol. 18, pp. 1–37.

116. Z.-H. Liu, C. Hu, J.-H. Huang, T.-Y. Chan, M.-C. Jeng, P. K. Ko, and Y. C. Cheng, "Threshold Voltage Model for Deep-Submicrometer MOSFET's," *IEEE Trans. Electron. Devices*, vol. 40, no. 1, pp. 86–94, Jan. 1993.

117. C.-H. Chen and M. J. Deen, "Channel Noise Modeling of Deep-Submicron MOSFETs," *IEEE Trans. Electron. Devices*, vol. 49, no. 8, pp. 1484–1487, Aug. 2002.

118. J. P. Nougier and M. Rolland, "Differential Relaxation Times and Diffusivities of Hot Carriers in Isotropic Semiconductors," *J. Appl. Phys.*, vol. 48, no. 4, pp. 1683–1687, Apr. 1977.

119. J. P. Nougier, "Noise and Diffusion of Hot Carriers," in *Physcis of Nonlinear Transport in Semi-conductors*, D. K. Ferry, J. R. Barker, and C. Jacobini, Eds. Plenum Press, 1980, pp. 415–477.

120. R. E. Robson, "Diffusivity of Charge Carriers in Semiconductors in Strong Electric Field," *Phys. Rev. Lett.*, vol. 31, no. 13, pp. 825–828, Sept. 1973.

121. J. P. Nougier and M. Rolland, "Mobility, Noise Temperature and Diffusivity of Hot Holes in Germanium," *Phys. Rev. B*, vol. 8, no. 12, pp. 5728–5737, Dec. 1973.

122. M. Lundstrom, *Fundamentals of Carrier Transport*, 2nd ed. Cambridge University Press, 2000.

123. K. Hess, "Phenomenological Physcis of Hot Carriers in Semiconductors," in *Physcis of Nonlinear Transport in Semiconductors*, D. K. Ferry, J. R. Barker, and C. Jacoboni, Eds. Plenum Press, 1980, pp. 1–43.

124. J. R. Barker and D. K. Ferry, "On the Physics and Modeling of Small Semi-Conductor Devices-I," *Solid-State Electron.*, vol. 23, no. 6, pp. 519–530, June 1980.

125. J.-S. Goo, C.-H. Choi, A. Abramo, J.-G. Ahn, Z. Yu, T. H. Lee, and R. W. Dutton, "Physical Origin of the Excess Thermal Noise in Short Channel MOSFETs," *IEEE Trans. Electron. Device Lett.*, vol. 22, no. 2, pp. 101–103, Feb. 2001.

126. J.-S. Goo, C.-H. Choi, F. Danneville, E. Morifuji, H. S. Momose, Z. Yu, H. Iwai, T. H. Lee, and R. W. Dutton, "An Accurate and Efficient High Frequency Noise Simulation Technique for Deep Submicron MOSFETs," *IEEE Trans. Electron. Devices*, vol. 47, no. 12, pp. 2410–2419, Dec. 2000.

127. A. Schenk, "Simulation of RF Noise in MOSFETs Using Different Transport Models," *IEICE Trans. Electron.*, vol. E86-C, pp. 481–489, Mar. 2003.

128. K. Han, H. Shin, and K. Lee, "Analytical Drain Thermal Noise Current Model Valid for Deep Submicron MOSFETs," *IEEE Trans. Electron. Devices*, vol. 51, no. 2, pp. 261–269, Feb. 2004.

129. K. Hess, *Advanced Theory of Semiconductor Devices*. IEEE Press, 2000.

130. A. J. Scholten, H. J. Tromp, L. F. Tiemeijer, R. v. Langevelde, R. J. Havens, P. W. H. d. Vreede, R. F. M. Roes, P. H. Woerlee, A. H. Montree, and D. B. M. Klaassen, "Accurate Thermal Noise Model for Deep-Submicron CMOS," in *Proc. Int. Electron. Device Meet.*, pp. 155–158, Dec. 1999.

131. A. A. Abidi, "High-Frequency Noise Measurements on FET's with Small Dimensions," *IEEE Trans. Electron. Devices*, vol. 33, no. 11, pp. 1801–1805, Nov. 1986.

132. K.-H. Oh, Z. Yu, and R. W. Dutton, "A Bias Dependent Source/Drain Resistance Model in LDD MOSFET Devices for Distortion Analysis," in *Int. Conf. VLSI CAD*, Seoul, pp. 190–193, Oct. 1999.

133. C.-H. Choi, J.-S. Goo, Z. Yu, and R. W. Dutton, "Shallow Source/Drain Extension Effects on External Resistance in Sub-0.1 m MOSFET's," *IEEE Trans. Electron. Devices*, vol. 47, no. 3, pp. 655–658, © 2000 IEEE.

134. A. J. Scholten, L. F. Tiemeijer, R. v. Langevelde, R. J. Havens, A. T. A. Z. v. Duijnhoven, and V. C. Venezia, "Noise Modeling for RF CMOS Circuit Simulation," *IEEE Trans. Electron. Devices*, vol. 50, no. 5, pp. 618–632, Mar. 2003.

135. B. Razavi, R.-H. Yan, and K. F. Lee, "Impact of Distributed Gate Resistance on the Performance of MOS Devices," *IEEE Trans. Circ. Syst. I*, vol. 41, no. 11, pp. 750–754, Nov. 1994.

136. I. C. Chen and W. Liu, "High-Speed or Low-Voltage, Low-Power Operations," in *ULSI Devices*, C. Y. Chang and S. M. Sze, Eds. John Wiley, 2000, pp. 547–630.

137. H. S. Momose, E. Morifuji, T. Yoshitomi, T. Ohguro, M. Saito, T. Morimoto, Y. Katsumata, and H. Iwai, "High-Frequency AC Characteristics of 1.5nm Gate Oxide MOSFETs," in *Proc. Int. Electron. Devices Meeting*, pp. 105–108, © 1996 IEEE.

138. H. Beneking, *High Speed Semiconductor Devices – Circuit aspects and fundamental Behaviour*. Chapman and Hall, 1994.

139. H. Cho and D. E. Burk, "A Three-Step Method for the De-embedding of High-Frequency S-Parameter Measurements," *IEEE Trans. Electron. Devices*, vol. 38, no. 6, pp. 1371–1375, June 1991.

140. M. C. A. M. Koolen, J. A. M. Geelen, and M. P. J. G. Versleijen, "An Improved De-Embedding Technique for On-Wafer High-Frequency Characterization," in *IEEE Proc. Bipolar Circ. Technol. Meet.*, pp. 188–191, 1991.

141. W. Liu, R. Gharpurey, M. C. Chang, U. Erdogan, R. Aggarwal, and J. P. Mattia, "R.F. MOSFET Modeling Accounting for Distributed Substrate and Channel Resistances with Emphasis on the BSIM3v3 SPICE Model," in *Proc. Int. Electron. Devices Meet.*, pp. 309–312, Dec. 1997.

142. Y. Cheng and M. Matloubian, "On the High-Frequency Characteristics of Substrate Resistance in RF MOSFETs," *IEEE Electron. Device Lett.*, vol. 21, no. 12, pp. 604–606, Dec. 2000.

143. L. F. Tiemeijer and D. B. M. Klassen, "Geometry Scaling of the Substrate Loss of RF MOSFETs," in *Proc. European Solid-State Dev. Res. Conf.*, pp. 481–483, Sept. 1998.

144. S. F. Tin and K. Mayaram, "Substrate Network Modeling for CMOS RF Circuit Simulation," in *Proc. IEEE Custom Integr. Circ. Conf.*, pp. 583–586, May 1999.

145. G. D. Vendelin, A. M. Pavio, and U. L. Rohde, *Microwave Circuit Design Using Linear and Nonlinear Techniques*. John Wiley, 1990.

146. G. Gonzalez, *Microwave Transistor Amplifiers – Analysis and Design*, 2nd ed. Upper Saddle River: Prentice-Hall, 1996.

147. S. H. Jen, C. Enz, D. R. Pehlke, M. Schroter, and B. J. Sheu, "Accurate MOS Transistor Modeling and Parameter Extraction Valid up to 10-GHz," *IEEE Trans. Electron. Devices*, vol. 46, no. 11, pp. 2217–2227, © 1999 IEEE.

148. S. H. Jen, C. Enz, D. R. Pehlke, M. Schroter, and B. J. Sheu, "Accurate MOS Transistor Modeling and Parameter Extraction Valid up to 10-GHz," in *Proc. European Solid-State Dev. Res. Conf.*, Bordeaux, pp. 484–487, Sept. 1998.

149. D. Schreurs, J. Verspecht, S. Vandenberghe, and E. Vandamme, "Straightforward and Accurate Non Linear Device Model Parameter Estimation Method Based on Vectorial Large-Signal Measurements," *IEEE Trans. Microw. Theory Tech.*, vol. 50, no. 10, pp. 2315–2319, Oct. 2002.

150. H. Johnson, "Noise in Field-Effect Transistors," in *Field-Effect Transistors*, J. T. Wallmark and H. Johnson, Eds. Englewood Cliffs, NJ: Prentice-Hall, 1966, pp. 160–186.

151. T. H. Lee, *The Design of CMOS Radio-Frequency Integrated Circuits*, 2nd ed. Cambridge: Cambridge University Press, 1998.

152. E. Morifuji, H. S. Momose, T. Ohguro, T. Yoshitomi, H. Kimijima, F. Matsuoka, M. Kinugawa, Y. Katsumata, and H. Iwai, "Future Perspective and Scaling Down Roadmap for RF CMOS," in *Dig. Tech. Pap. Symp. VLSI Technol.*, pp. 163–164, June 1999.

153. A.-S. Porret and C. C. Enz, "Non-Quasi-Static (NQS) Thermal Noise Modelling of the MOS Transistor," *IEE Proc.-Circ. Devices Syst.*, vol. 151, no. 2, pp. 155–166, Apr. 2004.

154. L.-J. Pu and Y. Tsividis, "Small-Signal Parameters and Thermal Noise of the Four-Terminal MOS-FET in Non-Quasistatic Operation," *Solid-State Electron.*, vol. 33, no. 5, pp. 513–521, May 1990.

155. C. H. Chen and M. J. Deen, "High Frequency Noise of MOSFETs – Part I: Modeling," *Solid-State Electron.*, vol. 42, no. 11, pp. 2069–2081, Nov. 1998.

156. A. S. Roy, C. C. Enz and J.-M. Sallese, "Noise modeling methodologies in the presence of mobility degradation and their equivalence," *IEEE Trans. on Electron Devices*, vol. 53, no. 2, pp. 348–355, Feb. 2006.

Index

Accumulation, 19, 216, 221
Active region, 9
Admittance
    gate-to-bulk, 68, 274
    gate-to-source, 68, 270, 274
    gate-to-drain, 68, 270, 274
    bulk-to-drain, 68, 270
    bulk-to-source, 68, 270
    substrate, 255
    correlation, 263, 261, 265

Band gap, 112
    extrapolated, 113
    widening, 156, 133, 157, 158, 160, 161
Bode theorem, 106–107, 92, 274n
Boltzmann constant, 10, 87

Capacitance
    gate oxide, 17, 214
    silicon, 18, 19, 224
    gate, 13, 18, 19, 30, 73, 233
    depletion, 19, 30
    intrinsic, 70, 72, 73, 75, 233
    overlap, 214, 220, 221, 222, 223, 232, 233
    junction, 54, 214, 224–226, 232, 244, 249, 288
    fringing-field, 220, 221, 222, 233,
    bottom-wall, 220, 226
    side-wall, 226
Capacitor
    overlap, 10, 66, 77
Carrier
    fluctuations, 96, 100
    heating, 167, 205–209, 211, 285
Cascode configuration, 226

Channel
    buried, 143, 144, 145
    homogeneous, 47–49
    length, 46, 53, 54, 60, 164–167, 176, 186–188,
        198, 208, 232, 285
    length modulation, 33, 48, 50, 52–54, 167,
        186–188, 208
    narrow, 50
    residual current, 45
    voltage, 34, 38, 43, 49, 52, 78, 83, 116
    width, 34, 46, 60, 164, 218
    gradual, 13, 14, 16, 167, 187, 189
    conductance, 86, 89, 90, 199, 201
    region, 14, 51, 81, 198, 218, 219, 222, 223, 231
Channel length modulation, 48, 50, 52–54, 167,
        186–188, 208
Charge
    elementary, 10, 189, 227
    fixed, 10, 11, 15–17, 150, 151, 155, 187
    inversion, 4, 10, 11, 20, 21, 24, 30, 65, 87, 96,
        143, 146, 171–176, 206, 211
    total, 11, 13, 17, 21, 27, 144
    concentration, 13, 14, 33, 187
    depletion, 11, 12, 15, 17, 21, 23, 26, 30, 47, 49,
        97, 134, 138, 157, 224
    specific, 13, 24, 32, 117, 133, 154, 161
Charge density, 11
    depletion, 11, 12, 15, 17, 19, 21, 26, 30, 47, 97,
        134, 138, 157, 224
    in silicon, 23, 27, 189
    inversion, 20, 21, 25, 30, 65, 87, 96, 143, 148,
        172–176, 192, 211
    specific, 13, 24, 32, 117, 133, 154, 159, 161
    fixed, 11, 17, 150, 151, 155, 187

Charge-Based MOS Transistor Modeling: The EKV Model for Low-Power and RF IC Design   C. Enz and E. Vittoz
© 2006 John Wiley & Sons, Ltd.

Transconductance-to-current ratio, 62, 183
Transducer gain, 237
Transfer parameter, 34, 40, 118, 121
Transistor
  bipolar, 12, 37, 45, 229
  extrinsic part, 9, 131, 213, 214, 227, 231
  intrinsic part, 9, 61, 66, 213, 214
  schematic cross section, 10
  symbols, 12
Transit frequency, 55, 66, 232, 233, 234, 235,
        239, 268, 287
  peak, 234
Transit time, 239, 240
  total, 239
  of intrinsic part, 239
  of extrinsic part, 239
Trapping, 96, 98, 100
Tunneling, 98

Unilateral power gain, 236, 237
Unity gain transit frequency, 233

Velocity
  Saturation, 86, 167, 168, 171, 177, 232, 235
  saturation region, 186, 187
  drift, 167, 168, 173

Velocity-field models, 169–171
Voltage
  channel length modulation, 60
  drain, 10, 33, 35, 36, 38, 42, 56, 60, 170, 171
  gain, 61, 64, 167, 197, 198
  gate, 10, 17, 18, 23, 25, 27, 32, 35, 40, 50, 143,
        150, 153
  pinch-off, 3, 25, 32, 35, 36, 43, 135, 180, 211
  intrinsic, 216
  saturation, 37, 44
  source, 10
  thermodynamic, 10, 83
  channel, 9, 14, 19, 23, 25, 26, 77, 83, 116
  flat-band, 17, 49, 50, 52, 114, 221
  band gap, 112
  extrapolated band gap, 113
  avalanche breakdown, 224

Weak inversion, 17, 18, 20, 26, 30, 36, 39
  current approximation, 43–45
Well
  perimeter, 226
  area, 226

Y-parameters, 234, 237, 243, 249, 251, 252, 253,
        255, 261, 263, 265, 273

Printed and bound by CPI Group (UK) Ltd, Croydon, CR0 4YY

16/04/2025

14658471-0004